Handbook of Modern Biophysics

Series Editor
Thomas Jue
University of California Davis
Davis, California

For other titles published in this series, go to
www.springer.com/series/7845

Roland Faller • Thomas Jue
Marjorie L. Longo • Subhash H. Risbud
Editors

Biomembrane Frontiers

Nanostructures, Models, and the Design of Life

Volume 2

Humana Press

Editors
Roland Faller
Department of Chemical Engineering
and Materials Science
University of California, Davis
One Shields Avenue
Davis, CA 95616
USA
rfaller@ucdavis.edu

Thomas Jue
Department of Biochemistry and Molecular Medicine
University of California, Davis
4418 Tupper Hall
One Shields Avenue
Davis, CA 95616
USA
tjue@ucdavis.com

Marjorie L. Longo
Department of Chemical Engineering
and Materials Science
University of California, Davis
One Shields Avenue
Davis, CA 95616
USA
mllongo@ucdavis.edu

Subhash H. Risbud
Department of Chemical Engineering
and Materials Science
University of California, Davis
One Shields Avenue
Davis, CA 95616
USA
shrisbud@ucdavis.edu

ISBN 978-1-60761-313-8 e-ISBN 978-1-60761-314-5
DOI 10.1007/978-1-60761-314-5
Springer Dordrecht Heidelberg London New York

Library of Congress Control Number: 2009927494

Printed on acid-free paper

Springer is part of Springer Science+Business Media (www.springer.com)

Preface

The book, *Biomembrane Frontiers: Nanostructures, Models, and the Design of Life*, a volume in the *Handbook of Modern Biophysics* series, is based on a workshop held on the 20th and 21st of March 2008 at the University of California Davis. Unlike other meeting monographs, the book presents the exciting frontiers of biomembrane research for both expert and student colleagues interested in research at the interface of biology and physics.

The idea of the workshop originated from discussions about how to create an effective outreach for the NSF-NIRT joint project "Aerogel and Nanoporous Materials for Biomolecular Applications" between the Longo, Faller, and Risbud groups at UC Davis and the groups of Curt Frank at Stanford and Joe Satcher at Lawrence Livermore National Laboratory. In the project we interacted with researchers from diverse backgrounds and hoped to create an opportunity to foster a multi- and interdisciplinary exchange of ideas. Thus, the workshop idea was conceived.

The workshop brought together experts working on many different aspects of biological membranes: from theory and simulation, to supported model bilayers, and to clinical applications. Several material scientists working on the interactions of biological membranes with biological or nonbiological materials also participated. Such a diverse set of experts in one meeting is unusual, as the different communities of theorists and experimentalists working on model membranes and real biological systems are typically quite distinct and do not often interact. Very few, if any, conferences take up the challenge of embracing a broad range of research interests. The chapters of the volume reflect the dynamic synergism of the diverse research interests in biomembrane research and present invaluable, leading ideas to a broad community of researchers and students.

At the workshop, the lively discussion made clear that everybody learned from this unique interaction with colleagues from several disciplines. It was obvious that many aspects of membranes cut across a variety of disciplines and that only research using a combination of ideas and techniques can facilitate real progress.

Several of the speakers invited their graduate students to the workshop. A large number of local graduate students and faculty also participated. Indeed, the graduate students benefited from examining the common theme of membranes from many different perspectives. These students also presented impressive work during poster sessions. The workshop had more than 70 participants.

The book is arranged topically. It leads from theory to applications. After an introduction by Harden McConnell on the history of lipid complexes over the last century, several chapters on theoretical and computational descriptions of membranes follow. Even within this subgroup there is already great diversity. Studying membranes at many different time and length scales is important and requires a wide variety of theoretical approaches to address them. The next block of chapters deals with techniques and applications in model membranes of increasing complexity. In this area there is always the compromise between the simplicity necessary to understand the system in as much detail as possible and the complexity to mimic real biological membranes as realistically as possible. The final chapters address questions of biological and clinical importance involving real membranes.

The workshop and ultimately this book would not have been possible without the dedicated support of a number of people: these are, of course, first and foremost, the speakers and graduate student poster presenters at the workshop and the authors of the chapters. They volunteered their time and effort to make this workshop a success. Specifically, we would like to thank our graduate students Allison Dickey, Emel Goksu, Clark Henderson, Matthew Hoopes, Monica Lozano, Barbie Nellis, Mike Skaug, Juan Vanegas, and Chenyue Xing for their help in organizing and running the workshop. In particular, we would like to acknowledge the organizational talent of Jenny McDonald. Finally, we would like to thank the NSF-NIRT program and the Graduate Group in Biophysics at UC Davis for their financial support, and Springer Science+Business Media for the opportunity to publish this book.

Roland Faller, Thomas Jue, Marjorie Longo, and Subhash Risbud

CONTENTS

3 Structure and Dynamics of Lipid Monolayers: Theory and Applications

Svetlana Baoukina, Siewert J. Marrink, and D. Peter Tieleman

4 Multiscale Modeling of Supported Lipid Bilayers

Matthew I. Hoopes, Chenyue Xing, and Roland Faller

5 Collective Dynamics in Lipid Membranes: From Pore Formation to Flip-Flops

Andrey A. Gurtovenko and Ilpo Vattulainen

6 Spatiotemporal Organization of Spin-Coated Supported Model Membranes

Adam Cohen Simonsen

7 Nanopore Analysis of Nucleic Acids: Single-Molecule Studies of Molecular Dynamics, Structure, and Base Sequence

Felix Olasagasti and David W. Deamer

8 Complex Applications of Simple FRAP on Membranes

Minchul Kang and Anne K. Kenworthy

9 Punching Holes in Membranes: How Oligomeric Pore-Forming Proteins and Lipids Cooperate to Form Aqueous Channels in Membranes

Cécile Fradin, Dmitri Satsoura, and David W. Andrews

10 Morphogens, Membranes and Mechanotransduction in Articular Cartilage

Shirley Motaung, Stephanie Chan, and A. Hari Reddi

11 Lifecycle of a Lipoprotein from a Biophysical Perspective

John C. Rutledge, Thomas Huser, John Voss, James Chan, and Atul Parikh

12 Targeting Apolipoproteins in Magnetic Resonance Imaging

Renuka Sriram, Jens O. Lagerstedt, Haris Samardzic, Ulrike Kreutzer, Jitka Petrolova, Hongtao Xie, George A. Kaysen, John C. Voss, Jean F. Desreux, and Thomas Jue

CONTRIBUTORS

David W. Andrews (chap. 9)
Department of Biochemistry
& Biomedical Sciences
McMaster University
1280 Main Street W
Hamilton, Ontario, L8S4M1, Canada

Svetlana Baoukina (chap. 3)
Department of Biological Sciences
University of Calgary
2500 University Drive NW
Calgary AB T2N1N4, Canada

James Chan (chap. 11)
Center for Biophotonics Science
& Technology
University of California Davis
2700 Stockton Boulevard
Sacramento, CA 95817, USA

Stephanie Chan (chap. 10)
Department of Orthopaedic Surgery
Lawrence Ellison Center for Tissue Regeneration
University of California Davis,
School of Medicine
Research Building 1, Room 2000
4635 Second Avenue
Sacramento, CA 95817, USA

David W. Deamer (chap. 7)
University of California Santa Cruz
1156 High Street
Santa Cruz, CA 95064-1077, USA
deamer@chemistry.ucsc.edu

Markus Deserno (chap. 2)
Department of Physics
Carnegie Mellon University
5000 Forbes Avenue
Pittsburgh, PA 15213, USA
deserno@andrew.cmu.edu

Jean-François Desreux (chap. 12)
Coordination and Radiochemistry
Sart Tilman-B16
University of Liège
Liège 4000, Belgium
jf.desreux@ulg.ac.be

Roland Faller (chap. 4)
Graduate Group in Biophysics
University of California Davis
Bainer Hall, One Shields Avenue
Davis, CA 95616, USA
rfaller@ucdavis.edu

Cécile Fradin (chap. 9)
Department of Physics & Astronomy
McMaster University
1280 Main Street W
Hamilton, Ontario, L8S4M1, Canada
fradin@physics.mcmaster.ca

Laura J. D. Frink (chap. 1)
Colder Insights
Saint Paul, MN 55126, USA
ljfrink@coldersinsights.com

Amalie L. Frischknecht (chap. 1)
Sandia National Laboratories
P.O. Box 5800 MS 1411
Albuquerque, NM 87185, USA
alfrisc@sandia.gov

Andrey A. Gurtovenko (chap. 5)
Computational Biophysics Laboratory
Institute of Pharmaceutical Innovation
University of Bradford
Bradford, West Yorkshire, BD7 1, UK
A.Gurtovenko@gmail.com

Matthew I. Hoopes (chap. 4)
Graduate Group in Biophysics
University of California Davis
Bainer Hall, One Shields Avenue
Davis, CA 95616, USA

Thomas Huser (chap. 11)
Department of Medicine
School of Medicine
University of California Davis
5404 GBSF
Davis, CA 95618, USA

Thomas Jue (chap. 12)
Department of Biochemistry & Molecular Medicine
University of California Davis
One Shields Avenue
Davis, CA 95616 USA
tjue@ucdavis.edu

Minchul Kang (chap. 8)
Department of Molecular Physiology & Biophysics
Vanderbilt University School of Medicine
718 Light Hall
Nashville, TN 37232, USA
minchul.kang@vanderbilt.edu

George A. Kaysen (chap. 12)
Department of Biochemistry & Molecular Medicine
University of California Davis
One Shields Avenue
Davis, CA 95616 USA
gakaysen@ucdavis.edu

Anne K. Kenworthy (chap. 8)
Department of Molecular Physiology & Biophysics
Vanderbilt University School of Medicine
718 Light Hall
Nashville, TN 37232, USA
anne.kenworthy@vanderbilt.edu

Ulrike Kreutzer (chap. 12)
Department of Biochemistry
& Molecular Medicine
University of California Davis
One Shields Avenue
Davis, CA 95616 USA
umkreutzer@ucdavis.edu

Jens O. Lagerstedt (chap. 12)
Department of Biochemistry & Molecular Medicine
University of California Davis
One Shields Avenue
Davis, CA 95616 USA
jlagerstedt@ucdavis.edu

Siewert J. Marrink (chap. 3)
Groningen Biomolecular Sciences
& Biotechnology Institute
University of Groningen, Nijenborgh 4
9747 AG Groningen, The Netherlands

Harden McConnell (intro)
Department of Chemistry
Stanford University
Stanford, CA 94305-5080, USA
harden@leland.stanford.edu

Shirley Motaung (chap. 10)
Faculty of Science
Department of Biomedical Sciences
Tshwane University of Technology
Private Bag X680
Pretoria, South Africa, 0001
smotaung@gmail.com

Felix Olasagasti (chap. 7)
Department of Chemistry & Biochemistry
University of California Santa Cruz
1156 High Street
Santa Cruz, CA 95064-1077, USA
olasagas@chemistry.ucsc.edu

Atul Parikh (chap. 11)
Department of Medicine
School of Medicine
University of California Davis
5404 GBSF
Davis, CA 95618, USA

Jitka Petrolova (chap. 12)
Department of Biochemistry
& Molecular Medicine
University of California Davis
One Shields Avenue
Davis, CA 95616 USA
jpetrolova@ucdavis.edu

A. Hari Reddi (chap. 10)
Department of Orthopaedic Surgery
Lawrence Ellison Center for Tissue Regeneration
University of California Davis,
School of Medicine
Research Building 1, Room 2000
4635 Second Avenue
Sacramento, CA 95817, USA
ahreddi@ucdavis.edu

John C. Rutledge (chap. 11)
Department of Medicine
School of Medicine
University of California Davis
5404 GBSF
Davis, CA 95618, USA
jcrutledge@ucdavis.edu
jcrutledge@gmail.com

Haris Samardzic (chap. 12)
Department of Biochemistry
& Molecular Medicine
University of California Davis
One Shields Avenue
Davis, CA 95616 USA
hsamardzic@ucdavis.edu

Dmitri Satsoura (chap. 9)
Department of Biochemistry
& Biomedical Sciences
McMaster University
1280 Main Street W
Hamilton, Ontario, L8S4M1, Canada

Adam Cohen Simonsen (chap. 6)
MEMPHYS —Center for Biomembrane Physics
Department of Physics and Chemistry (IFK)
University of Southern Denmark
Campusvej 55
DK-5230 Odense M, Denmark
http://memphys.sdu.dk/~adam
e-mail:adam@memphys.sdu.dk

Renuka Sriram (chap. 12)
Department of Biochemistry
& Molecular Medicine
University of California Davis
One Shields Avenue
Davis, CA 95616 USA
rsriram@ucdavis.edu

D. Peter Tieleman (chap. 3)
Department of Biological Sciences
University of Calgary
2500 University Drive NW
Calgary AB T2N1N4, Canada
http://moose.bio.ucalgary.ca
tieleman@ucalgary.ca

Ilpo Vattulainen (chap. 5)
Department of Physics
Tampere University of Technology
P.O. Box 692
FI-33101 Tampere, Finland
Ilpo.Vattulainen@tut.fi

John C. Voss (chaps. 11, 12)
Department of Biochemistry
& Molecular Medicine
University of California Davis
One Shields Avenue
Davis, CA 95616 USA
jcvoss@ucdavis.edu

Hongtao Xie (chap. 12)
Department of Biochemistry
& Molecular Medicine
University of California Davis
One Shields Avenue
Davis, CA 95616 USA
hxie@ucdavis.edu

Chenyue Xing (chap. 4)
Department of Chemical Engineering
& Materials Science
University of California Davis
Bainer Hall, One Shields Avenue
Davis, CA 95616, USA
Cxing@ucdavis.edu

Perspectives: Complexes in Liquids, 1900–2008

Harden McConnell
Department of Chemistry, Stanford University,
Stanford, California

INTRODUCTION

The interplay of chemical and phase equilibria is one of the classical problems in physical chemistry, having literature citations as far back as the early 1900s. Phase diagrams of particular interest have been those for ternary liquid mixtures (and alloys) that form closed loops. These loops correspond to composition regions in which two liquids are formed. These diagrams were sometimes interpreted in terms of the reversible formation of compounds, or complexes between components of the mixtures. This subject is now of renewed interest in connection with the discovery of liquid–liquid immiscibility in monolayer and bilayer membranes composed of phospholipids and cholesterol. Specific ternary mixtures of phospholipids and cholesterol in bilayers have closed-loop phase diagrams, and have also been interpreted in terms of the formation of complexes. Immiscibility is found in model mixtures of defined composition, and also in lipids extracted from cell membranes and in lipid blebs from animal cells. The observed consolute critical temperatures of some of these mixtures are not far from temperatures characteristic of animal cells, raising the possibility of biophysical significance.

The purpose of this discussion will be to describe simple models of intermolecular interactions that have been used to model liquid–liquid immiscibility in monolayers and bilayers.

COMPLEXES AND THE ORIGIN OF LIQUID–LIQUID IMMISCIBILITY

Our interest in complex formation between cholesterol and phospholipids originated with attempts to model phase diagrams observed for monolayer mixtures at the air–water interface

Address correspondence to Harden McConnell, Department of Chemistry, Stanford University, Stanford, CA 94305-5080, USA, 650 723-4571, <harden@leland.stanford.edu>.

[1,2]. In our calculations we followed the earlier general theoretical studies of Corrales and Wheeler [3]. Much earlier investigators were also interested in nonideal concentrated liquid mixtures [4]. Of particular interest were closed-loop phase diagrams that describe liquid–liquid immiscibility in ternary mixtures [5]. One example is the phase diagram for a ternary mixture of phenol, acetone, and water. At temperatures between 65 and 92°C any pair of these substances is completely miscible, but mixtures containing all three components simultaneously give rise to two liquid phases. An explanation, doubtless oversimplified, is that (a) acetone and phenol form a complex, and (b) this complex is immiscible with water.

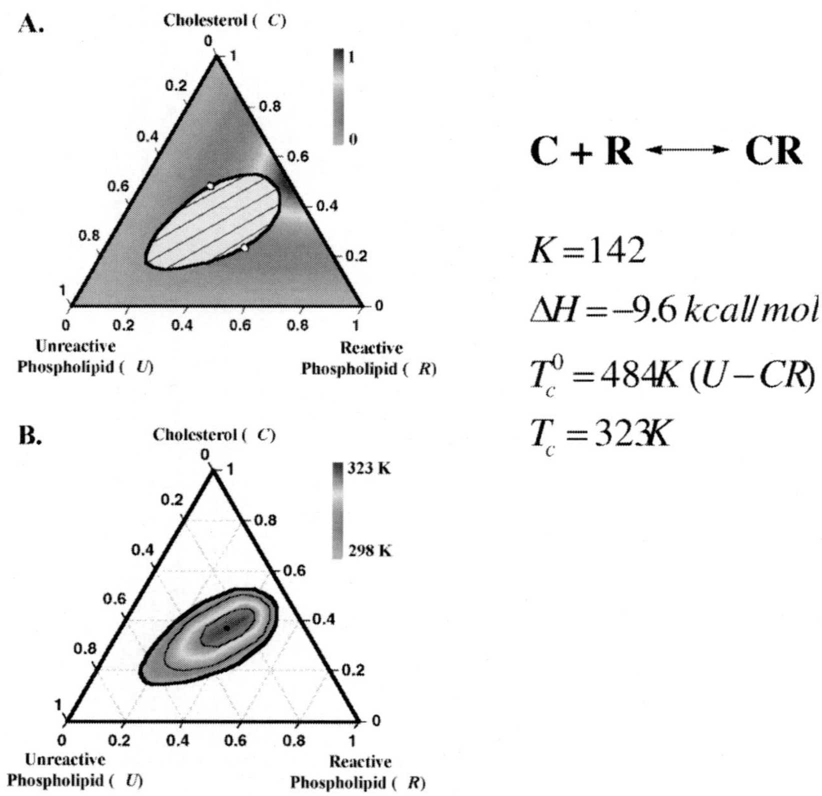

Figure 1. Calculated phase diagram for a ternary lipid mixture in which two components — C (cholesterol) and R (saturated phospholipid) — react reversibly to form a 1:1 molecular complex, and in which this complex undergoes repulsive interactions with the third component, U (dioleoylphosphatidylcholine). Please visit http://www.springer.com/series/7845 to view a high-resolution full-color version of this illustration.

A number of ternary lipid mixtures in bilayers also show closed-loop phase diagrams, as shown in the work of Veatch and Keller [6]. The ternary lipids usually consist of cholesterol, a saturated phospholipid, and an unsaturated phospholipid. We proposed that this immiscibility is related to the formation of "condensed complexes" between cholesterol and the saturated phospholipid [7], by analogy with the model used earlier for monolayers [1,2]. Figure 1 gives the result of a theoretical calculation of the phase diagram of a ternary lipid mixture: cholesterol

(C), dipalmitoylphosphatidylcholine (DPPC, or "reactive" phospholipid (R)), and dioleoylphosphatidylcholine (DOPC, or "unreactive" phospholipid (U)) [8]. In constructing this diagram it was assumed that cholesterol and DPPC form a complex, and that this complex tends to be immiscible with DOPC. The diagram in Figure 1 seeks to model the experimental results of Veatch and Keller for the phase diagrams of ternary lipid mixtures containing cholesterol and phospholipids [6]. Of course, over the years there have been many earlier qualitative proposals for the formation of cholesterol–phospholipid complexes, but there have been very few attempts at quantitative studies. (See [2] for extensive references to these earlier proposals.)

One can summarize some of the theoretical methods that can be used to calculate the thermodynamic phase diagrams yielding liquid–liquid immiscibility with the following formulas.

Two components, mean field repulsion:

$$F = -TS + |a| X_A X_B . \tag{1}$$

Three components, mean field attraction:

$$F = -TS - |a| X_A X_B , \tag{2}$$

Two, three, or more components, mean field plus complexes:

$$F = -TS - kTX_{\text{cpx}} \ln K + |a| X_{\text{cpx}} X_{\text{DOPC}} . \tag{3}$$

Ising lattice model:

$$E = -\Sigma_{i<j} J_{ij} (s_i s_j - 1) . \tag{4}$$

In Eqs. (1)–(3), $-TS$ gives the contribution of the entropy of mixing to the free energy, and $|a|$ is a parameter that represents a mean-field repulsion (Eqs. (1) and (3)) or attraction (Eq. (2)) between molecules. Equation (3) was used in constructing the phase diagram in Figure1; here K is the equilibrium constant describing the formation of a complex between cholesterol and DPPC, and the parameter $|a|$ describes the putative repulsion between the complex and DOPC. The quantities X represent the mole fractions of components.

Equation (1) is the most elementary equation that can be used to model liquid–liquid immiscibility, in this case between two components A and B. In this model there is a competition between the entropy that favors random mixing and the energy term proportional to $|a|$ that favors a separation of the A and B molecules. The mean field term $|a| X_A X_B$ has the form of a long-range interaction between A and B molecules. There are usually no long-range interactions in bilayers. The Ising model, described in Eq. (4), can also be discussed in terms of A and B molecules, and does not involve a mean-field free energy.

In the Ising model, spins with possible values of $S_i = \pm 1$ are placed on a square lattice, where the nearest neighbor interactions are repulsive when neighboring spins have opposite values, $S_i = -S_j$. Otherwise, there is no interaction. One can also describe this Ising problem in terms of the interactions between molecules A and B. That is, equal numbers of molecules A and B are placed on a square lattice. Molecules A and B repel one another. This gives rise to the temperature-dependent "liquid–liquid" phase separation sketched in Figure 2, for which the critical transition temperature is T_C.

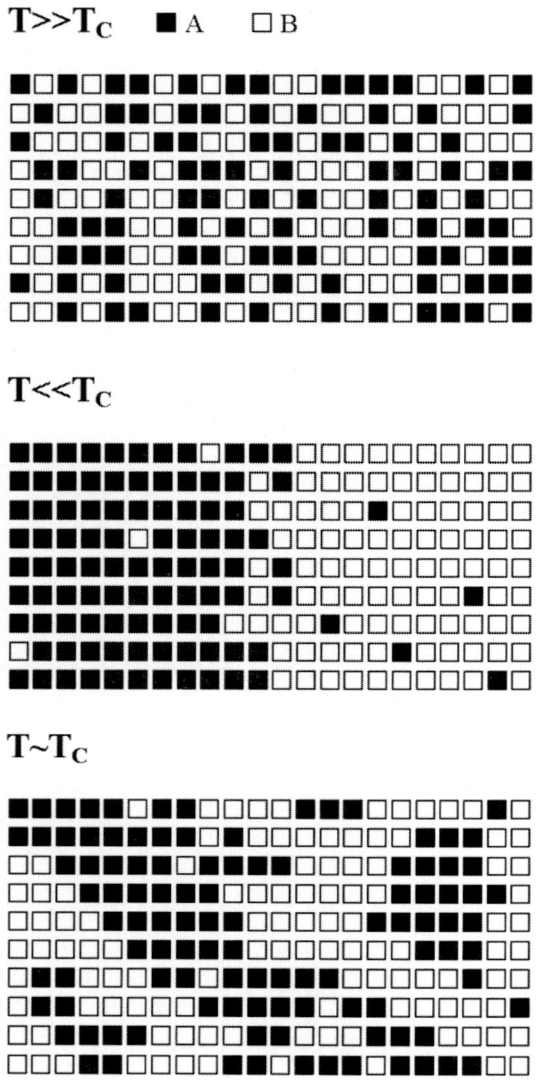

Figure 2. Schematic representation of the Ising problem in terms of molecules A and B where A and B repel one another. At high temperatures the distribution of molecules is dominated by the entropy of mixing, whereas at lower temperatures the phase separation into a liquid rich in A and a liquid rich in B is dominated by the repulsive AB interaction. At the critical transition temperature, T_c, there is long-range composition correlation in the sense that molecules of one type are most likely to be surrounded by molecules of the same type, that is, A (B) molecules are most likely to be surrounded by A (B) molecules.

It has recently been shown that ternary lipids mixtures containing cholesterol, as well as animal cell lipids, have properties (critical exponents) that conform to this 2D Ising model [9,10]. A lattice model can also be used in describing complex formation, as illustrated in Figure 3. Here molecules A and B react reversibly to form a complex C, and this complex undergoes repulsions with both A and B. Some models, termed "decorated lattice models," can describe complex formation and can be mapped mathematically on the Ising model [3].

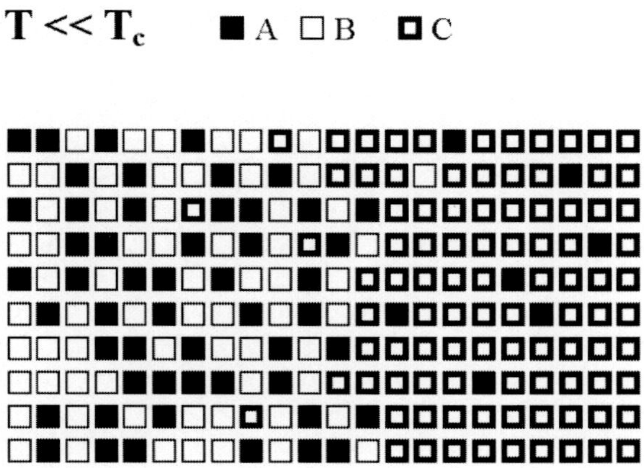

Figure 3. Phase separation can also be described by means of a lattice model, including complex formation. Here phase separation is brought about by the reaction of A and B to form a complex C, where C undergoes a repulsive interaction with A and B.

Clearly, a number of thermodynamic models can be used to describe liquid–liquid immiscibility in membranes containing phospholipids and cholesterol. Some of these models explicitly postulate the formation of molecular complexes. In these concentrated, condensed liquids molecular pairs may preferentially associate with one another due to repulsive interactions among other molecules. (Note the preferred association of the A molecules with one another in Fig. 2 due to the repulsion between the A and B molecules.) In general we use the term "complex" formation between molecules X and Y to describe molecular liquids in which the probability of finding neighboring XY molecules exceeds that expected for random mixing, irrespective of whether or not isolated pairs of X and Y "attract" one another.

We find the idea of complexes to be chemically intuitive, and to facilitate the calculations of the shapes of phase diagrams, chemical activities, heat capacities, and NMR spectra. The NMR spectra of bilayers containing cholesterol and phospholipids may reflect the kinetics of formation and dissociation of complexes (for leading references, see [11]).

OUTLOOK

Our objective in using complexes to interpret the physical chemical properties of phospholipid-cholesterol mixtures is twofold. First, as briefly summarized above, this is a convenient intuitive first-approximation. Second, the idea of complex formation is likely to be appropriate to other weak but relatively specific reversible lipid–protein and protein–protein interactions in cell membranes. References [1,2,6–11] provide many references to earlier as well as more recent work in this field.

REFERENCES

1. Radhakrishnan A, McConnell HM. 1999. Condensed complexes of cholesterol and phospholipids. *Biophys J* **77:**1507–1517.
2. McConnell HM, Radhakrishnan A. 2003. Condensed complexes of cholesterol and phospholipids. *Biochim Biophys Acta* **1610**:159–173.
3. Corrales LR, Wheeler JC. 1989. Chemical reaction-driven phase transitions and critical points. *J. Chem. Phys.* **91:**7097–7112.
4. Dolezalek F. 1908. Zur theorie binaren gemische und konzentrieten losungen. *Phys Chem* **64:**727–747.
5. Schreinemakers FAH. 1900. *Z Phys Chem* **33:**78–79; **39**:489 (1902).
6. Veatch SL, Keller SL. 2003. Separation of liquid phases in giant vesicles of ternary mixtures of phospholipids and cholesterol. *Biophys J* **85**:3074–3083.
7. McConnell H. 2005. Complexes in ternary cholesterol-phospholipid mixtures. *Biophys J* **88**: L23–L25.
8. Radhakrishnan A, McConnell HM. 2002. Critical points in charged membranes containing cholesterol. *Proc Nat Acad Sci USA* **102**:13391–13396.
9. Honerkamp-Smith AR, Cicuta P, Collins MD, Veatch SL, den Nijs M, Schick M, Keller SL. 2008. Line tensions, correlation lengths, and critical exponents in lipid membranes near critical points. *Biophys J* 95(1):236–246.
10. Veatch SL, Cicuta P, Sengupta P, Honerkamp-Smith A, Holowka D, Baird B. 2008. Critical fluctuations in plasma membrane vesicles. *ACS Chem Biol* **3**:287–293.
11. McConnell H, Radhakrishnan A. 2008. Molecular motion at the critical point in lipid membranes. *Biophys J* **95**(2):L22–L24.

1

MOLECULAR THEORY APPLIED TO LIPID BILAYERS AND LIPID–PROTEIN INTERACTIONS

Amalie L. Frischknecht
Sandia National Laboratories, Albuquerque, New Mexico, USA

Laura J. D. Frink
Colder Insights, Saint Paul, Minnesota, USA

1.1. INTRODUCTION

The fundamental, necessary components of biomembranes are the lipids that form the membrane bilayer. Although a typical plasma membrane is composed of about 50% percent proteins by mass [1], the lipids provide the structure of the membrane through their self-assembly into a bilayer. The self-assembly is driven by the amphiphilic nature of the lipids: they contain a headgroup that is hydrophilic and a tailgroup that is hydrophobic. The basic principles of lipid self-assembly are well understood [2].

The physical properties and behavior of the lipid bilayer have been studied theoretically using a variety of models and techniques. With increasing computational resources available, a popular approach has been to study membranes using atomistic molecular dynamics (MD) simulations. In such simulations, all the atoms are represented explicitly. The interactions between atoms are described by effective potentials which must be obtained through either quantum mechanical calculations or by fitting various properties to experiment. Such simulations give detailed information about both the structure and dynamics of the bilayer. Recently atomistic MD simulations have been able to treat patches of membrane up to a few tens of nanometers in lateral extent, over timescales of a few tens of nanoseconds.

Address all correspondence to Amalie L. Frischknecht, Sandia National Laboratories, Post Office Box 5800 MS 1411, Albuquerque, NM 87185, USA, 505 284-8585, 505 844-9781 (fax), <alfrisc@sandia.gov>.

R. Faller et al. (eds.), *Biomembrane Frontiers: Nanostructures, Models, and the Design of Life*,
Handbook of Modern Biophysics, DOI: 10.1007/978-1-60761-314-5_1,

However, many processes in biomembranes occur on time and length scales that are too large to access by atomistic simulations. This has led researchers to study *coarse-grained* models of lipid bilayers. Typically these models remove some of the degrees of freedom in the molecule. One way to do this is to lump several atoms together into an effective, or "united" atom, and to formulate new effective potentials for the interactions between these coarse-grained atoms [3–7]. Another approach is to develop very simple, minimalist models of the lipid molecules, such as modeling a lipid as a rigid rod with appropriate hydrophilic and hydrophobic regions. Particle-based simulations, using, for example, molecular dynamics or dissipative particle dynamics simulations, of these coarse-grained lipid models run considerably faster than atomistic simulations. Coarse-grained simulations of lipid bilayers are an active area of research; for recent reviews see [8–10]. Since these kinds of simulations are discussed by Faller and by Tieleman elsewhere in this volume, we will not discuss them further here.

Instead, we turn to theoretical approaches to understanding membrane behavior. Theoretical methods provide a conceptual framework as well as a mathematical description of the system of interest. Using theoretical methods to describe membranes is attractive for several reasons. Often a theoretical approach can highlight which ingredients are essential in controlling a particular behavior. Very simple theories may be solved analytically, and can be used to quickly analyze experimental or simulation results. More complicated theories that must be implemented numerically can probe long-time equilibrium behavior not easily accessible with simulation methods.

An early and important approach is to think of a biomembrane literally as a thin membrane, a two-dimensional surface with certain properties [11,12]. At this level, the most important features of the membrane are its elastic properties. An assembly of lipids has a preferred curvature, termed the intrinsic curvature, that depends on the geometry of the lipid molecules. Additionally, bending the membrane costs free energy. Various phenomenological theories based on a curvature-dependent free energy have been used to study the elastic properties of biomembranes and the effects of perturbations on the membranes, such as those caused by intrinsic membrane proteins [13]. These theories are continuum-level theories, which do not take into account the detailed molecular structure of the lipids.

In contrast, this chapter focuses primarily on theories based on minimal molecular models of lipid-based membranes. In these kinds of models, a very simple model of the lipid molecules is constructed, and additional features are added as necessary. This minimal model must then be coupled with an appropriate theory that can describe the statistical mechanics of the chosen model. A variety of theoretical approaches based on minimal molecular models have been taken in the literature. Section 1.2 presents a brief overview of theoretical methods that have been developed in the context of membranes. Section 1.3 presents a more detailed description of one particular class of theory known as fluids density functional theory (Fluids-DFT). Finally, results from studies that used the Fluids-DFT approach to investigate the structure and properties of lipid bilayers [14,15], the effects of added alcohols [16], and pore-forming model peptide assemblies [17] are presented in Section 1.4 and thereafter.

This chapter is by no means a comprehensive review of theoretical studies of biomembranes, and we note that the literature in this area is active and growing. For a more complete treatment of the basic thermodynamics of lipid self-assembly see [2,12]. For a recent review of membrane elasticity theory see [13]. For a discussion of coarse-grained approaches to membranes, with a focus on self-consistent field theories see [10]. Choosing an optimal numerical or theoretical approach depends on what aspects of the system are under investigation as well as on available computational resources.

1.2. ESSENTIAL ELEMENTS FOR MOLECULAR MEMBRANE THEORIES

The focus of this chapter is molecular-level theories of membranes. In these theories, the basic element is a representation of the individual lipid molecules that form the membrane. The task of the theorist is to construct a statistical mechanical theory that describes the behavior of this collection of model lipid molecules, and then to use the theory to obtain properties of interest. To date, most such theories are limited to the equilibrium properties of membranes, i.e., they do not include any dynamics. Dynamical molecular theories have been developed for various fluid systems, such as simple fluids and polymer melts [18,19], but these are much less developed than equilibrium theories and have not yet been applied to model membranes.

There are two main approaches to deriving an equilibrium, statistical mechanical theory of a molecular fluid. The most common is to construct an approximate free energy for the system of interest. The form of the free energy will depend on the chosen lipid model. It will typically incorporate various kinds of physics, such as the translational entropy of the molecules, the energy of interactions between different parts of the molecules, and for a model with internal degrees of freedom, the configurational entropy. The free energy F is related to the partition function Z by [20]

$$F = -kT \ln Z, \tag{1.1}$$

where k is Boltzmann's constant and T is the temperature. The task in statistical mechanics is to calculate the partition function Z. For most models it is not possible to evaluate Z analytically, and so approximations must be made. These approximations lead to a variety of different kinds of theories. A common approximation is the mean-field approximation. The idea is that, instead of attempting to solve the many-body problem of many interacting molecules, one focuses on one particular molecule in the system and assumes that all the other molecules contribute to an average (mean) field that acts on the molecule of interest. This approach neglects many-body fluctuations (by construction) but can be quite accurate, especially when the system is not near a critical point [20].

A second approach to a statistical mechanical theory is to start with the structural properties of the model fluid. These fluid state theories are based on approximate expressions for correlation functions [21]. One such correlation function is the radial distribution function or intermolecular correlation function $g(r)$, which, given a fluid particle at the origin, gives the probability of another fluid particle being a distance r away. Fluid state theories typically lead to nonlinear integral equations that must be solved numerically. Both free energy–based and correlation function–based theories have been applied to model lipid membranes.

There are several ways one could build a model lipid, depending on the aspect of membrane behavior of interest. The molecular structure of a common phospholipid found in plasma membranes, a phosphatidylcholine (PC), is shown in Figure 1.1. The headgroup is hydrophilic, due to its polar nature and ability to form hydrogen bonds with water. At physiological pH, PCs are

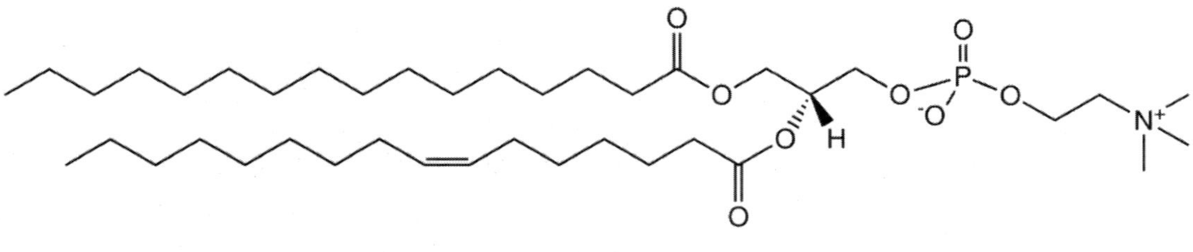

Figure 1.1. Molecular structure of a phosphatidylcholine, 1-hexadecanoyl-2-(7Z-hexadecenoyl)-sn-glycero-3-phosphocholine.

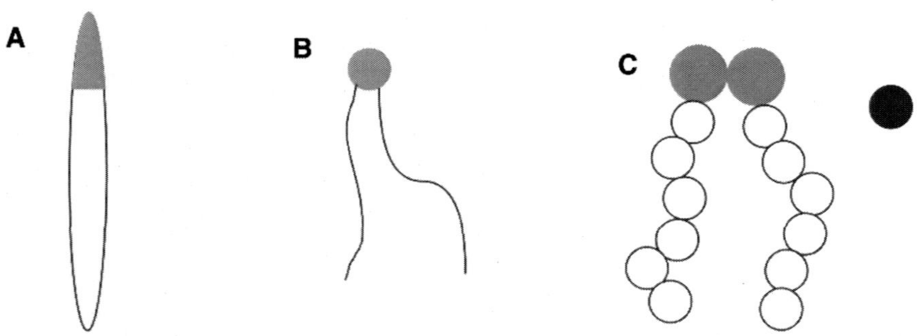

Figure 1.2. Sketches of coarse-grained lipid models. (A) An anisotropic, rigid model lipid with hydrophilic (gray) and hydrophobic (open) regions. (B) A "thread" model commonly used in SCF theories, where the lipid tails are represented by flexible curves in space. (C) A more atomistic model, with head beads (gray), tail beads (open), and an explicit solvent (black).

zwitterionic, meaning the headgroup is overall neutral but with a positive and negative charge, as shown. Other lipids, such as phosphatidylserines, are charged. The fatty acid tailgroups consist of hydrocarbon chains, typically consisting of between 14 and 24 carbon atoms. The tails are hydrophobic and also flexible. The degree of flexibility depends on the number of unsaturated (double) bonds, which varies for different lipids. Lipids have a considerable amount of configurational entropy, since the tails can adopt many different conformations, especially in the liquid state. The overall shape of phospholipids is cylindrical due to the double tails, so they will spontaneously self-assemble into bilayer structures in water with no intrinsic curvature [2].

To study the self-assembly behavior, the most important features are the geometry and the amphiphilic nature of lipids. A very simple model of a lipid as a rigid rod, in which one end is hydrophilic and one end is hydrophobic, as sketched in Figure 1.2A, is sufficient to obtain a bilayer structure. In a rigid model, all the internal degrees of freedom of the lipid molecule are ignored. Such a model thus ignores the configurational entropy of the tails. The interactions in such a model include the hydrophilic and hydrophobic interactions, and the anisotropic excluded volume since the rods cannot interpenetrate. In early work, Somoza and coworkers [22] studied such a model using a Fluids-DFT approach (see §1.3 below). They found distinct solutions to the theory corresponding to planar lipid bilayers, spherical micelles, and spherical vesicles. Such a simple model thus exhibits the self-assembly characteristic of real lipid systems. Brannigan and Brown [23] performed Monte Carlo simulations on a similar rigid lipid model that included only three interaction parameters, and were able to obtain a variety of self-assembled structures including bilayers. Neither of these models included the water molecules explicitly—they are examples of *implicit solvent* models.

A variety of flexible lipid models have also been introduced. In early work, many of these focused on the behavior of the lipid tails. The conformational entropy of the lipid tails contributes to the free energy of any lipid assembly. When, for example, a lipid bilayer is perturbed by the presence of a surface, such as a substrate or a membrane protein, it can cause a change in the lipid tail entropy. This entropy can be calculated using various mean-field theories [24]. Several groups have studied membrane systems in which the bilayer geometry is assumed (it is an input to the theory, rather than an output), but the lipid conformations and packing properties are allowed to vary. For example, May and Ben-Shaul [25,26] studied lipid–protein interactions with a theory that described the lipids with a free energy that included headgroup repulsions, a water/membrane interfacial energy, and the contribution from the configurational entropy of the tails.

Along similar lines, Lagüe and coworkers [27–29] developed an integral equation theory based on the laterally averaged response of the lipid hydrocarbon tails (obtained from atomistic MD simulations). They used the theory to calculate the potential of mean force between transmembrane proteins, modeled as rigid cylinders, that were embedded in the bilayer. However, the theory neglects the lipid headgroups, the entropy of the tails, and any fluctuations normal to the bilayer.

A more complete theory should include the self-assembly of the lipids in water. Several groups have used a self-consistent field (SCF) theory for this purpose. SCF theory is a mean-field theory that was originally developed in the field of polymer physics [30]. Typically in SCF theories, the molecule (polymer) is modeled as an infinitely thin, continuous "thread," as illustrated in Figure 1.2B. Repulsive interactions (excluded volume interactions) are included in the theory by assuming that the fluid is incompressible. To model lipids with SCF, one typically includes van der Waals-type interactions between various species such as the head, tail, and solvent groups, and the configurational entropy of the lipid tails. This approach has been extensively applied to lattice models, in which the lipids and solvent occupy sites on a regular lattice. Leermakers and coworkers [31–33] have developed a lattice SCF theory for lipids that includes chain stiffness and anisotropic interactions between tail segments. They are able to predict a fluid-to-gel phase transition for their bilayers. Similar lattice SCF theories are used by other groups [34–35].

Lattice models are not well suited to systems with arbitrary geometries, or for, e.g., the study of phase transitions between phases with different symmetries. To overcome these issues, Schick and coworkers applied off-lattice SCF theories to lipid assemblies [36–39]. Most recently they have applied the theory to studies of vesicle fusion [10]. We note that in both lattice and off-lattice SCF calculations it is possible to include the effects of charges on the lipid headgroups by also solving the Poisson-Boltzmann equation for the electrostatic potential, along with the SCF equations that result from minimizing the free energy [32,40,37,39,41].

Even more molecular detail can be introduced into a lipid model by adopting a united atom-type approach that is similar to that used in coarse-grained simulations. A simple example is the model shown in Figure 1.2C. The model consists of a flexible chain of connected spherical "beads" or sites. Different kinds of sites represent the headgroup region of the lipid, the tails, and the water (or solvent). The beads are connected together by bonds that can either be rigid or have some kind of spring potential between them. Additionally, one could impose a preferred angle between three neighboring beads and a preferred torsion angle between four neighboring beads with additional potentials. This kind of coarse-grained model was first used for lipids by Goetz and Lipowsky in MD simulations [42]. Bead-spring models have a long history in the field of polymer simulations [43], where often the most important feature is simply the topology of the long polymer molecule. The minimal molecular model shown in Figure 1.2C can also be treated using Fluids-DFT approaches. These particular approaches will be the focus of the remainder of the chapter.

1.3. FLUIDS DENSITY FUNCTIONAL THEORIES

Fluids density functional theories are statistical mechanical theories that typically describe the equilibrium, thermodynamic properties of inhomogeneous fluids. The inhomogeneity can come from different sources. One possibility occurs when the fluid is adjacent to a surface of some type, such as a fluid film sitting on a substrate. Alternatively, the fluid could have interfaces due to phase separation or self-assembly among components of the fluid, which is the case for free membranes in solution.

These theories were originally developed to describe simple atomic fluids or even more coarse-grained descriptions of fluids, and were then extended to treat molecular systems. Some density functional theories are based on molecular models, while others are based on phenomenological fluid properties. Some density functional theories utilize bulk liquid state information (known as correlation functions) calculated from an integral equation theory as part of the DFT approach.

The basic starting point for Fluids-DFTs is an expression for the free energy of the system. Given a model of the fluid, the free energy expression then includes the important physics in the model. Density functional theory is based on a mathematical theorem proved by Hohenberg and Kohn in 1964 [44] and extended by Mermin [45], in the context of an inhomogeneous gas of electrons. Originally DFT was applied to calculations of the quantum mechanical, electronic structure of atoms and molecules, and this electronic DFT has become a major technique in computational chemistry. The application of the ideas of DFT to classical fluids began in 1976 [46]. Reviews include those by Evans [47] and the books by Davis [48] and that edited by Henderson [49]. Recently Fluids-DFTs have been applied to interfacial and phase behavior in a wide variety of systems, including colloidal fluids, polymeric fluids, liquid crystals, and biological systems [50–52].

The basic idea is as follows. Consider a system in an open ensemble specified by the temperature T, volume V, and chemical potentials of all the species μ_α, and which is subject to any static external potential $V(\mathbf{r})$. This external potential could arise from the presence of a surface, or from some other kind of field such as an applied electric field. The theorem states that the total external potential for all the chemical species, $\psi_\alpha(\mathbf{r}) = V(\mathbf{r}) - \mu_\alpha$, is uniquely determined by the spatial distribution of the fluid species given by the equilibrium fluid density profiles $\rho_\alpha^0(\mathbf{r})$. Thus only one function $\psi_\alpha(\mathbf{r})$ can result in a particular density distribution $\rho_\alpha^0(\mathbf{r})$. As a consequence of this, it can be shown that the Helmholtz free energy F can be expressed as a unique functional[1] of the density profiles, which is independent of the external potential $V(\mathbf{r})$:

$$F = F[\rho_\alpha(\mathbf{r})]. \tag{1.2}$$

Additionally, the grand potential free energy can also be written as a functional of the density profiles:

$$\Omega[\rho_\alpha(\mathbf{r})] = F[\rho_\alpha(\mathbf{r})] - \int d\mathbf{r}\left(\mu_\alpha - V(\mathbf{r})\right)\rho_\alpha(\mathbf{r}). \tag{1.3}$$

This functional obeys a variational principle. The equilibrium density profile $\rho_\alpha^0(\mathbf{r})$ minimizes the functional $\Omega[\rho_\alpha(\mathbf{r})]$:

$$\left.\frac{\delta\Omega[\rho_\alpha(\mathbf{r})]}{\delta\rho_\alpha(\mathbf{r})}\right|_{\rho_\alpha^0} = 0, \tag{1.4}$$

and results in the grand potential free energy of the system, $\Omega = \Omega[\rho_\alpha^0(\mathbf{r})]$. Thus, given an expression for the intrinsic Helmholtz free energy functional and an external potential, one can solve Eq. (1.4) for the equilibrium density profiles and the grand potential free energy. Other thermodynamic quantities can then be calculated from these two basic outputs of the theory. From thermodynamics, the bulk pressure p is directly related to the grand free energy of a homogeneous system:

$$p = -\frac{1}{kTV}\Omega_{\text{bulk}}. \tag{1.5}$$

In an inhomogeneous system, the force exerted on a surface by the fluid can be calculated from derivatives of $\Omega[\rho^0(\mathbf{r})]$. The total number of molecules of species α present in the system, the *adsorption* Γ_α, is calculated from the density profile:

$$\Gamma_\alpha = \int d\mathbf{r}\rho_\alpha^0(\mathbf{r}). \tag{1.6}$$

In the following we will drop the superscript 0 to denote the equilibrium solutions to Eq. (1.4), as this will be clear from context.

To construct a Fluids-DFT for a particular model fluid, we need to determine the form of $F[\rho_\alpha(\mathbf{r})]$. The most basic contribution is the translational entropy of the fluid molecules in the absence of any interactions among them, in other words, the free energy of an ideal gas. This term is usually known exactly; for a simple one-component fluid it is

$$F^{\text{id}}[\rho] = kT\int d\mathbf{r}\rho(\mathbf{r})\left(\ln(\lambda^3\rho(\mathbf{r})) - 1\right), \tag{1.7}$$

where k is Boltzmann's constant and $\lambda = \left(h^2/2\pi mkT\right)^{1/2}$ is the thermal de Broglie wavelength. Additional terms in the free energy account for the interactions between the fluid components. If the fluid includes a connected molecule, as is the case for polymers or lipids, the connectivity of the molecule (the intramolecular interactions) must also be taken into account. There have been distinct efforts in the Fluids-DFT community to develop accurate theories for spherical model fluids, simple molecular fluids, and polymer fluids. Of these, spherical model fluids and polymer fluids have received the most attention. The spherical model fluids can include atomistic systems (such as argon), coarse-grained models of small molecules, or larger colloidal particles. In any of these cases, the Fluids-DFT approach is often constructed as a perturbation theory where a hard sphere model (essentially billiard balls that only interact at contact and cannot overlap) is the reference fluid, and other physical effects (charge, attractions, polarization, even bonds) are layered on top of the reference fluid model. For these approaches in any application domain, it is essential that the hard sphere reference fluid be treated accurately. Over the last 20 years, research into high-accuracy hard sphere models has resulted in a class of Fluids-DFTs known as the Fundamental Measure Theories (FMT) [53,54]. These nonlocal theories exhibit good accuracy both for bulk fluid properties and for interfacial fluid structure.

Fluids-DFTs for polymeric fluids mostly fall in two categories. In both kinds of DFT, a chain-like molecule is typically treated as a connected series of spherical interaction sites as sketched in Figure 1.2C. The free energy can be built on a reference state of hard spheres, with bonding added as a perturbation, as described above. Alternatively, the free energy can be based on bulk liquid state correlation functions. Below we will describe one of each of these types of Fluids-DFT, which can be used to treat model lipids as shown in Figure 1.2C.

1.3.1. CMS-DFT

As mentioned above, one approach to molecular DFTs is to use a bulk liquid state theory as input. Here we give an overview of this kind of DFT, originally derived by Chandler, McCoy, and Singer (CMS) [55–57] for molecular systems and later extended to polymers by McCoy and coworkers [58–60]. The particular version of the CMS-DFT used here was described previously by Donley et al. [61] and by Frischknecht et al. [62,63].

The fluid system of interest is characterized by its inhomogeneous site density profile, given by $\rho(\mathbf{r})=\sum_{\alpha}\rho_{\alpha}(\mathbf{r})$, where $\rho_{\alpha}(\mathbf{r})$ is the density of site type α at $\mathbf{r}$. In, for example, the lipid model in Figure 1.2C above, there are three types of sites, consisting of tail beads, head beads, and solvent beads. Each density $\rho_{\alpha}(\mathbf{r})$ gives the spatial distribution of the sites of type α.

In CMS-DFT, the free energy is measured relative to a reference system. This reference system is a homogeneous, bulk system in which all species are uniformly mixed. The reference system serves as the reservoir fluid, since the Fluids-DFT is formulated in an open ensemble. The density of each species in the reference system $\rho_{\alpha,r}$ indirectly sets the species chemical potentials. To derive the theory, we start with the difference in the grand canonical free energy between the inhomogeneous system and the reference system:

$$\Delta\Omega[T,V,\mu_{\alpha}]=\Omega[T,V,\mu_{\alpha}]-\Omega_{r}, \tag{1.8}$$

where μ_{α} is the chemical potential acting on site α. In density functional theory, the grand potential is generalized to also be a functional of the density profile, $\Omega[T,V,\mu_{\alpha};\rho_{\alpha}(\mathbf{r})]$. As described above, the density profile of the equilibrium system is determined by minimizing $\Delta\Omega$ with respect to the density:

$$\frac{\delta\Delta\Omega[T,V,\mu;\rho(\mathbf{r})]}{\delta\rho(\mathbf{r})}=0. \tag{1.9}$$

To solve Eq. (1.9) we need an approximate form for Ω. In the CMS-DFT this is obtained by expanding the free energy in a functional Taylor series about the reference state to second order:

$$\Delta\Omega = \frac{1}{2}\int\int d\mathbf{r}d\mathbf{r}r \left.\frac{\delta^2\Omega}{\delta\rho(\mathbf{r})\delta\rho(\mathbf{r}')}\right|_r \Delta\rho(\mathbf{r})\Delta\rho(\mathbf{r}') + E[\rho(\mathbf{r})], \tag{1.10}$$

where $\Delta\rho(\mathbf{r}) = \rho(\mathbf{r}) - \rho_r$, the first-order term is zero to satisfy the minimization in Eq. (1.9), and $E[\rho(\mathbf{r})]$ includes all higher-order terms in the density. Neglecting the terms higher than second order results in a linear response expression for the density in a weak field. This is inadequate for systems with strong interactions and leads to unphysical results [61]. In the CMS formalism, the higher-order terms are accounted for approximately by introducing an "ideal" system of noninteracting molecules. The free energy for the ideal system is also expanded in a functional Taylor series to second order about the reference state, and one approximates the higher-order terms in $\Delta\Omega$ of the real system by those of the noninteracting ideal system:

$$E[\rho(\mathbf{r})] \approx E^I[\rho(\mathbf{r})], \tag{1.11}$$

where the superscript I indicates the ideal system. Since the ideal system is noninteracting, this is a mean-field approximation, in which we neglect the effects of intermolecular interactions beyond second order, while retaining an exact calculation of the intramolecular correlations.

In order for this scheme to make sense, the ideal system must be constrained to have the same density profile as the real system of interest. The constraint is enforced by the mean field $U_\alpha^I(\mathbf{r})$, which acts on each site α, and incorporates in an average way the effects of interactions among all the sites in the fully interacting system. We also need to include the intramolecular correlations in a molecule. In general, the probability of finding a particular conformation of the molecule can be written as a function $S(\mathbf{r}_1,\cdots,\mathbf{r}_N)$, which gives the probability that the first segment is at $\mathbf{r}_1$, the second is at $\mathbf{r}_2$, etc. Then the density of a site α on the molecule is given by an integral over all possible conformations of the molecule, weighted by a "Boltzmann" weight given by the mean field [62]:

$$\rho_\alpha(\mathbf{r}) = \int\cdots\int d\mathbf{r}_1\cdots d\mathbf{r}_{\alpha-1}d\mathbf{r}_{\alpha+1}\cdots d\mathbf{r}_N \exp[\beta\sum_\gamma(\mu_\gamma^I - U_\gamma^I(\mathbf{r}_\gamma))]S(\mathbf{r}_1,\cdots,\mathbf{r}_N). \tag{1.12}$$

The sum over γ gives the total potential for a single conformation and the integrals are over all possible positions of all segments in the molecule $\neq\alpha$. Even though this ideal system is a simplified version of the real system, the fundamental difficulty of performing DFT on chain-like molecules is that the density at a point r depends not only on the field at that point but also on the field experienced by all other segments of the same molecule in all possible conformations.

In order to obtain an expression for $U_\alpha^I(\mathbf{r})$, the free energy $\Delta\Omega$ must be minimized with respect to $\rho_\alpha(\mathbf{r})$, subject to the constraint of Eq. (1.12). Subtracting the expanded free energy of the ideal system from that of the real system, we obtain [59]:

$$\Delta\Omega[\rho_\alpha(\mathbf{r})] = \Delta\Omega^I - \sum_\alpha\int d\mathbf{r}\left[\mu_\alpha - V_\alpha(\mathbf{r}) - \mu_{\alpha,r}\right]\rho_\alpha(\mathbf{r}) + \sum_\alpha\int d\mathbf{r}\left[\mu_\alpha^I - U_\alpha^I(\mathbf{r}) - \mu_{\alpha,r}^I\right]\rho_\alpha(\mathbf{r})$$
$$+\frac{1}{2}\sum_{\alpha,\beta}\int\int d\mathbf{r}d\mathbf{r}' \left.\frac{\delta^2(F-F^I)}{\delta\rho_\alpha(\mathbf{r})\delta\rho_\beta(\mathbf{r}')}\right|_r \Delta\rho_\alpha(\mathbf{r})\Delta\rho_\beta(\mathbf{r}'), \tag{1.13}$$

where $F - F^I$ is the excess Helmholtz free energy, and its second derivative is calculated in the homogeneous reference fluid. Minimizing Eq. (1.13) gives [59]

$$\mu_\alpha^I - U_\alpha^I(\mathbf{r}) - \mu_{\alpha,r}^I = \mu_\alpha - V_\alpha(\mathbf{r}) - \mu_{\alpha,r} + \sum_\beta \int d\mathbf{r}\mathbf{r} \left.\frac{\delta^2(F - F^I)}{\delta\rho_\alpha(\mathbf{r})\delta\rho_\beta(\mathbf{r}')}\right|_r \Delta\rho_\beta(\mathbf{r}')$$

$$= \mu_\alpha - V_\alpha(\mathbf{r}) - \mu_{\alpha,r} + \sum_\beta \int d\mathbf{r}' c_{\alpha\beta}(\mathbf{r}-\mathbf{r}')\Delta\rho_\beta(\mathbf{r}'), \tag{1.14}$$

where $c_{\alpha\beta}(\mathbf{r}-\mathbf{r}')$ is called the *direct correlation function* between sites α and β in the reference fluid. This function describes, roughly, the effective interactions and packing between the sites on the molecule. It is an input to the Fluids-DFT and must be obtained separately, either from a simulation of the model fluid or from a fluid state theory. In the CMS-DFT, the direct correlation function is obtained from an integral equation theory of polymers known as PRISM theory [64–66].

A condition of thermodynamic equilibrium is that the chemical potential is constant throughout the system. Thus, we can set the chemical potential of the real inhomogeneous system equal to the reference chemical potential, $\mu_\alpha = \mu_{\alpha,r}$ in Eq. (1.14). After some rearrangement (and a small change in the definition of the mean field; for details see [62]), we arrive at two coupled expressions for the mean field and the density profiles:

$$U_\alpha(\mathbf{r}) = V_\alpha(\mathbf{r}) - \sum_\beta \int d\mathbf{r}' c_{\alpha\beta}(\mathbf{r}-\mathbf{r}')\Delta\rho_\beta(\mathbf{r}'), \tag{1.15}$$

$$\rho_\alpha(\mathbf{r}) = \frac{V}{NZ_{\text{mol}}}\rho_r \int \prod_k d\mathbf{r}_k \delta(\mathbf{r}-\mathbf{r}_\alpha)\exp\left[-\beta\sum_\beta U_\beta(\mathbf{r}_\beta)\right] S(\mathbf{r}_1,\cdots,\mathbf{r}_N)$$

$$= \frac{V}{N}\rho_r \left\langle \delta(\mathbf{r}-\mathbf{r}_\alpha)\exp\left[-\beta\sum_\beta U_\beta(\mathbf{r}_\beta)\right]\right\rangle_S, \tag{1.16}$$

where Z_{mol} is the conformational partition function of a single molecule and serves as a normalization factor in the first line of Eq. (1.16):

$$Z_{\text{mol}} = \int \prod_\alpha d\mathbf{r}_\alpha S(\mathbf{r}_1,\cdots,\mathbf{r}_N), \tag{1.17}$$

and the second line of Eq. (1.16) indicates a statistical average over all the molecular conformations. The excess grand free energy is found by substituting the result in Eq. (1.14) into Eq. (1.13):

$$\Delta\Omega = \Delta\Omega^I + \frac{1}{2}\sum_{\alpha\beta}\int\int d\mathbf{r}d\mathbf{r}' c_{\alpha\beta}(\mathbf{r}-\mathbf{r}')\left[\rho_\alpha(\mathbf{r})\rho_\beta(\mathbf{r}') - \rho_{\alpha,r}\rho_{\beta,r}\right]$$

$$= -\frac{1}{N}\sum_\alpha \int d\mathbf{r}\Delta\rho_\alpha(\mathbf{r}) + \frac{1}{2}\sum_{\alpha\beta}\int\int d\mathbf{r}d\mathbf{r}' c_{\alpha\beta}(\mathbf{r}-\mathbf{r}')\left[\rho_\alpha(\mathbf{r})\rho_\beta(\mathbf{r}') - \rho_{\alpha,r}\rho_{\beta,r}\right], \tag{1.18}$$

where the grand potential difference $\Delta\Omega^I$ can be shown to be simply related to the excess adsorption $\Gamma = \sum_\alpha \int d\mathbf{r}\Delta\rho_\alpha(\mathbf{r})$ through a thermodynamic identity [59]. The important results here are Eq. (1.15) for the mean field, Eq. (1.16) for the density, and Eq. (1.18) for the free energy. These are the equations that must be solved to obtain the density profiles and grand free energy of the system.

We can now make a further simplification, and apply the CMS-DFT to a particular model lipid/solvent system. We specialize to model lipid chains that are freely jointed. Thus the bond lengths between neighboring sites are kept fixed at a length $\sigma_{\alpha\beta}$, but the sites are allowed to rotate freely in any direction. In this case, the configurational probability for the lipid chain is known analytically:

$$S(\mathbf{r}_1,\cdots,\mathbf{r}_{N_L}) = \prod_{s=2}^{N_L}\left[\omega_{\alpha\beta}(\mathbf{r}_s - \mathbf{r}_{s-1})\right], \tag{1.19}$$

where $\omega_{\alpha\beta}(\mathbf{r}-\mathbf{r}')$ is the probability of a bond of a certain length between a site α at $\mathbf{r}$ and a site β at $\mathbf{r}'$. For freely jointed chains with bond lengths $\sigma_{\alpha\beta}$, this is

$$\omega_{\alpha\beta}(\mathbf{r}) = \frac{1}{4\pi\sigma_{\alpha\beta}^2}\delta(|\mathbf{r}| - \sigma_{\alpha\beta}). \tag{1.20}$$

Given the definition for $S\{\mathbf{r}\}$ in Eq. (1.19), the density distribution for sites of type α on a linear chain lipid becomes

$$\rho_\alpha(\mathbf{r}) = \frac{\rho_{\alpha,r}}{N_\alpha}\sum_{s\in\alpha}\frac{G_s(\mathbf{r})G_s^{\text{inv}}(\mathbf{r})}{\exp[-\beta U_\alpha(\mathbf{r})]}, \tag{1.21}$$

where the sum over s is a sum over all the sites of type α in the molecule, N_α is the total number of sites of type α, and the G and G^{inv} are propagator functions that describe the configurational probability based on chain connectivity. The propagator functions in Eq. (1.21) obey the recursion relations

$$G_s(\mathbf{r}) = \exp[-\beta U_{\alpha(s)}(\mathbf{r})]\int dr'\omega_{\alpha\beta}(\mathbf{r}-\mathbf{r}')G_{s-1}(\mathbf{r}'), \tag{1.22}$$

$$G_s^{\text{inv}}(\mathbf{r}) = \exp[-\beta U_{\alpha(s)}(\mathbf{r})]\int d\mathbf{r}'\omega_{\alpha\beta}(\mathbf{r}-\mathbf{r}')G_{s+1}^{i}(\mathbf{r}'), \tag{1.23}$$

for the 2 … N_α sites in the case of Eq. (1.22) and the $N_\alpha - 1 \ldots 1$ sites in the case of Eq. (1.23). They satisfy the "initial" conditions $G_1(\mathbf{r}) = \exp[-\beta U_{\alpha(1)}(\mathbf{r})]$ and $G_{N_s}^{\text{inv}}(\mathbf{r}) = \exp[-\beta U_{\alpha(N_L)}(\mathbf{r})]$. Here $\alpha(s)$ indicates the site type of bead s. These initial conditions state that the probability of the end sites being at a position $\mathbf{r}$ is just given by the Boltzmann distribution in the mean field $U(\mathbf{r})$. The propagator functions then "propagate" this probability distribution to all the other connected sites. For example, site 2 is a fixed distance σ_{12} away from site 1, which is accounted for by the integral over the delta-function weight in $G_2(\mathbf{r})$ as calculated from Eq. (1.22). For a single-site liquid, such as the solvent bead (see Fig. 1.2c), Eq. (1.21) reduces to

$$\rho_s(\mathbf{r}) = \rho_{s,r}\exp[-\beta U_s(\mathbf{r})], \tag{1.24}$$

which is simply the Boltzmann distribution for a single-component fluid in the field U_s.

The CMS-DFT equations for the lipid/solvent model thus consist of Eqs. (1.15), (1.21), (1.22), and (1.23) for the lipid and Eqs. (1.15) and (1.24) for the solvent. They form a set of nonlinear, nonlocal integral equations to solve self-consistently for the fields and the densities.

The propagator equations for $G_s(\mathbf{r})$ and $G_s^{\text{inv}}(\mathbf{r})$ ensure that the chain is connected, and are the part of the calculation that incorporates the conformational entropy. The interactions between sites (on either the same or different molecules) are embedded in the direct correlation functions that appear in the expression for $U_\alpha(\mathbf{r})$.

Note that in the CMS-DFT, the free energy is only calculated relative to the free energy of the reservoir fluid. All thermodynamic properties are therefore relative to this bulk reference state.

1.3.2. Modified i-SAFT

Here we outline a second type of Fluids-DFT that can be applied to a model lipid. The so-called modified i-SAFT theory was recently introduced by Jain et al. [67]. The theory is also based on spherical interaction sites, but in modified i-SAFT the reference fluid is different than in the CMS-DFT. Instead of noninteracting chains, in modified i-SAFT the reference fluid consists of a simple fluid of hard spheres, which are the sites that make up the molecules of interest. The bonding constraints are treated as a perturbation to this reference hard sphere fluid. One advantage to this approach is that very accurate DFTs have been developed for hard spheres and can be directly used in the modified i-SAFT theory. Attractions between the sites can also be included as a perturbation.

The intrinsic Helmholtz free energy in the modified i-SAFT theory consists of a series of terms:

$$\beta F[\rho_i(\mathbf{r})] = \beta F_{\text{id}}[\rho_i(\mathbf{r})] + \beta F_{\text{hs}}[\rho_i(\mathbf{r})] + \beta F_{\text{att}}[\rho_i(\mathbf{r})] + \beta F_{\text{ch}}[\rho_i(\mathbf{r})], \tag{1.25}$$

where $\beta = 1/kT$ and the terms on the right-hand side represent the intrinsic Helmholtz free energies for the ideal gas, hard sphere, attractive interactions, and chain constraints. The segment density profiles $\rho_i(\mathbf{r})$ are the densities of each segment i in the model fluid. The ideal gas contribution for a lipid molecule is the generalization of Eq. (1.7) to the case of a molecule of N segments:

$$\beta F_{\text{id}} = \sum_{i=1}^{N} \int d\mathbf{r} \rho_i(\mathbf{r}) \left[\ln\left(\Lambda_i^3 \rho_i(\mathbf{r})\right) - 1 \right]. \tag{1.26}$$

The hard sphere term F_{hs} describes the packing between the hard spheres that make up the chain. Its general form is

$$\beta F_{\text{hs}}[\rho_i(\mathbf{r})] = \int dr \Phi[n_i(\mathbf{r})]. \tag{1.27}$$

where the energy density for the hard sphere system, Φ, is a functional of a set of nonlocal, weighted densities, $n_i(\mathbf{r})$, which in turn depend on the fluid densities $\rho_i(\mathbf{r})$. Any one of several expressions for Φ that are available in the literature [53,68,69,54] can be used. These free energies have been constructed to result in accurate equations of state in the bulk liquid. Fluids-DFTs based on these fundamental measure theories are also *thermodynamically consistent*, meaning that they obey certain exact expressions (known as sum rules) that can be derived from the laws of statistical mechanics [49].

Attractions are included in Fluids-DFTs of this type typically through a simple mean-field expression:

$$\beta F_{\text{att}} = \frac{1}{2} \sum_i \sum_j \int_{|\mathbf{r}_2 - \mathbf{r}_1| > \sigma_{ij}} d\mathbf{r}_1 d\mathbf{r}_2 u_{ij}(|\mathbf{r}_2 - \mathbf{r}_1|) \rho_i(\mathbf{r}_1) \rho_j(\mathbf{r}_2), \tag{1.28}$$

where $u_{ij}(|\mathbf{r}_2-\mathbf{r}_1|)$ is a pair potential between two sites, such as the standard Lennard-Jones interaction.

Finally, an expression is needed that captures the free energy of bonding the sites together. This expression is based on Wertheim's first-order perturbation theory [70]. It is rather complicated, and so we just give a high-level description of it here. For details, see [71,72,67]. The modified i-SAFT theory treats the chains as a sequence of N tangentially bonded monomers enforced by giving each segment a label and allowing segments to bond exclusively to their specific matching segments. The resulting free energy is

$$\beta F_{\text{ch}}\left[\rho_i(\mathbf{r})\right]=\int d\mathbf{r}'\sum_{i=1}^{N}\rho_i(\mathbf{r}')\sum_{i'}^{\{i\}}\left(\ln X_A^i(\mathbf{r}')-\frac{X_A^i(\mathbf{r}')}{2}+\frac{1}{2}\right). \tag{1.29}$$

The first summation is over all the sites in one molecule, while the second summation is over all the possible bonding sites on segment i, which are bonded to other sites in the chain. The quantity $X_A^i(\mathbf{r}')$ is the fraction of segments of type i that are not bonded. Since we are interested in a fully bonded molecule, we take the limit that all sites are bonded. Due to the structure of the theory, $X_A^i(\mathbf{r}')$ does not go to zero as one might expect, but instead leads to an expression that involves the bonding potential,

$$\exp[-\beta v_{ij}^{\text{bond}}(\mathbf{r}_1,\mathbf{r}_2)]=\frac{\delta\left(|\mathbf{r}_2-\mathbf{r}_1|-\sigma_{ij}\right)}{4\pi\sigma_{ij}^2}, \tag{1.30}$$

for tangentially bonded spheres i and j (note the similarity to Eq. (1.20)), and the so-called cavity correlation function $y_{ij}(|\mathbf{r}_2-\mathbf{r}_1|)$. In the modified i-SAFT theory, one only needs to know an expression for $y_{ij}(|\mathbf{r}_2-\mathbf{r}_1|)$ when the spheres are in contact. However, we do not know the form of the cavity correlation function in the inhomogeneous fluid, so it is instead approximated by its form in the bulk homogeneous fluid, evaluated at a weighted density.

Minimization of the grand free energy according to Eq. (1.4) leads to the set of equations that must be solved,

$$\frac{\delta F[\rho_i(\mathbf{r})]}{\delta\rho_i(\mathbf{r})}=\mu_i-V_i, \tag{1.31}$$

for each segment i in the fluid, where F is the total Helmholtz free energy from Eq. (1.25). The equations obtained from Eq. (1.31) are again a set of nonlocal integral equations that must be solved numerically. Because of the bonding constraints, it is easiest to write them in terms of a set of propagator equations for the chains that are very similar to the propagators G and G^{inv} in the CMS-DFT (see Eqs. (1.22) and (1.23)). For the modified i-SAFT theory, the propagators can be defined as

$$G_i(r)=\Xi_i\int d\mathbf{r}'y_{i,i-1}(\mathbf{r},\mathbf{r}')\frac{\delta(|r-r'|-\sigma_{i,i-1})}{4\pi\sigma_{i,i-1}^2}G_{i-1}(\mathbf{r}'),$$

$$G_i^{\text{inv}}(\mathbf{r})=\Xi_i\int d\mathbf{r}'y_{i,i+1}(\mathbf{r},\mathbf{r}')\frac{\delta(|\mathbf{r}-\mathbf{r}'|-\sigma_{i,i+1})}{4\pi\sigma_{i,i+1}^2}G_{i+1}(\mathbf{r}'). \tag{1.32}$$

with $G_1(\mathbf{r}) = \exp[-\Xi_1(\mathbf{r})]$ and $G_N^{\text{inv}}(\mathbf{r}) = \exp[-\Xi_N(\mathbf{r})]$. Note that now, in addition to integrating over the bonding weight function,

$$w_{ij}^{\text{bond}}(\mathbf{r},\mathbf{r}') = \frac{\delta(|\mathbf{r}-\mathbf{r}'|-\sigma_{ij})}{4\pi\sigma_{ij}^2}. \tag{1.33}$$

we also must include the cavity correlation functions y_{ij} in the integrand. The function Ξ_i, while more complicated, plays a similar role as the CMS-DFT mean-field U_α. It is given by

$$\ln \Xi_i(\mathbf{r}) = D_i(\mathbf{r}) - \beta V_i(\mathbf{r}), \tag{1.34}$$

where

$$D_i(\mathbf{r}) = \frac{1}{2}\sum_{j=1}^{m}\sum_{k}^{\{k=j\pm1\}}\int \rho_j(\mathbf{r}')\frac{1}{y^{\alpha(j)\alpha(k)}(\mathbf{r}')}\sum_{\varepsilon=2,3}\frac{\partial y^{\alpha(j)\alpha(k)}}{\partial \xi_\varepsilon}(\mathbf{r}')\omega_i^{(\varepsilon)}(|\mathbf{r}-\mathbf{r}'|)d\mathbf{r}'$$

$$-\int\sum_{\varepsilon}\frac{\delta\Phi(\mathbf{r}')}{\delta\rho_\varepsilon(\mathbf{r})d\mathbf{r}'} - \sum_j\int\rho_j(\mathbf{r}')u_{ij}(|\mathbf{r}-\mathbf{r}'|)d\mathbf{r}'. \tag{1.35}$$

Note that $D_i(\mathbf{r})$ (and hence $\Xi_i(\mathbf{r})$) is a functional of the densities $\rho_i(\mathbf{r})$. It also depends on the cavity correlation functions *y*, the hard sphere free energy density Φ, and the pair potentials between sites $u_{ij}(|\mathbf{r}-\mathbf{r}'|)$. Thus Ξ_i, just like the CMS-DFT mean field, contains information about all the site–site interactions in the theory. Finally, the density of site type α is found by summing over all segments *i* of type α:

$$\rho_\alpha(\mathbf{r}) = \sum_{i;\alpha(i)}\rho_i(\mathbf{r}) = \Xi_\alpha^{-1}\sum_{i;\alpha(i)} e^{\beta\mu_M} G_i G_i^{\text{inv}}, \tag{1.36}$$

where $\mu_M = \sum_{i=1}^{N}\mu_i$ is the chemical potential of the molecule. For the modified-iSAFT we must solve Eqs. (1.32), (1.34), and (1.36). These are very similar in form to the CMS-DFT Eqs. (1.22)–(1.23), (1.15), and (1.21), respectively.

Since the modified i-SAFT is based on a known equation of state and a reference fluid of individual spheres, it is possible in this theory to calculate the chemical potentials μ_i given the input bulk densities. This allows determination of the free energy of the homogeneous bulk state as well as of the inhomogeneous system. Thus the equilibrium free energy in the modified i-SAFT theory is not relative to the bulk free energy as for the CMS-DFT, but instead is based on the underlying equation of state for the model fluid.

1.4. A SIMPLE LIPID MODEL

Consider the coarse-grained lipid model shown in Figure 1.3. This model lipid has two tails which are 8 beads long each This maps roughly two CH_2 groups per bead. In this model, the headgroups are lumped into two larger beads in the center of the chain. Finally a single-site solvent is included in the model. The polar/apolar characteristics of the head- and tailgroups are retained in the model by assigning different energetic properties to the different coarse-grained groups. More specifically, the different sites in the lipid are assumed to interact with standard Lennard-Jones (LJ) potentials,

$$u_{\alpha\beta}(r) = u_{\alpha\beta}^{\mathrm{LJ}}(r) - u_{\alpha\beta}^{\mathrm{LJ}}(r_c), \tag{1.37}$$

$$u_{\alpha\beta}^{\mathrm{LJ}}(r) = \frac{4\varepsilon_{\alpha\beta}}{kT}\left[\left(\frac{\sigma_{\alpha\beta}}{r}\right)^{12} - \left(\frac{\sigma_{\alpha\beta}}{r}\right)^{6}\right], \tag{1.38}$$

where r_c is the cutoff distance where the potential goes to zero, k is Boltzmann's constant, and T is the temperature.

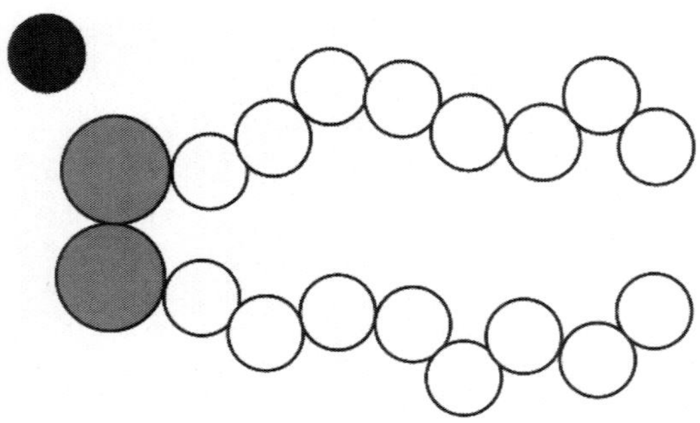

Figure 1.3. A sketch of the model lipid-solvent system. The tail beads (white) and the solvent (black) are the same size. The headgroup beads (grey) are larger by a factor of 1.44.

The ratio of head to tail bead diameters should be chosen to preserve a bilayer morphology. In our calculations, we set $\sigma_h / \sigma_t = 1.44$, giving an overall headgroup volume fraction of 0.27. On the basis of simple packing arguments and previous theoretical work on similar lipid models [39], this ratio results in lamellar/bilayer-forming lipids. The solvent diameter is set to $\sigma_s = \sigma_t \equiv \sigma$. Finally, the cross-terms in the bead diameters are set using the Berthelot scaling rules, so that $\sigma_{\alpha\beta} = 0.5(\sigma_\alpha + \sigma_\beta)$.

The energetics of this coarse-grained model are set as follows. The tail–solvent and tail–head interactions are chosen to be purely repulsive with $r_c = 2^{1/6}\sigma_{\mathrm{ts}}$ and $r_c = 2^{1/6}\sigma_{\mathrm{th}}$, respectively. Solvent–solvent, solvent–head, head–head, and tail–tail interactions are all uniformly attractive with a cutoff of $r_c = 3.5\sigma$. Finally, $\varepsilon_{\alpha\beta} \equiv \varepsilon = 1$. This combination of parameters allows for a self-assembling bilayer to form. In the following sections, lengths are reported in units of σ and energies in units of kT unless otherwise specified.

1.5. PERFORMING FLUIDS-DFT CALCULATIONS FOR MODEL LIPID BILAYERS

The Fluids-DFT equations are a set of nonlinear integral equations that must be solved numerically. Both the CMS-DFT and modified i-SAFT equations have been implemented into a general Fluids-DFT code called Tramonto [73]. The equations are solved in real space on a regular Cartesian mesh. Our numerical methods have been enumerated extensively elsewhere [62,50,74]. We use a Newton's method approach with update dampening, as required for good convergence. In the rest of this chapter, all DFT calculations will refer to the CMS-DFT, with the exception of the preliminary results reported in Section 1.6.2. The calculation methods are, however, quite general, and can be applied to other fluid DFTs.

Recall that Fluids-DFTs are formulated in the grand canonical ensemble. A given thermodynamic state, or state point, is specified in this ensemble by the values of μ_s, μ_L, V, and T, where μ_s is the solvent chemical potential, μ_L is the lipid chemical potential, V is the system volume, and T is the temperature. The μ variables are defined indirectly in the CMS-DFT by specifying the site number densities of solvent and lipid in a mixed bulk reservoir fluid.

There is no way in this grand canonical ensemble to guess a priori how the state variables should be chosen in order to obtain a bilayer structure. In our work, we chose to keep the value of the total site densities, $\rho_b = \rho_L + \rho_s$, fixed at a constant value. This leaves three state variables, V, T, and one other chemical potential or bulk density variable. We characterize this second variable by the fraction of solvent in the bulk state, defined by $x_s = \rho_s / \rho_b$.

A flat lipid bilayer has planar symmetry, in which the fluid densities are nonuniform only perpendicular to the bilayer. To find a bilayer solution to the DFT equations it is sufficient to allow variables to depend only on the direction perpendicular to the bilayer. We perform analytical integration of the equations in the two dimensions parallel to the bilayer. Note that a homogeneous solution of mixed lipids and solvent at constant bulk density is always a solution of the DFT equations, i.e., $\rho_\alpha(r) = \rho_{b,\alpha}$ is a valid solution. However, this is clearly not the solution of interest. In order to obtain bilayer-like solutions, suitable initial guesses are required. We used initial guesses consisting of step-function density profiles (with some trial and error) to generate a nonuniform bilayer-like solution. Obtaining these solutions was easiest with reflecting boundary conditions, which assume $\rho_{B+i} = \rho_{B-i}$ where B is the node at the boundary, i is an integer, and ρ is any variable in the calculation. For single bilayer solutions, we often locate the center of the bilayer at one of the reflecting boundaries. This ensures symmetric bilayer solutions, and removes any numerical instabilities due to drift of bilayer–solvent interfaces in the computational domain.

There are several properties of interest that can be calculated directly from a converged solution to the DFT equations. In general, the properties of interfacial fluids are defined by the excess surface free energy. The appropriate definition of this free energy for an assembled bilayer in solvent is the free energy difference between the bilayer–solvent system of interest and a state that contains only the pure solvent, at the same density as the solvent density in the bilayer system far from the bilayer. This pure solvent density is not known a priori, but is observed as the solvent density far from the bilayer in any inhomogeneous solution. The excess surface free energy is given by

$$\Omega^{ex} = \frac{(\Delta\Omega[\rho(\mathbf{r})] - \Delta\Omega^{s})}{A}, \tag{1.39}$$

where $\Delta\Omega$ is calculated from Eq. (1.18), $\Delta\Omega^{s}$ is the difference in free energy between the pure solvent and the bulk homogeneous reservoir, and A is the total area.

Interestingly, the excess free energy, Ω^{ex}, is also exactly the net surface tension across the bilayer:

$$\gamma = \Omega^{ex}. \tag{1.40}$$

In biological membranes, the net surface tension across the bilayer is zero. If the surface tension were positive, the bilayer would shrink, whereas if negative the bilayer would expand. The stable point is thus when $\gamma = \Omega^{ex} = 0$.

The basic structural quantities of interest for a fluid bilayer are the bilayer thickness t and the density of lipids in the bilayers, usually reported as the area per lipid A_L. Reported thicknesses are given as the distance between headgroup density peaks, while A_L is calculated from

$$A_L = (\Gamma_L / A)^{-1} \equiv \left(N_\alpha^{-1} \int \rho_\alpha(x)dx\right)^{-1}, \tag{1.41}$$

where $\rho_\alpha(x)$ is the density profile of any single site type α on the lipid and Γ_L is the total number of lipid molecules in the system.

In a self-assembling system such as our lipid/solvent model, it is possible to have multiple solutions to the DFT equations. To explore these solutions, we used a pseudo-arclength continuation algorithm of Keller [75] as implemented in the LOCA library of Salinger et al. [76]. Given one solution to the equations, this algorithm will follow a curve of solutions as a function of some parameter in the model, for example, as a function of temperature T. The algorithm is capable of following metastable branches of solutions, finding unstable (spinodal points), and turning around these points. We can thus find both metastable branches of the solution curve and also the unstable branches that often connect two stable or metastable branches [62]. This algorithm was instrumental in understanding the thermodynamic behavior of our system [14], since a given initial guess does not necessarily lead to the lowest free-energy solution. Additionally, the algorithm can be used to find the coexistence points between bilayers with a large excess pure solvent region and a pure solvent solution, which are the points where $\gamma = 0$. This allows us to find all zero net tension bilayers as a function of temperature in an automated way.

1.6. LIPID BILAYER STRUCTURE

One of the principal results of a Fluids-DFT calculation are the equilibrium fluid density profiles $\rho_\alpha(r)$ for all the species in the system. The density profile for a lipid bilayer using the model system of Figure 1.3 at the state point $kT/\varepsilon = 1.3$ and $\rho_b\sigma^3 = 0.59$ is shown in Figure 1.4 [14]. The state point was chosen so that there is zero net tension across the bilayer, $\gamma = 0$. The density profile has the qualitative shape one would expect for the lipid bilayer. The tail beads are in the inner part of the bilayer, the headgroups are clustered at the bilayer–solvent interface, and the solvent is excluded from the bilayer. The density profile does not exhibit

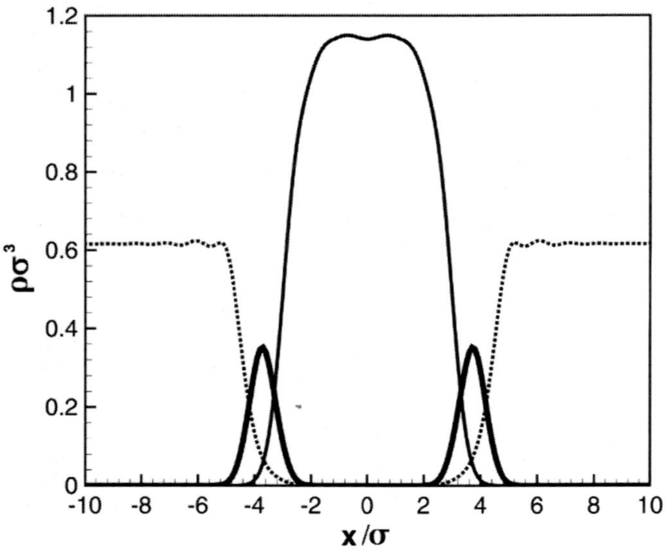

Figure 1.4. Density profile for a zero tension bilayer, showing the headgroups (bold solid curve), tail groups (solid curve), and solvent (dotted curve). The distance between headgroup peaks on either side of the bilayer is found to be 7.4σ. The state point is $\{x_s = 0.415, \rho_r\sigma^3 = 0.59, kT/\varepsilon = 1.3\}$.

significant packing structure, indicating that the bilayer is in a fluid state. The Fluids-DFT density profile compares well with molecular dynamics simulation results on the same model lipid system [15].

Recall that in Fluids-DFT, because we are in an open ensemble, the chemical potential is the input variable and the number of molecules in the system is an output of the calculation. The number of lipids in the bilayer, as well as the solvent density outside it, can be immediately calculated from the density profile. Figure 1.5A shows the area per lipid, calculated from Eq. (1.41), as a function of temperature, while Figure 1.5B shows the bilayer thickness. Along the physically realistic $\gamma = 0$ line, we see that, as the bilayer thickness increases, the area per lipid decreases. The area per lipid decreases roughly linearly with temperature, in agreement with experiment [77] and previous simulations [78].

We can compare the structure of our model bilayers to experiment by a rough mapping of the coarse-grained model. The ratio of headgroup volume to tail volume in our lipid is similar to that in dipalmitoylphosphatidylcholine (DPPC); if we assume that the volume of one of our tail beads is equivalent to two CH_2 groups in DPPC, we can identify $\sigma = 4.75$ Å. Here we have used a volume of 28.1 Å^3 for the CH_2 groups in DPPC [79]. The thicknesses and the areas per lipid can then be compared with values in the literature. For fully hydrated DPPC at 50°C, the head-to-head distance $t = 38.3$ Å and the area per lipid $A_L = 64$ Å^2 [80]. These conditions should apply to our fluid bilayers at zero tension. Using our mapping for σ, we find that the thickness for bilayers in this model varies from 38 to 35.2 Å, and the area per lipid from 77.4 to 110.3 Å^2 for zero tension bilayers where the temperature ranges from $kT/\varepsilon = 0.92$ to 1.3. The best fit to the DPPC data is thus at low temperatures in the model.

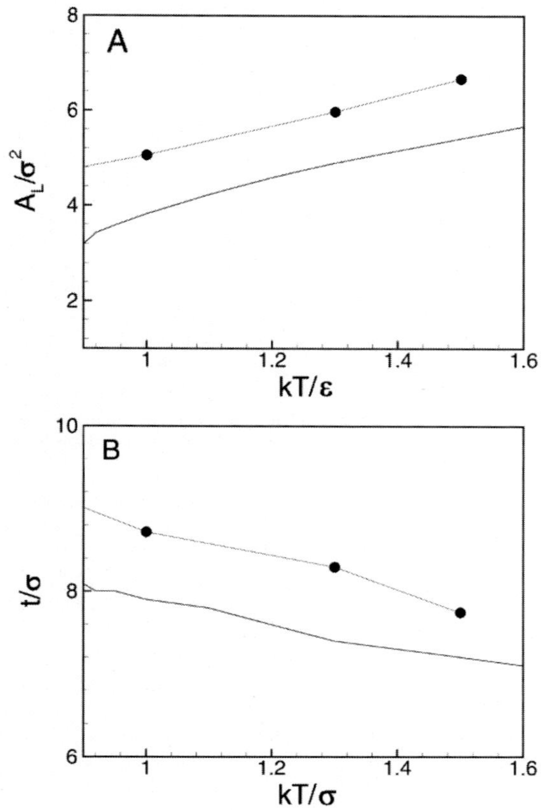

Figure 1.5. Bilayer properties for the coarse-grained lipid models: (A) the area per lipid, and (B) the bilayer thickness. The thick solid line shows the Fluids-DFT results along the γ = 0 line. These results are compared to those from MD simulations, shown in the filled circles. Modified from [15].

1.6.1. Variations on the Model

Since the model system introduced in Section 1.4 is rather arbitrary, one might be able to obtain a closer agreement in bilayer structure between DFT and experiment by modifying the parameters in the model. We have explored this possibility to some extent by calculating bilayer properties for two additional models.

Model II

Consider lipid chains still consisting of tangent spheres, with $\sigma_h = 1.44\sigma$ and $\sigma_t = \sigma$, but with a more realistic size for the solvent spheres. For DPPC, the total volume of the headgroup is 320.9 Å^3, while the volume of a water molecule is about 30 Å^3, so the ratio of water volume to headgroup volume for this case is about 0.093 [79]. If we enforce the same ratio in our model lipid, we obtain $\sigma_s = 0.82\sigma$. For simplicity, we explore this model keeping the total site density fixed at $\rho_b\sigma^3 = 0.59$ as before. This means that at the same value of x_s, the actual volume of solvent will be smaller than for the larger solvent we used previously.

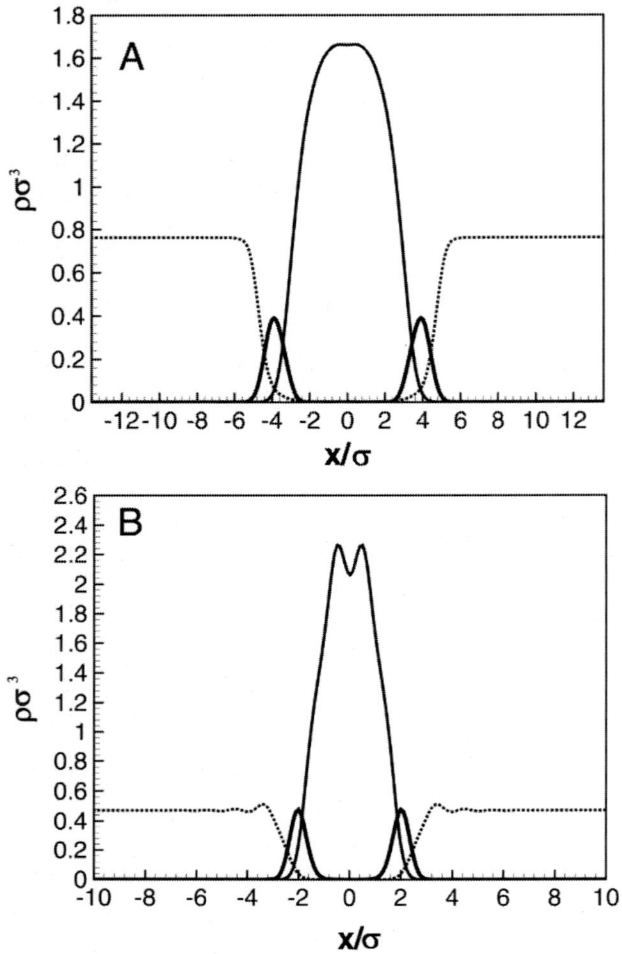

Figure 1.6. (A) Model II: density profile for $\gamma = 0$ at the state point $\{kT/\varepsilon = 1.73, x_s = .5439\}$. (B) Model III: density profile for $\gamma = 0$ at the state point $\{kT/\varepsilon = 2.0, x_s = .3037\}$.

A density profile for Model II is shown in Figure 1.6A, and is qualitatively similar to profiles obtained using the original model, such as the one in Figure 1.4. The area per lipid and bilayer thickness for this model are shown in Figure 1.7. The discontinuous jumps in t are due to inaccuracies in determining the exact distance between headgroup peaks from the density profiles (calculated with a mesh size of 0.1σ). The temperature dependence is different than was found from the original model in Figure 1.5, so changes in the model do result in somewhat different behavior, although the trend of increasing area per lipid with increasing temperature remains.

Model III

Here we consider lipid chains in which the monomers are allowed to overlap. This should give a more realistic representation of the lipid molecule, since generally bond lengths are considerably shorter than van der Waals radii in real molecules. The bond length between CH_2 groups is approximately 1.54 Å, so the effective bond length between sets of $2CH_2$ groups is about 2.58 Å. Assuming that each tail bead represents two CH_2 groups, we can identify $\sigma = 4.75$ Å by equating volumes. A bond length of 2.58 Å then corresponds to $l_{tt} = 0.54\sigma$. We arbitrarily set the head–head and head–tail bond lengths to have the same ratio of $l_{tt}/\sigma = 0.54$, so that $l_{hh} = 0.78\sigma$ and $l_{th} = 0.66\sigma$.

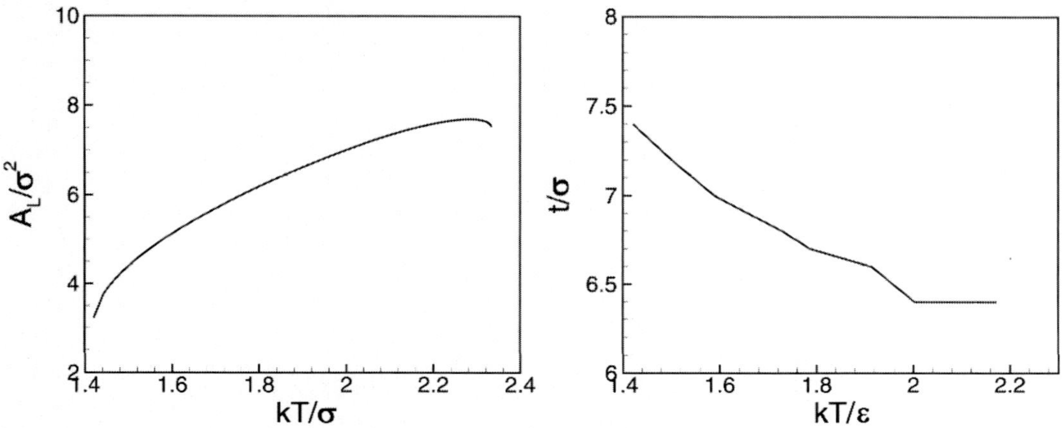

Figure 1.7. Model II: area per lipid A_L and thickness t along the $\gamma = 0$ line for a single bilayer as a function of temperature.

Since the sites in the lipid molecules overlap, the lipids occupy a smaller total volume than before. The volume can be calculated following work by Abrams and Kremer [81]. The lipid now has a volume of about $8.26\sigma^3$, compared to the tangent site lipids that have a volume of about $11.5\sigma^3$. One way to choose the bulk density for this model is to maintain the same lipid packing fraction as at the bilayer–lamellar–solvent triple point in the tangent site model, which gives a total site density of $\rho_b\sigma^3 = 0.728$. However, the Fluids-DFT code had numerical difficulties at this density, and so here we give results at a higher bulk site density of $\rho_b\sigma^3 = 0.825$.

A typical density profile for Model III at $\rho_b\sigma^3 = 0.825$ is shown in Figure 1.6B. The bilayer is considerably more narrow than before and exhibits a pronounced depletion in tail density in the center of the bilayer. The densities in the tail regions are high because the bilayers are so narrow. We anticipated that Model III might work better for comparing bilayer properties to those of real lipids, since in this model the volumes and bond lengths in the lipid molecule are consistent with reasonable values. For zero-tension bilayers we find areas per lipid in the range of $A_L/\sigma^2 = 3 - 4\sigma^2$, which corresponds to areas of 67.7 to 90.2 Å^2. The thickness of the bilayers is in the range of 5σ, which corresponds to a thickness of 23.8 Å. The areas per lipid are well within the experimental range for phospholipids such as DPPC, but the thickness is too small. This is presumably due to the excess flexibility of the model lipid tails.

We conclude that simultaneously obtaining a physically realistic area per lipid and bilayer thickness requires introduction of some chain stiffness into the model. The random-walk chains are too coiled and lead to overly thin bilayers. Nevertheless, the simple model of Section 1.4, or small variations on it, does predict bilayers with properties reasonably close to those of experimental lipid bilayers.

1.6.2. Variations in the Theory: Lipids with i-SAFT

Jain et al. [67] have performed some preliminary calculations using their i-SAFT DFT and the lipid model described in Section 1.4. They first calculated the density profiles for a melt of the lipid molecules (without solvent) and a lipid/solvent solution near a flat surface, in order to compare with previous MD simulations of these systems [15]. The agreement with the MD simulations is very good, and is better than the results for the same calculation done with the CMS-DFT. The improvement is due at least in part to the more accurate treatment of the solvent in i-SAFT compared to the CMS-DFT. Essentially, the CMS-DFT was not constructed to treat single sites very accurately, and does not give the correct equation of state for the solvent [15], whereas the i-SAFT theory reduces to the hard sphere DFT for the single-site solvent, which is known to be very accurate.

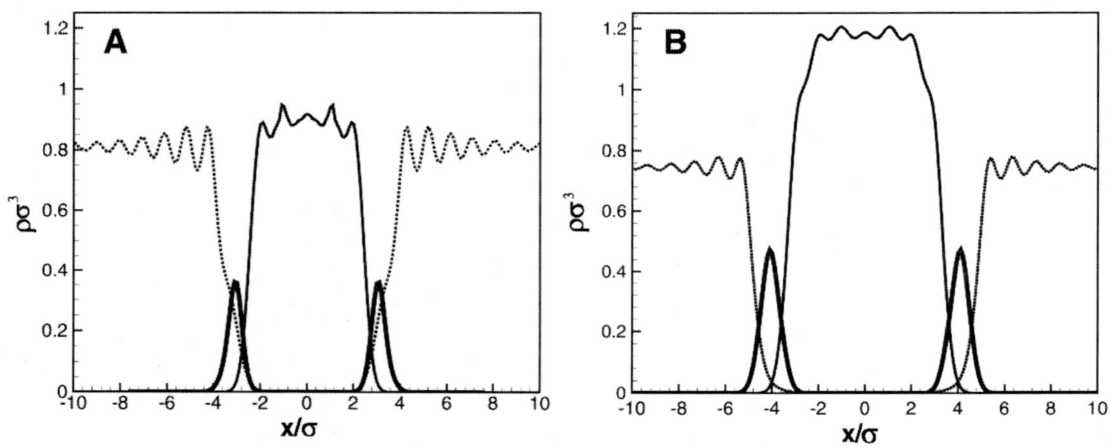

Figure 1.8. Bilayer density profiles from (A) the i-SAFT DFT [67] and (B) the CMS-DFT [15] for the state point { $\rho_b\sigma^3$ = 0.68, x_s = 0.442, T = 1}.

A prediction for the lipid bilayer structure obtained from i-SAFT at the state point { $\rho_b\sigma^3$ = 0.68, x_s = 0.442, T = 1} is shown in Figure 1.8, along with the bilayer density profile for the same state point as obtained from the CMS-DFT. The i-SAFT DFT predicts a somewhat different bilayer profile than the CMS-DFT. This is not surprising given the different approximations in the two functionals. There is more ordering in both profiles than in the profile shown in Figure 1.4, due to the lower temperature. The solvent penetrates the entire headgroup region in the i-SAFT profile, in contrast to the CMS profile. Additionally the i-SAFT profile is thinner and has a considerably lower tailgroup density. Further research is required to elucidate the relative merits of these two different DFTs for studying membranes.

1.6.3. Effects of Added Alcohols on Lipid Structure

When additional components are added to the lipid bilayer, they perturb the structure. One direct use of the Fluids-DFT is to make predictions of how the bilayer structure changes with various added components. All biological membranes are composed of a mixture of different lipids, along with a variety of other molecules, such as cholesterol and, of course, membrane-associated proteins.

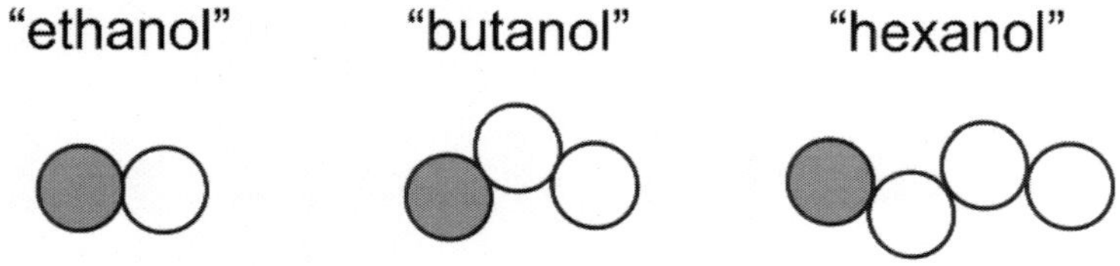

Figure 1.9. Sketch of model alcohols, with OH groups in gray and tail beads in white.

As a first step to investigating the effects of mixtures of surfactant-like molecules, we investigated the effects of adding alcohols to a lipid bilayer [16]. Since short-chain alcohols are amphiphilic, perhaps not surprisingly they have been found in both experiment and atomistic simulation to localize primarily in the headgroup region of the lipid bilayer [82–85]. Figure 1.9 gives a sketch of three model coarse-grained alcohol molecules — ethanol, butanol, and hexanol. Each alcohol consists of a headgroup (the OH group) and a number of tail beads, with the sizes of both beads equal to σ. The alcohol headgroup bead is thus somewhat smaller than that of the lipid headgroup, while the tail beads are the same size. The tail and headgroup interactions are the same as for the lipid model described in Section 1.4.

Using this simple model, we found that the alcohols are indeed located preferentially near the headgroup-tailgroup interface, as shown in the density profiles in Figure 1.10. As the tailgroup gets longer going from ethanol to butanol to hexanol, the alcohol tails penetrate further into the tail region of the bilayer. In Figure 1.10, the aqueous ethanol concentration x_a is 3.4%, and thus it has a significant contribution to the density in the aqueous solution outside the bilayer, whereas the other aqueous concentrations are much smaller — 0.12% for butanol and only 0.00048% for hexanol. These densities are so low because most of the longer alcohols adsorb into the bilayer. The longer the tail, the less alcohol could be accommodated in the bilayer; as the alcohol concentration was increased, eventually there were no more bilayer-like solutions to the Fluids-DFT equations, indicating that the bilayers were not stable above a certain concentration of alcohol. This effect is also seen experimentally. The density profiles in Figure 1.10 are at the upper limits of alcohol concentration. Density profiles at lower alcohol concentrations are similar, with lower peaks in the alcohol density profile as x_a decreases.

Since the alcohols are adsorbed in the interfacial region, one might expect that the area per lipid A_L would increase as alcohol is adsorbed in the bilayer. This is in fact what Longo et al. infer from their data on alcohol adsorption in SOPC vesicles [86]. The Fluids-DFT also predicts an increase in A_L as alcohol is added in qualitative agreement with experiment. It is difficult to compare number densities directly between theory and experiment since the coarse-grained model does not map completely onto the physical system. However, experimentally there is a maximum concentration for which the vesicles are stable enough for micropipette aspiration. For comparison purposes, we compare values between this maximum experimental concentra-

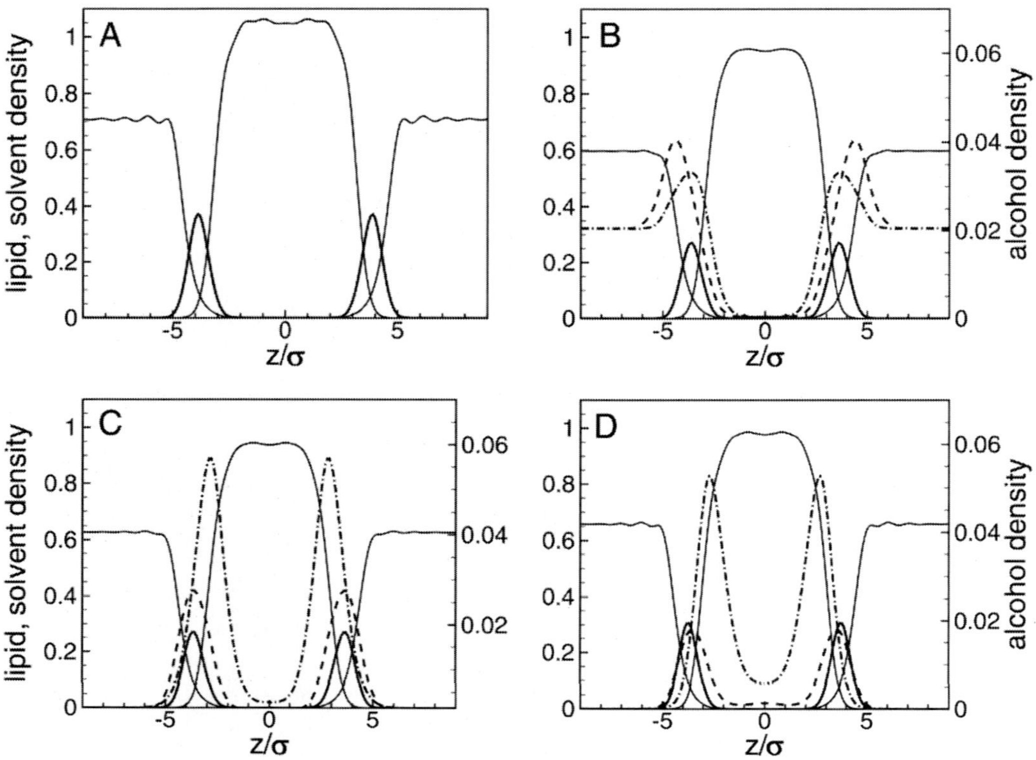

Figure 1.10. Density profiles for bilayers with added alcohols: (A) pure lipid bilayer; (B) at 3.4% ethanol; (C) at 0.12% butanol; and (D) at 0.00048% hexanol. The curves show the lipid tailgroups (solid black curves), lipid headgroups (thick gray curves), solvent (thin gray curves), OH groups (dashed curves), and alcohol tails (dash-dotted curves). Reprinted with permission from [16]. Copyright © 2006, The Biophysical Society.

tion and the maximum concentration at which we also obtain stable bilayer solutions to the DFT. Thus we see changes in the area per lipid, relative to the area per lipid with no alcohol A_{L0}, of $\Delta A_L/A_{L0}$ between 16 and 24% at the upper limits of bilayer stability [15], which is similar to Ly and Longo's results of about a 16% change for ethanol and a 27% change for butanol [86]. The shape of the $\Delta A_L/A_{L0}$ vs. concentration curves is also qualitatively similar to Figure 12 in [86].

Interestingly, the thickness of the lipid bilayer did not decrease as much as expected from the increase in A_L. We found a decrease in thickness of roughly 0.4σ at the maximum ethanol concentration and somewhat less for the butanol and hexanol. The thickness of the pure bilayer at $x_a = 0$ is 7.7σ, so the change in thickness is at most about 5%, which is considerably less than the change in A_L [15]. We are not aware of any independent experimental data on the bilayer thickness as a function of alcohol concentration, so currently this issue is unresolved.

1.6.4. Effects of Peptides on Lipid Structure

The Fluids-DFT can also be used to investigate the effect of added transmembrane proteins or peptides to the lipid bilayer. Any such study will need to begin with a suitable coarse-grained model of the protein or peptide of interest. To date, most coarse-grained peptide models have treated the peptides as rigid geometrical shapes, and studied how such an object perturbs the lipid bilayer. Recently, more atomistic coarse-grained models of proteins/peptides have been developed [5,87], and it will be an interesting future research endeavor to see if these can be incorporated into molecular theories.

Here we discuss one particular kind of peptide assembly with biological relevance, namely peptides that can assemble and form pores in membranes. Often pore-forming peptides are amphipathic cationic α-helices. The pores formed by these peptide assemblies are the basis for toxicity for a wide range of antimicrobial peptides (AMPs), and are part of the mechanism used by some toxins and viruses in penetrating cell membranes. Many experiments have been performed to investigate the structure of a lipid bilayer in the vicinity of a transmembrane pore formed from an assembly of α-helices [88–90]. The results of these experiments suggest that many AMP assemblies form a toroidal (or wormhole) membrane geometry where the headgroups of the lipid chains are found lining the pore in order to form an interface between the solvent-filled pore and the hydrocarbon lipid tails [90]. In this geometry, the bilayer bends around the peptide assembly. A second hypothesized geometry is a "barrel-stave" pore, in which the membrane does not deform much and just surrounds the peptide assembly.

To investigate the possible geometries, we performed calculations [17] for an assembly of model peptides embedded in our model lipid bilayer, at $kT/\varepsilon = 1$ and at zero net tension $\gamma = 0$. Like previous researchers, we modeled the peptide fragments as cylinders. The cylinders have a diameter of $d = 2\sigma_t \approx 9.5$ Å, where we use the mapping discussed in Section 1.6 of $\sigma_t \approx 4.75$ Å. We considered a hexagonal arrangement of six cylinders, all oriented normal to the bilayer. The distance a between the centers of neighboring cylinders ranged from $a = d$ to $a = 3.25d$. The cylinders are treated in the CMS-DFT as external surfaces, rather than fluid components. They thus exert an external field $V(\mathbf{r})$ on the lipid and solvent species that consisted of purely hard (volume exclusion) interactions.

Figures 1.11 and 1.12 show density profiles for each of the three fluid species (tailgroups, headgroups, and solvent) in the system, in two slices through the three-dimensional computational volume. The first set of slices (Fig. 1.11) are taken through the center of the bilayer, parallel to the lipid headgroup–solvent interface, but perpendicular to the assembly of peptides. The second set of slices (Fig. 1.12) are taken perpendicular to the lipid headgroup–solvent interfaces and parallel to the long axis of the peptide assembly. The various rows of panels in each figure show peptide assemblies with varying lattice constants. In all cases, significant perturbation to the uniform membrane is observed.

The rows in each of these figures show distinctly different morphologies for the lipid bilayer in the vicinity of the peptides as the assembly size is varied. When $a = 30.875$ Å, there is no membrane spanning pore although the structures of both the bilayer and the solvent in the vicinity of the peptide assembly are clearly perturbed. In contrast, pores are found for both $a = 16.625$ Å and $a = 9.5$ Å. However, the specific structure of the membrane–pore interface is markedly different in the two cases.

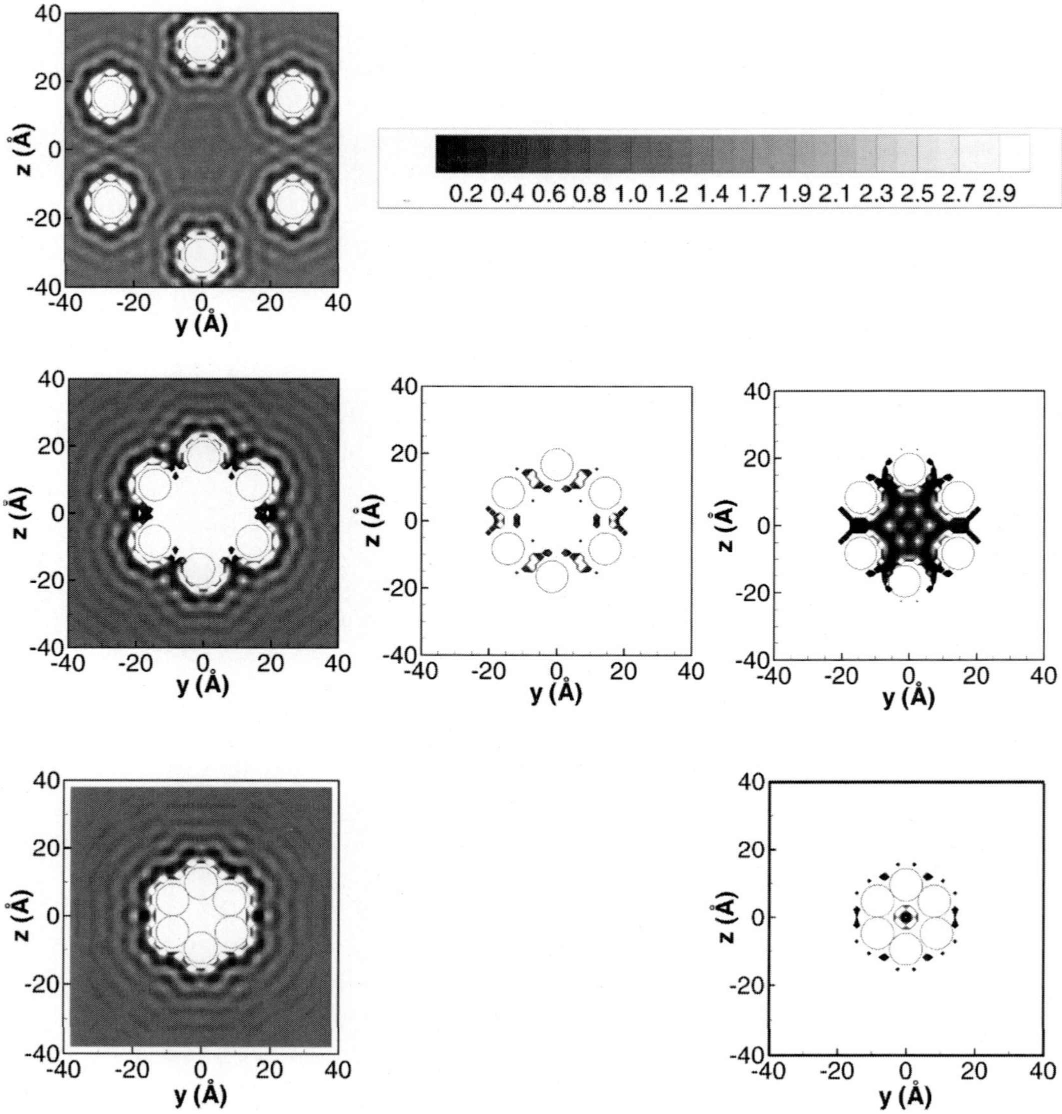

Figure 1.11. Density contours (in reduced units of $\rho\sigma^3$) for slices perpendicular to the long axis of the peptides. The slice is taken through the center of the bilayer. The top figure has a lattice constant of a = 30.875 Å. the second row has a = 16.625 Å, and the third row has a = 9.5 Å. The first column shows density distributions for the lipid tailgroups, the second column shows lipid headgroups, and the third column shows the solvent densities. Note that the three missing panels had zero densities in this plane. The contour legend applies to all plots. Regions where the density is smaller than $\rho\sigma_i^3 = 0.1$ for heads and tails or $\rho\sigma_i^3 = 0.001$ for solvent are blanked to white to show clearly the regions where there is no fluid (i.e., in the space occupied by the peptides). The computational domain for all cases was 1/4 of the displayed figure with reflective boundaries applied at $y = 0$ and $z = 0$. The total size of the computational domains was {x = 42.75 Å, $y = x$ = 57 Å} for the a = 30.875 Å assembly, {$x = y$ = 42.75 Å} for the a = 16.625 Å assembly, and {x = 42.75 Å, $y = z$ = 38 Å} for the a = 9.5 Å assembly. Modified from [17].

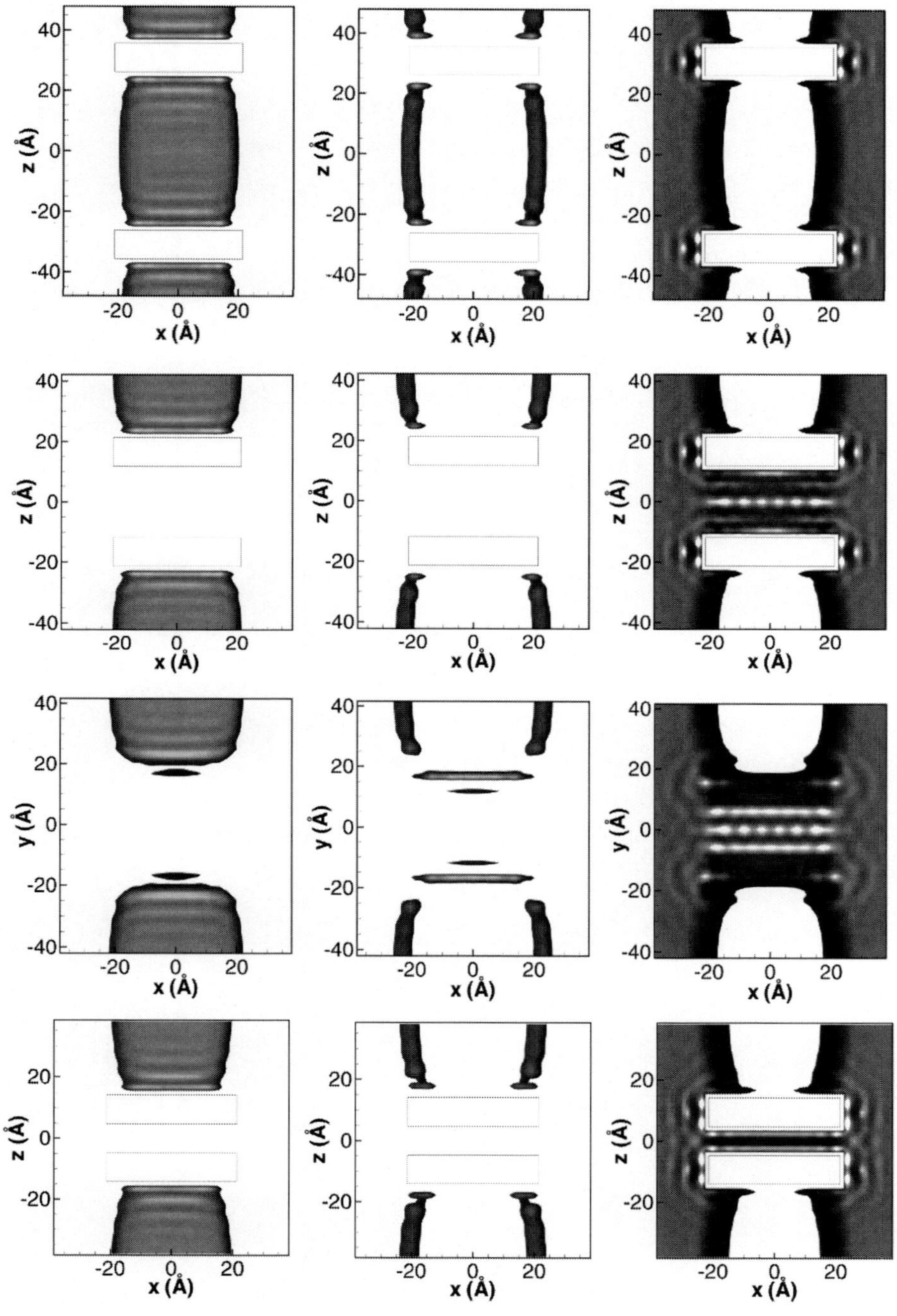

Figure 1.12. Density contours for slices parallel to the long axis of the peptides. The slices in the first, second, and fourth rows were taken in the plane $y = 0$ where the plane intersects two of the cylinders of the assembly (see Fig. 1.1). The slice in the third row is taken in the plane $z = 0$ between the cylinders in order to demonstrate the toroidal nature of the headgroup distribution. The top row of figures has a lattice constant of $a = 30.875$ Å, the second and third rows have $a = 16.625$ Å, and the fourth row has $a = 9.5$ Å. Column ordering and the color contour is identical to Fig. 1.11. Modified from [17].

For the a = 9.5 Å assembly, a barrel-stave type pore is found. No headgroups are found within the pore volume, and solvent completely fills the pore. The solvent that fills this very narrow pore is at a significantly higher density than the bulk fluid. The pore size essentially allows for a single-file arrangement of solvent particles in the pore, although the hard boundary used to describe the peptides results in the most probable solvent positions to be in a ring defined by the closest approach of the solvent particles to the surface.

The intermediate assembly (a = 16.625 Å) is perhaps the most interesting case since the peptides are far enough apart that a barrel-stave type assembly would require tailgroup–solvent contact. These unfavorable interactions are avoided by rearrangement of the lipids in the vicinity of the peptide assembly. This rearrangement allows headgroups to line the pore in the gaps between the peptides and form a barrier between the tailgroup beads and the solvent. The continuity of the headgroup–solvent interface through the pore is seen in Figure 1.12 (third row, center column). This is essentially a toroidal pore morphology.

In order to assess the stability of various peptide assemblies, we computed the surface excess free energy, Ω^{ex} of the peptide–membrane assembles as the difference between the grand free energy of the system with the peptide assembly, Ω, and the grand free energy of the uniform membrane at zero tension, Ω_{zt} :

$$\Omega^{ex} = \Omega - \Omega_{zt} . \qquad (1.42)$$

The solid line in Figure 1.13 shows this excess free energy per cylinder (Ω^{ex}/N) as a function of the lattice spacing of the assembly. The free energy minimum is found at a = 16.625 Å, where the toroidal pore morphology was found. The barrel-stave pores are only found for very tightly packed assemblies, and they have a high energy relative to isolated cylinders (the $a \to \infty$ limit).

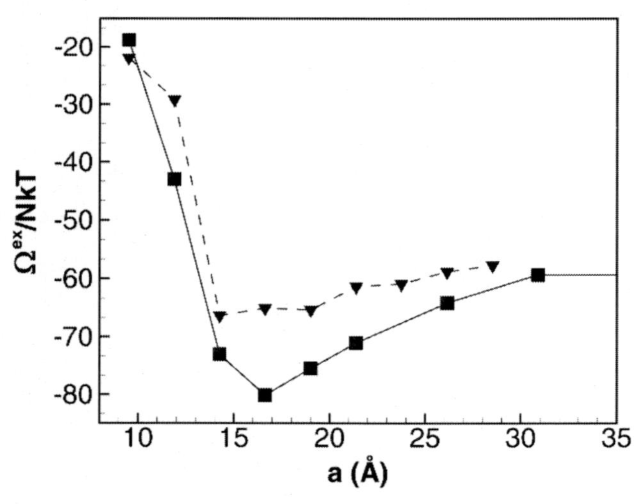

Figure 1.13. The excess surface free energy of a hexagonal assembly of six peptides (N = 6) as a function of the lattice constant of the assembly. The solid curve (squares) is the multibody free energy, while the dashed line (triangles) is an approximate free energy curve based on a superposition approximation for pairwise solvation interactions. The lines are simply a guide to the eye. Modified from [17].

The surface excess free energy reported here is closely related to a *multibody* potential of mean force (PMF), W_N, defined as

$$W_N(r) = \Omega^{ex} - \Omega^{ex}(\infty), \tag{1.43}$$

where $\Omega^{ex}(\infty) = N\Omega_1^{ex}$ is the free energy of the peptide–membrane assembly when the peptides are infinitely far apart, but still in the membrane. This limit is computed based on the free energy Ω_1^{ex} of a single peptide multiplied by the number of peptides in the assembly, $N = 6$.

One strategy for simulation of many peptides in a membrane would involve constructing a pairwise additive potential of mean force, W_{ij}. Here we consider the accuracy of a PMF constructed from the interaction of just two peptides in the membrane for comparison with the multibody PMF as defined above. Note that these calculations also require a series of 3D-DFT calculations. Given the pairwise PMF, the *approximate* energy of an N-mer assembly would be

$$\Omega^{ex}(r) \approx N\Omega_1^{ex} + \sum_{i<j}^{N} W_{ij}(r) \tag{1.44}$$

using a superposition approximation. Figure 1.13 also includes this pair potential–based assembly energy for an $N = 6$ hexagonal assembly in the dashed line. The pair potential–based assembly energy underpredicts the deep free energy minimum associated with the formation of the membrane spanning pore, but it does have a similar shape to the full multibody energy.

1.7. MECHANICAL PROPERTIES OF BILAYERS

In addition to structure, an important result of Fluids-DFT calculations is the free energy of the equilibrium system. In the context of bilayers, several mechanical properties can be extracted from either the free energy itself or from the free energy density $\Omega(x)/V$. The mechanical quantity that is the most straightforward to calculate is the area compressibility modulus K_A of the membrane. This modulus can be obtained from the dependence of the surface tension (excess surface free energy) on the area per lipid near the zero tension point,

$$\gamma \approx K_A\left(A_L - A_{L0}\right)/A_{L0}, \tag{1.45}$$

for small deviations of A_L from its value A_{L0} at $\gamma = 0$. In the CMS-DFT predictions for single-component bilayers, the solvent density far from the bilayer is nearly constant as we change the area per lipid by changing x_s [14]. Thus, $\gamma(A_L)$ can be obtained simply from arc-length continuations done at constant temperature. For larger magnitudes of A_L, γ becomes nonlinear and saturates at large values of A_L. Near the equilibrium area per lipid we find that γ is linear in A_L and can extract $K_A = 9.72\varepsilon/\sigma^2$ for $kT/\varepsilon = 1.0$ and $K_A = 5.31\varepsilon/\sigma^2$ for $kT/\varepsilon = 1.3$. As expected, the compressibility is lower (higher modulus) at lower temperatures.

Further elastic constants could be calculated from the Fluids-DFT. These include the bending modulus κ and the splay modulus $\bar{\kappa}$. However, calculating these moduli from Fluids-DFT would require solutions in curved geometries [12,91], which we have not yet performed. These moduli are the principal ingredients of membrane elasticity theories. A multiscale approach to modeling a particular membrane system could thus entail calculating all the elastic moduli from DFT using a molecular model, and then passing these parameters to a mesoscale elasticity theory.

A final mechanical quantity of interest is the lateral pressure profile, or stress profile, across the membrane. The bilayer is an inhomogeneous environment, with different interactions among the molecules at different positions across the bilayer. If one imagines inserting a surface across the membrane, the lateral forces (parallel to the bilayer) on that surface would be different at different positions in the bilayer. These differing forces can in fact influence the behavior of transmembrane proteins. For example, certain ion channels are *mechanosensitive* and are believed to be gated by changes in the lateral pressure exerted on the protein by the lipid bilayer [92,93]. More generally, it is thought that changes in membrane composition and properties alter the shape of the lateral pressure profile, which then alters the amount of mechanical work associated with conformational changes in membrane proteins [94,95,34,96]. As of yet there are no direct experimental probes of lateral pressure profiles in membranes [97]. This makes calculation by molecular models such as Fluids-DFT attractive.

Normally the lateral pressure profile $\pi(x)$ is defined as [98]

$$-\gamma = \int_{-\infty}^{\infty} dx \pi(x) = \int_{-\infty}^{\infty} dx \left[P_T(x) - P_N(x) \right], \tag{1.46}$$

where the lateral pressure profile across the bilayer is $\pi(x) = P_T(x) - P_N(x)$, x is the direction normal to the bilayer, $P_N(x)$ is the normal component of the pressure tensor (a constant for all x as dictated by mechanical equilibrium), and $P_T(x)$ is the tangential component of the pressure tensor. (Note that one can also define a stress profile $s(x)$ as $s(x) = -\pi(x)$). The pressure tensor can be calculated in molecular dynamics simulations directly from the virial expression for the pressure, and recently several groups have calculated the lateral pressure profile from atomistic simulations of lipid bilayers [99,93].

Thermodynamically, the grand free energy density $\Omega(x)/V$ can also be identified with the lateral pressure $P_T(x)$ [100,101], and thus a stress profile is rather straightforward to compute with a Fluids-DFT. (The stress profile is also related to the curvature elasticity of the membrane [12,91].) Using the energetic route in the CMS-DFT based on Eqs. (1.39), (1.40), and (1.18), we can identify the lateral pressure profile with the excess surface free energy density, i.e., $\pi(x) = -\Omega^{ex}(x)$, so that

$$\pi(x) = \left(\rho_s(x) - \rho_{b,s} \right) + \frac{1}{N_L} \sum_{\alpha\{L\}} \left(\rho_\alpha(x) - \rho_{b,\alpha} \right)$$
$$-\frac{1}{2} \sum_{\alpha\beta} \int dx' c_{\alpha\beta}(|x - x'|) \left[\rho_\alpha(x) \rho_\beta(x') - \rho_{b,\alpha} \rho_{b,\beta} \right] + S, \tag{1.47}$$

where S is a constant arising from the pure solvent contribution, and the first sum is taken over the beads in the lipid. In a pure solvent $P_T \to 0$, and so S is the isotropic pressure in a pure solvent where

$$S = -\left(\rho_{ps,s} - \rho_{b,s} \right) + \frac{\rho_{s,ps}^2}{2} \int dx' c_{ss}(|x - x'|) + \frac{1}{2} \sum_{\alpha\beta} \int dx' c_{\alpha\beta}(|x - x'|) \rho_{b,\alpha} \rho_{b,\beta}, \tag{1.48}$$

and $\rho_{ps,s}$ is the density of the solvent in a solution that has a pure uniform solvent.

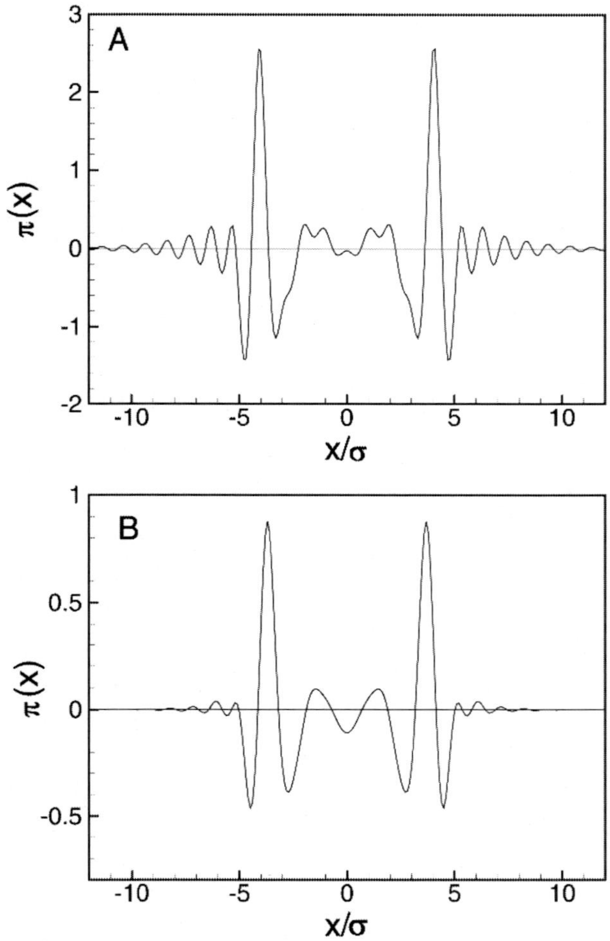

Figure 1.14. Lateral pressure profiles calculated from the CMS-DFT at (A) $\{kT/\varepsilon = 1,\ \sigma^3 = 0.68,\ \gamma = 0\}$ and (B) $\{kT/\varepsilon = 1.3,\ \rho_b\sigma^3 = 0.59\}$.

Figure 1.14 shows lateral pressure profiles at two different temperatures calculated from Eq. (1.47) and shifted by a constant so that $\pi(x) = 0$ in the bulk solvent region. The lateral pressure profiles have several different peaks. The physics underlying the spatial variations in $\pi(x)$ is fairly clear. As it must be, the total lateral pressure is zero in the solvent region, and reaches a negative peak at the solvent/headgroup interface corresponding to the solvent–head (SH) interfacial tension. The adjacent positive (compressive) region is associated with the attractive head–head (HH) potential, and is followed by a second negative peak at the head–tail (HT) interface corresponding to the HT interfacial tension. There is a fair amount of smaller-scale structure in the lateral pressure profiles, due to the density oscillations found in the density profiles. The integral under the $\pi(x)$ curve is zero since the net tension across the bilayer is zero.

Several previous authors have also calculated lateral pressure profiles from mean-field theories of lipid bilayers [102,103,34]. All of these theories have assumed that the lipid tails form an incompressible liquid, and furthermore they calculate only the contribution of the tails to the lateral pressure profile. The general result is that there is a peak in the lateral pressure, $\pi(x)$ in the middle of each leaflet of the bilayer, with the lateral pressure falling in the middle of the bilayer and near the tail/head interface. This is also roughly what we see in the tail region of the bilayer. In the previous theoretical work, the lateral pressure was found to be positive throughout the tail region, as it must be since it is calculated as the constraint that must be applied to impose incompressibility and thus a constant density through the tail region. In contrast, we find a dip below zero in $\pi(x)$ in the tail region in the Fluids-DFT for $kT/\varepsilon = 1.3$, indicating that the ends of the tails are under tension in our model for this state point.

1.7.1. Effects of Alcohols on Mechanical Properties

We saw above that short-chain alcohols sit near the headgroups in the lipid bilayer. This disrupts the packing in the lipid bilayer and leads to a variety of changes, among them experimentally observed increases in membrane fluidity [104], in membrane permeability [105], and in lipid lateral mobility [106]. Ly and Longo found that alcohol partitioning into SOPC lipid bilayers follows Traube's rule: for each additional alcohol CH_2 group, the concentration required to reach the same interfacial tension is reduced by a factor of three [107]. As discussed above, this reduction in interfacial tension could have a substantial impact on intrinsic membrane proteins. In fact, a recent experimental study of the dissociation of the tetrameric potassium channel KcsA by small alcohols suggested that the alcohols interacted with the protein via changes in the lateral pressure profile [108].

In Figure 1.15 we show the effect of the alcohols on the area compressibility modulus K_A of the bilayer. K_A decreases with increasing alcohol concentration and also with increasing alcohol chain length. Decreases in K_A at the highest x_a values were about 44% for ethanol and butanol, and about 29% for hexanol. The magnitude of the decrease is similar to the 35% reduction in K_A measured by Ly and Longo (see Fig. 5 of [86]).

The alcohols also affect the lateral pressure profiles $\pi(z)$, where now z is the direction normal to the bilayer. The profiles obtained upon varying the alcohol concentrations are shown in Figure 1.16. All of these bilayers are still at zero net tension, $\gamma = 0$, so the area under all of the $\pi(z)$ curves integrates to zero. Although the shape of $\pi(z)$ remains similar, the alcohols reduce the magnitudes of all the peaks, both the pressures and the tensions, with increasing alcohol concentration. Similar effects of ethanol on the lateral pressure were recently seen in atomistic MD simulations [109]. Because the thickness of the bilayer decreases somewhat, the profiles become somewhat more narrow so the peaks shift toward the center of the bilayer. The alcohols have almost no effect in the very center of the bilayer.

Note that the changes in $\pi(z)$ induced by alcohol adsorption are asymmetric (in each leaflet), in that the change is greater in some regions of the bilayer than in others. These changes in the pressure profile could potentially influence the conformational state of intrinsic membrane proteins, and hence provide a nonspecific mechanism for the biological effects of alcohol.

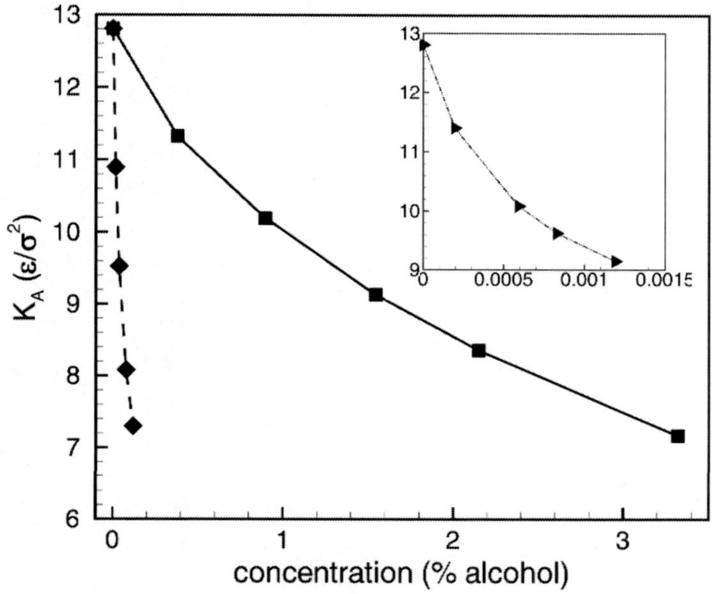

Figure 1.15. Area compressibility modulus as a function of the alcohol concentration in the bulk for ethanol (squares), butanol (diamonds), and hexanol (triangles, inset). Reprinted with permission from [16]. Copyright © 2006, The Biophysical Society.

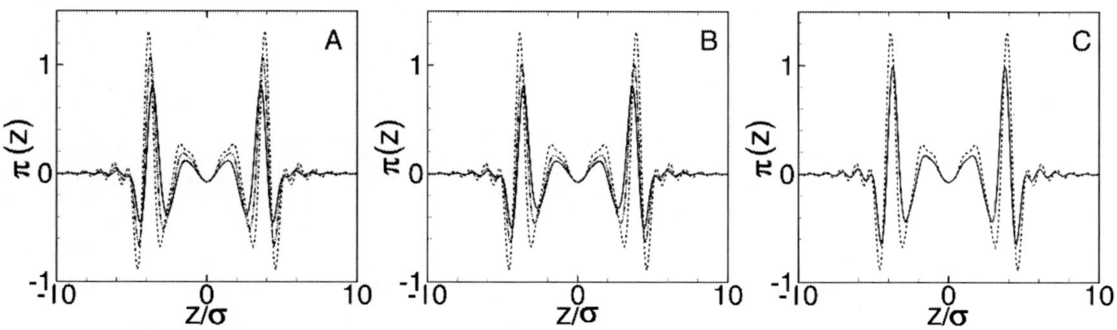

Figure 1.16. Lateral pressure profiles for bilayers with (A) ethanol at x_a = 0% (dashed), 0.89% (dash-dot-dot), and 3.32% (solid); (B) butanol at x_a = 0% (dashed), 0.037% (dash-dot-dot), and 0.12% (solid); and (C) hexanol at x_a = 0% (dashed) and 0.00119% (solid). Reprinted with permission from [16]. Copyright © 2006, The Biophysical Society.

1.8. CONCLUSIONS

In this chapter we have described the application of molecular theories to study various physical properties of lipid bilayers. We feel that Fluids-DFTs are a very promising route to modeling biomembranes. Since it is possible to treat all the species in a common theoretical framework in Fluids-DFTs, the self-assembled structure of a lipid bilayer is an output of the theory rather than an input, as in many phenomenological theories.

We have shown that perturbations to a single-component bilayer structure due to added components can be calculated with the Fluids-DFT. In the case of adding model alcohols, we find, as expected, that the alcohols are located preferentially near the headgroup–tailgroup interface, with longer-chain alcohols penetrating further into the bilayer interior. The theory predicts the correct qualitative trends for the changes in the area per lipid, the adsorption of alcohol into the bilayer, and the decreases in the area compressibility modulus and headgroup-tailgroup interfacial tension. For quantities that could be compared with experiment we generally obtained the right order of magnitude from the theory and in some cases even semiquantitative agreement.

Our study of a model peptide assembly inserted into the bilayer required fully three-dimensional calculations. We found large changes in the structure of the lipid bilayer due to the presence of peptides, including the spontaneous formation of membrane spanning pores. We find that toroidal pores are favored over either barrel-stave pores or isolated peptides in the membrane. The results suggest that barrel-stave pores are only likely to be found when peptides are tightly packed. We demonstrated the importance of considering multibody potentials of mean force rather than a superposition of pairwise additive potentials of mean force in understanding the formation of membrane spanning pores.

Future work will no doubt lead to improvements in both the fidelity of the coarse-grained models and in theoretical techniques. The results described here were for a very minimal lipid model with rather arbitrary interactions. A more systematic approach to developing a coarse-grained model could potentially lead to more predictive results. Additional features could also be added to the minimal model in order to explore other effects. Two obvious examples are the addition of charges in order to study electrostatic phenomena, and the addition of bond angle constraints, which would lead to stiffer lipid tails and enable the calculation of, for example, phase behavior in mixtures of two lipids with different degrees of tail stiffness.

There is a spectrum of Fluids-DFTs available in the literature, utilizing different approximations for the free energy functional. Further development of these theories is ongoing. Application of more accurate Fluids-DFTs to membranes should also result in more predictive results.

ACKNOWLEDGMENTS

We thank Mike Heroux, Andy Salinger, Walter Chapman, Mark Stevens, Julie Heffernan, and John Curro for collaborations and discussions over the course of this work, and Walter Chapman for the data shown in Figure 1.8. Sandia is a multiprogram laboratory operated by Sandia Corporation, a Lockheed Martin Company, for the United States Department of Energy under Contract DE-AC04-94AL85000. This work was supported by the Sandia LDRD program.

NOTE

1. A *functional* is a function whose value depends, not on a number, but on an entire function. For an introduction to functionals, see, for example, Chapter 9 in [48].

PROBLEMS

1.1. As we have discussed, a Fluids-DFT is typically formulated in an open ensemble in which the state variables are the temperature *T*, volume *V*, and chemical potentials of all the species, μ_α. Thus for a single-component membrane system consisting of lipids and solvent, we must specify four variables: μ_s, μ_L, *V*, and *T*. In the CMS-DFT, the chemical potentials are determined by the densities in the bulk reservoir. One way to set these is to fix the total bulk site density $\rho_b = \rho_L + \rho_s$ and then to consider bilayer properties as a function of the solvent number fraction $x_s = \rho_s / \rho_b$. Note that at fixed *T*, *V*, and ρ_b, the additional constraint of $\gamma = 0$ uniquely determines x_s, so that there is only one bilayer for this set of conditions. We identify this unique bilayer solution as the physical one for a bilayer at that particular temperature (this assumes our choice of the overall density is a realistic one, i.e., resulting in realistic overall fluid densities).

What happens if we now add a second component to the bilayer, such as the alcohols discussed in the text? We now have three chemical potentials: μ_s, μ_L, and μ_a for the alcohol. We still need to maintain the constraint $\gamma = 0$. What is the dimensionality of the set of possible membrane solutions at zero tension and fixed temperature and volume? How might we choose the chemical potential variables to obtain physically realistic results?

1.2. Consider the CMS-DFT described in the text (§1.3.1). The free energy is given by a functional Taylor expansion about the bulk reference state up to second order in the density as expressed in Eq. (10). Explain why this implies that the CMS-DFT cannot support a fluid/vapor phase transition or interface.

FURTHER READING

Davis HT. 1996. *Statistical mechanics of phases, interfaces, and thin films*. New York: Wiley-VCH.

Evans R. 1979. The nature of the liquid–vapor interface and other topics in the statistical mechanics of non-uniform classical fluids. *Adv Phys* **28**:143–200.

Frink LJD, Salinger AG, Sears MP, Weinhold JD, Frischknecht AL. 2002. Numerical challenges in the application of density functional theory to biology and nanotechnology. *J Phys Cond Matter* **14**:12167–12187.

Henderson D, ed. 1992. *Fundamentals of inhomogeneous fluids*. New York: Marcel Dekker.

Müller M, Katsov K, Schick M. 2006. Biological and synthetic membranes: what can be learned from a coarse-grained description? *Phys Rep* **434**:113–176.

Safran SA. 1994. *Statistical thermodynamics of surfaces, interfaces, and membranes*. Reading, MA: Addison-Wesley.

Wu J. 2006. Density functional theory for chemical engineering: from capillarity to soft materials. *AIChE J* **52**:1169–1193.

REFERENCES

1. Alberts B, Johnson A, Lewis J, Raff M, Roberts K, Walter P. 2002. *Molecular biology of the cell*. New York: Garland Science, Taylor & Francis Group.
2. Israelachvili J. 1992. *Intermolecular and surface forces*. Academic Press.
3. Shelley JC, Shelley MY, Reeder RC, Bandyopadhyay S, Klein ML. 2001. A coarse-grain model for phospholipid simulations. *J Phys Chem B* **105**:4464–4470.
4. Marrink SJ, Vries AH, Mark AE. 2004. Coarse-grained model for semiquantitative lipid simulations. *J Phys Chem B* **108**:750–760.
5. Marrink SJ, Risselada HJ, Yefimov S, Tieleman DP, Vries AH. 2007. The MARTINI force field: coarse grained model for biomolecular simulations. *J Phys Chem B* **111**:7812–7824.
6. Bond PJ, Holyoake J, Ivetac A, Khalid S, Sansom MSP. 2007. Coarse-grained molecular dynamics simulations of membrane proteins and peptides. *J Struct Bio* **157**:593–605.
7. Orsi M, Haubertin DY, Sanderson WE, Essex JW. 2008. A quantitative coarse-grain model for lipid bilayers. *J Phys Chem B* **112**:802–815.
8. Venturoli M, Sperotto MM, Kranenburg M, Smit B. 2006. Mesoscopic models of biological membranes. *Phys Rep—Rev Sect Phys Lett* **437**:1–54.
9. Shillcock JC, Lipowsky R. 2006. The computational route from bilayer membranes to vesicle fusion. *J Phys Cond Matter* **18**:S1191–S1219.
10. Müller M, Katsov K, Schick M. 2006. Biological and synthetic membranes: what can be learned from a coarse-grained description?. *Phys Rep—Rev Sect Phys Lett* **434**:113–176.
11. Helfrich W. 1973. Elastic properties of lipid bilayers—theory and possible experiments. *Z Naturforsch* **28c**:693–703.
12. Safran SA. 1994. *Statistical thermodynamics of surfaces, interfaces, and membranes*. Reading, MA: Addison-Wesley.
13. May S. 2000. Theories on structural perturbations of lipid bilayers. *Curr Opin Colloid Interface Sci* **5**:244–249.
14. Frink LJD, Frischknecht AL. 2005. Density functional theory approach for coarse-grained lipid bilayers. *Phys Rev E* **72**:041923.
15. Frischknecht AL, Frink LJD. 2005. Comparison of density functional theory and simulation of fluid bilayers. *Phys Rev E* **72**:041924.
16. Frischknecht AL, Frink LJD. 2006. Alcohols reduce lateral membrane pressures: predictions from molecular theory. *Biophys J* **91**:4081–4090.
17. Frink LJD, Frischknecht AL. 2006. Computational investigations of pore forming peptide assemblies in lipid bilayers. *Phys Rev Lett* **97**:208701.
18. Marconi UMB, Tarazona P. 1999. Dynamic density functional theory of fluids. *J Chem Phys* **110**:8032–8044.
19. Fraaije JGEM, Van Vlimmeren BAC, Maurits NM, Postma M, Evers OA, Hoffmann C, Altevogt P, Goldbeck-Wood G. 1997. The dynamic mean-field density functional method and its application to the mesoscopic dynamics of quenched block copoymer melts. *J Chem Phys* **106**:4260–4269.
20. Chandler D. 1987. *Introduction to modern statistical mechanics*. Oxford: Oxford UP.
21. Hansen J.-P, McDonald IR. 1986. *Theory of simple liquids*. Academic Press.
22. Somoza AM, Chacon E, Mederos L, Tarazona P. 1995. A model for membranes, vesicles and micelles in amphiphilic systems. *J Phys Cond Matter* **7**:5753–5776.
23. Brannigan G, Brown FLH. 2004. Solvent-free simulations of fluid membrane bilayers. *J Chem Phys* **120**:1059–1071.
24. Marcelja S. 1974. Chain ordering in liquid crystals, II: structure of bilayer membranes. *Biochim Biophys Acta* **1024**:139–151.
25. May S, Ben-Shaul A. 1999. Molecular theory of lipid–protein interaction and the La–HII transition. *Biophys J* **76**:751–767.
26. May S, Ben-Shaul A. 2000. A molecular model for lipid-mediated interaction between proteins in membranes. *Phys Chem Chem Phys* **2**:4494–4502.
27. Lague P, Zuckermann MJ, Roux B. 1998. Protein inclusion in lipid membranes: a theory based on the hypernetted chain integral equation. *Faraday Discuss* **111**:165–172.
28. Lague P, Zuckermann MJ, Roux B. 2000. Lipid-mediated interactions between intrinsic membrane proteins: a theoretical study based on integral equations. *Biophys J* **79**:2867–2879.

29. Lague P, Zuckermann MJ, Roux B. 2001. Lipid-mediated interactions between intrinsic membrane proteins: dependence on protein size and lipid composition. *Biophys J* **81**:276–284.
30. Fredrickson GH. 2006. *The equilibrium theory of inhomogeneous polymers*. Oxford: Clarendon Press.
31. Leermakers FAM, Scheutjens JMHM. 1988. Statistical thermodynamics of association colloids, III: the gel to liquid phase transition of lipid bilayer membranes. *J Chem Phys* **89**:6912–6924.
32. Meijer LA, Leermakers FAM, Lyklema J. 1994. Headgroup conformations in lipid bilayer-membranes. *Recl Trav Chim Pays-Bas* **113**:167–175.
33. Leermakers FAM, Rabinovich AL, Balabaev NK. 2003. Self-consistent-field modeling of hydrated unsaturated lipid bilayers in the liquid–crystal phase and comparison to molecular dynamics simulations. *Phys Rev E* **67**:011910–1910.
34. Cantor RS. 1999. Lipid composition and the lateral pressure profile in bilayers. *Biophys J* **76**:2625–2639.
35. Whitmore MD, Whitehead JP. 1998. Self-consistent field theory of compressible phospholipid membranes at finite pressure. *Can J Phys* **76**:883–898.
36. Muller M, Schick M. 1998. Calculation of the phase behavior of lipids. *Phys Rev E* **57**:6973–6978.
37. Li XJ, Schick M. 2000. Fluctuations in mixtures of lamellar- and nonlamellar-forming lipids. *J Chem Phys* **112**:10599–10607.
38. Li XJ, Schick M. 2000. Distribution of lipids in nonlamellar phases of their mixtures. *J Chem Phys* **112**:6063–6072.
39. Li XJ, Schick M. 2000. Theory of lipid polymorphism: application to phosphatidylethanolamine and phosphatidylserine. *Biophys J* **78**:34–46.
40. Meijer LA, Leermakers FAM, Nelson A. 1994. Modeling of the electrolyte ion phospholipid layer interaction. *Langmuir* **10**:1199–1206.
41. Li XJ, Schick M. 2001. Theory of tunable pH-sensitive vesicles of anionic and cationic lipids or anionic and neutral lipids. *Biophys J* **80**:1703–1711.
42. Goetz R, Lipowsky R. 1998. Computer simulations of bilayer membranes: self-assembly and interfacial tension. *J Chem Phys* **108**:7397–7409.
43. Kremer K, Grest GS. 1990. Dynamics of entangled linear polymer melts: a molecular-dynamics simulation. *J Chem Phys* **92**:5057.
44. Hohenberg P, Kohn W. 1964. Inhomogeneous electron gas. *Phys Rev* **136**:B864–B871.
45. Mermin ND. 1965. Thermal properties of the inhomogeneous electron gas. *Phys Rev* **137**:A1441–A1443.
46. Ebner C, Saam WF, Stroud D. 1976. Density-functional theory of simple classical fluids, I: surfaces. *Phys Rev A* **14**:2264–2273.
47. Evans R. 1979. The nature of the liquid–vapor interface and other topics in the statistical mechanics of nonuniform classical fluids. *Adv Phys* **28**:143–200.
48. Davis TH. 1996. Statistical mechanics of phases, interfaces, and thin films. New York: VCH.
49. Henderson D, ed. 1992. *Fundamentals of inhomogeneous fluids*. New York: Marcel Dekker.
50. Frink LJD, Salinger AG, Sears MP, Weinhold JD, Frischknecht AL. 2002. Numerical challenges in the application of density functional theory to biology and nanotechnology. *J Phys Cond Matter* **14**:12167–12187.
51. Wu J. 2006. Density functional theory for chemical engineering: from capillarity to soft materials. *AIChE J* **52**:1169–1193.
52. Wu J. 2007. Density-functional theory for complex fluids. *Annu Rev Phys Chem* **58**:85–112.
53. Rosenfeld Y. 1989. Free-energy model for the inhomogeneous hard-sphere fluid mixture and density-functional theory of freezing. *Phys Rev Lett* **63**:980–983.
54. Roth R, Evans R, Lang A, Kahl G. 2002. Fundamental measure theory for hard-sphere mixtures revisited: the White Bear version. *J Phys Cond Matter* **14**:12063–12078.
55. Chandler D, McCoy JD, Singer SJ. 1986. Density functional theory of nonuniform polyatomic systems, I: general formulation. *J Chem Phys* **85**:5971.
56. Chandler D, McCoy JD, Singer SJ. 1986. Density functional theory of nonuniform polyatomic systems, II: rational closures for integral equations. *J Chem Phys* **85**:5977–5982.
57. McCoy JD, Singer SJ, Chandler D. 1987. A density functional treatment of the hard dumbbell freezing transition. *J Chem Phys* **87**:4853–4858.
58. Hooper JB, McCoy JD, Curro JD. 2000. Density functional theory of simple polymers in a slit pore, I: theory and efficient algorithm. *J Chem Phys* **112**:3090.

59. Hooper JB, Pileggi MT, McCoy JD, Curro JD, Weinhold JD. 2000. Density functional theory of simple polymers in a slit pore, II: the role of compressibility and field type. *J Chem Phys* **112**:3094.
60. Hooper JB, McCoy JD, Curro JD, van Swol F. 2000. Density functional theory of simple polymers in a slit pore, III: surface tension. *J Chem Phys* **113**:2021–2024.
61. Donley JP, Curro JD, McCoy JD. 1994. A density functional theory for pair correlation funcitons in molecular liquids. *J Chem Phys* **101**:3205.
62. Frischknecht AL, Weinhold JD, Salinger AG, Curro JG, Frink LJD, McCoy JD. 2002. Density functional theory for inhomogeneous polymer systems, I: numerical methods. *J Chem Phys* **117**:10385–10397.
63. Frischknecht AL, Curro JG. 2004. Comparison of random-walk density functional theory to simulation for bead-spring homopolymer melts. *J Chem Phys* **121**:2788–2797.
64. Schweizer KS, Curro JD. 1989. Integral equation theory of the structure and thermodynamcis of polymer blends. *J Chem Phys* **91**:5059–5081.
65. Schweizer KS, Curro JG. 1997. Integral equation theories of the structure, thermodynamics and phase transitions of polymer fluids. *Adv Chem Phys* **98**:1.
66. Heine DR, Grest GS, Curro JG. 2005. Structure of polymer melts and blends: comparison of integral equation theory and computer simulations. *Adv Polym Sci* **173**:209–252.
67. Jain S, Dominik A, Chapman WG. 2007. Modified interfacial statistical associating fluid theory: a perturbation density funcitonal theory for inhomogeneous complex fluids. *J Chem Phys* **127**:244904.
68. Rosenfeld Y, Schmidt M, Löwen H, Tarazona P. 1996. Dimensional crossover and the freezing transition in density functional theory. *J Phys Cond Matter* **8**:L577.
69. Rosenfeld Y, Schmidt M, Löwen H, Tarazona P. 1997. Fundamental-measure free-energy density functional for hard spheres: dimensional crossover and freezing. *Phys Rev E* **55**:4245–4263.
70. Wertheim MS. 1984. Fluids with highly directional attractive forces, II: thermodynamic perturbation theory and integral equations. *J Stat Phys* **35**:35–47.
71. Tripathi S, Chapman WG. 2005. Microstructure of inhomogeneous polyatomic mixtures from a density functional formalism for atomic mixtures. *J Chem Phys* **122**:094506.
72. Tripathi S, Chapman WG. 2005. Microstructure and thermodynamics of inhomogeneous polymer blends and solutions. *Phys Rev Lett* **94**:087801.
73. <http://software.sandia.gov/tramonto>.
74. Heroux MA, Salinger AG, Frink LJD. 2007. Parallel segregated Schur complement methods for fluid density functional theories. *SIAM J Sci Comp* **29**:2059–2077.
75. Keller HB. 1977. Numerical solution of bifurcation and nonlinear eigenvalue problems. In *Applications of bifurcation theory*, pp. 359–384. Ed PH Rabinowitz. New York, London: Academic Press.
76. Salinger AG, Bou-Rabee NM, Pawlowski RP, Wilkes ED, Burroughs EA, Lehoucq RB, Romero LA. 2002. *LOCA 1.0, library of continuation algorithms: theory and implementation manual.* Technical Report SAND2002-0396. Albuquerque, NM: Sandia National Laboratories.
77. Petrache HI, Dodd SW, Brown MF. 2000. Area per lipid and acyl length distributions in fluid phosphatidylcholines determined by ^{2}H NMR spectroscopy. *Biophys J* **79**:3172–3192.
78. Stevens MJ. 2004. Coarse-grained simulations of lipid bilayers. *J Chem Phys* **121**:11942–11948.
79. Armen RS, Uitto OD, Feller SE. 1998. Phospholipid component volumes: determination and application to bilayer structure calculations. *Biophys J* **75**:734–744.
80. Nagle JF, TristramNagle S. 2000. Structure of lipid bilayers. *Biochim Biophys Acta, Rev Biomembr* **1469**:159–195.
81. Abrams CF, Kremer K. 2001. The effect of bond length on the structure of dense bead-spring polymer melts. *J Chem Phys* **115**:2776–2785.
82. Barry JA, Gawrisch K. 1994. Direct NMR evidence for ethanol binding to the lipid–water interface of phospholipid bilayers. *Biochemistry* **33**:8082–8088.
83. Holte LL, Gawrisch K. 1997. Determining ethanol distribution in phospholipid multilayers with MAS-NOESY spectra. *Biochemistry* **36**:4669–4674.
84. Feller SE, Brown CA, Nizza DT, Gawrisch K. 2002. Nuclear Overhauser enhancement spectroscopy cross-relaxation rates and ethanol distribution across membranes. *Biophys J* **82**:1396–1404.
85. Patra M, Salonen E, Terama E, Vattulainen R, Faller B, Lee W, Holopainen J, Karttunen M. 2006. Under the influence of alcohol: the effect of ethanol and methanol on lipid bilayers. *Biophys J* **90**:1121–1135.

86. Ly HV, Longo ML. 2004. The influence of short-chain alcohols on interfacial tension, mechanical properties, area/molecule, and permeability of fluid lipid bilayers. *Biophys J* **87**:1013–1033.
87. Monticelli L, Kandasamy S, Periole X, Larson R, Tieleman DP, Marrink SJ. 2008. The MARTINI coarse-grained force field: extension to proteins. *J Chem Theory Comput* **4**:819–834.
88. Bechinger B. 1999. The strucutre, dynamics and orientation of antimicrobial peptides in membranes by multi-dimensional solid-state NMR spectroscopy. *Biochim Biophys Acta, Rev Biomembr* **1462**:157–183.
89. Brogden KA. 2005. Antimicrobial peptides: pore formers or metabolic inhibitors in bacteria? *Nature Rev Microbiol* **3**:238–250.
90. Huang HW. 2006. Molecular mechanism of antimicrobial peptides: the origin of cooperativity. *Biochim Biophys Acta* **1758**:1292–1302.
91. Oversteegen SM, Barneveld PA, Male J, Leermakers FAM, Lyklema J. 1999. Thermodynamic derivation of mechanical expressions for interfacial parameters. *Phys Chem Chem Phys* **1**:4987–4994.
92. Gullingsrud J, Schulten K. 2003. Gating of MscL studied by steered molecular dynamics. *Biophys J* **85**:2087–2099.
93. Gullingsrud J, Schulten K. 2004. Lipid bilayer pressure profiles and mechanosensitive channel gating. *Biophys J* **86**:3496–3509.
94. Cantor RS. 1997. The lateral pressure profile in membranes: a physical mechanism of general anesthesia. *Biochemistry* **36**:2339–2344.
95. Cantor RS. 1999. The influence of membrane lateral pressures on simple geometric models of protein conformational equilibria. *Chem Phys Lipids* **101**:45–56.
96. Attard GS, Templer RH, Smith WS, Hunt AN, Jackowski S. 2000. Modulation of CTP:phosphocholine cytidylyltransferase by membrane curvature elastic stress. *Proc Natl Acad Sci USA* **97**:9032–9036.
97. Marsh D. 1996. Lateral pressure in membranes. *Biochim Biophys Acta* **1286**:183–223.
98. Rowlinson J, Widom B. 1982. *Molecular theory of capillarity*. New York: Oxford UP.
99. Lindahl E, Edholm O. 2000. Spatial and energetic-entropic decoposition of surface tension in lipid bilayers from molecular dynamics simulations. *J Chem Phys* **113**:3882–3893.
100. Yang AJM, Flemming III PD, Gibbs JH. 1976. Molecular theory of surface tension. *J Chem Phys* **64**:3732–3747.
101. Henderson JR. 1992. Statistical mechanical sum rules. In *Fundamentals of inhomogeneous fluids*, pp. 23–84. Ed D Henderson. New York: Dekker.
102. Szleifer I, Kramer D, Ben-Shaul A, Gelbart WM, Safran S. 1990. Molecular theory of curvature elasticity in surfactant films. *J Chem Phys* **92**:6800–6817.
103. Harries D, Ben-Shaul A. 1997. Conformational chain statistics in a model lipid bilayer: comparison between mean field and Monte Carlo simulations. *J Chem Phys* **106**:1609–1619.
104. Chin JH, Goldstein DB. 1977. Effects of low concentrations of ethanol on fluidity of spin-labeled erythrocyte and brain membranes. *Mol Pharmacol* **13**:435–441.
105. Komatsu H, Okada S. 1997. Effects of ethanol on permeability of phosphatidylcholine/cholesterol mixed liposomal membranes. *Chem Phys Lipids* **85**:67–74.
106. Chen SY, Yang B, Jacobson K, Sulik KK. 1996. The membrane disordering effect of ethanol on neural crest cells in vitro and the protective role of GM1 ganglioside. *Alcohol* **13**:589–595.
107. Traube I. 1891. Über die capillaritatsconstanten organischer stoffe in wassriger losung. *Liebigs Ann Chem* **265**:27–55.
108. Laan E, Chupin V, Killian JA, Kruijff B. 2004. Small alcohols destabilize the KcsA tetramer via their effect on the membrane lateral pressure. *Biochemistry* **43**:5937–5942.
109. Terama E, Ollila OH Samuli, Salonen E, Rowat AC, Trandum C, Westh P, Patra M, Karttunen M, Vattulainen I. 2008. Influence of ethanol on lipid membranes: from lateral pressure profiles to dynamics and partitioning. *J Phys Chem B* **112**:4131–4139.

2

Membrane Elasticity and Mediated Interactions in Continuum Theory: A Differential Geometric Approach

Markus Deserno
Department of Physics, Carnegie Mellon University,
Pittsburgh, Pennsylvania 15213, USA

2.1. INTRODUCTION: BIOPHYSICS AND SCALE SEPARATION

Biomembranes are fantastically complex systems [1–4]: hundreds of different lipid and protein species self-assemble into a large two-dimensional aggregate of locally complex and laterally inhomogeneous structure, and a globally potentially daunting topology. Thermal motion of this elastically soft system contributes prominently to its properties, and active processes constantly drive it away from equilibrium. How can we ever hope to learn something quantitative about such a complicated thing?

The route to success lies — as so often in physics — in the observation that we can frequently study different aspects of the problem largely independently. For instance, if a system spans many orders of magnitude in spatial scales, chances are that on the large scales it is perfectly well described by an effective Hamiltonian, whose small number of phenomenological parameters depend on — and can in principle be determined from — the underlying microscopic physics. Yet, establishing this micro–macro relation is not prerequisite to a successful and productive effective macroscopic description, since these phenomenological parameters can alternatively be measured experimentally on the macroscopic level. Often several such coarsening steps can be performed, thus constructing a hierarchy of scales with much beautiful physics in the different tiers. For instance, the laws of quantum mechanics explain everything about water that we need to know — e.g., how water structure and hydrogen bonds give rise to many of water's anomalies — but we can often just describe it effectively as a substance with some

Address correspondence to Markus Deserno, Department of Physics, Carnegie Mellon University, 5000 Forbes Avenue, Pittsburgh, PA 15213, USA, 412 268-4401, 412 681-0648 (fax), <deserno@andrew.cmu.edu>.

R. Faller et al. (eds.), *Biomembrane Frontiers: Nanostructures, Models, and the Design of Life*,
Handbook of Modern Biophysics, DOI: 10.1007/978-1-60761-314-5_2,

measurable material parameters, such as density, heat of vaporization, melting point, and compressibility. And even of these parameters many become irrelevant if we're only interested in large-scale fluid motion, for which density and viscosity are often the only relevant properties. Having gotten so used to this separability, we sometimes even forget that the success of physics as a science rests entirely on it. If phenomena on different scales could not be disentangled, we would for instance not be able to describe the motion of the liquid in a stirred cup of coffee without a thorough appreciation of its atomic structure. Or, maybe we'd even need to understand quarks? Or strings? The fact that for all intents and purposes we can master our surrounding world quite well, *without* knowing what the ultimate structure of matter and the form of a Grand Unified Theory is, provides a vivid proof for the power of scale separation.

However, the ease with which this program can be implemented depends crucially on how well scales actually separate. For instance, the size of a water molecule (~Å) is about 10 orders of magnitude smaller than ordinary human scales (m) and 16 orders of magnitude smaller than scales characterizing, say, the Gulf stream (1000 km); and its vibration frequencies (~100 THz) are 14 orders of magnitude away from human scales (~s) and 24 orders of magnitude away from the times characterizing deep see ocean currents (~1000 yr). Hence, scale separation for water works extremely well. In contrast, many systems studied in soft matter physics show relaxation times that easily extend beyond hours, days, and years — even if their constituents are only two orders of magnitude larger than water molecules! In these cases more effort is required to understand such complex matter, the techniques are more specialized, and tend to be tailored to the specific system in question (e.g., liquid crystals, polymers, colloids, ...) [5]. But the benefits to be reaped are the same: a quantitative phenomenology based on a small set of measurable parameters describes much of the meso- or macroscopic behavior.

Is scale separation possible in biology? The success of the medical sciences suggests a cautious "yes." We are evidently able to understand enough about how organisms function to be able to "fix" them when certain (simple) things go wrong.[1] Nevertheless, biology is more complicated in many ways. First of all, most of the materials that nature uses are soft, so "trivial" scale separation hardly ever works and we must resort to the more sophisticated techniques developed in soft matter physics. However, biology is more than soft matter. It is never in equilibrium while alive. No living system is isolated from its outside world. Every living system is here for a purpose (namely, to carry its genes into the future), for which it has to compete or cooperate with other organisms, owing to limited resources. Every organism is evolutionarily optimized to do just that. Such concepts don't resonate easily with our classical physical approach to the world, but this should not aggrieve us to the point of giving up the program. The complexity of nature is sometimes used to argue that simplifications are impossible. If that were true, biological science would not be possible either. Quite to the contrary, given the complexity of nature and our inability to grasp it all at once, we depend more than ever on techniques that promise a reliable, systematic, and quantitative phenomenological approach. Moreover, despite its astounding complexity, biology is full of simple laws. Here's one: the metabolic rate of an organism scales like its weight to the power of ¾ — a law that holds over *more than 20 orders of magnitude in weight* [6]. Such relations must be explicable on very fundamental general principles, and indeed convincing explanations for this law (and others) have been proposed [7].

In this book we are concerned with only a small aspect of nature: biomembranes. Still, they are already sufficiently complex such as to necessitate a hierarchical treatment — as nicely illustrated by the scope spanned by the different chapters. In the present chapter we will set out to look at membranes from a comparatively large scale, where continuum elastic treatments become meaningful. Admittedly, membrane elasticity as such might not at first be perceived as a frontier in biosciences, and indeed the basic equations are more than three decades old by now [8,9]. However, it still fits within the context for the following reasons:

- Whether old or not, it is the mesoscopic scale on which a lot of cellular physics happens, and it therefore remains a crucial contribution in the mix of techniques useful to elucidate cellular behavior.
- Continuum elasticity theory provides a close and very natural link to stresses transmitted in membranes, and as such it constitutes an ideal framework for all situations in which such stresses play an important role, e.g., in morphology changes, protein interactions, or locomotion.
- Recent theoretical developments have provided new tools with which membrane stresses can be described in very geometrical (and thus often very *intuitive*) ways.
- Years of careful simulation work have begun to establish a systematic coarse-graining sequence through which microscopic membrane properties can be quantitatively linked to the phenomenological parameters on larger scales (see, e.g., the chapters by Hoopes et al., Vattulainen, and Tieleman in this volume). Continuum theory is thus bound even more tightly to the finer levels. In particular, its phenomenological parameters have become *predictable*.

2.2. CONTINUUM THEORY FOR MEMBRANES: A FIRST LOOK

Lipids are the major structural component of most biomembranes. These are typically two-tailed amphiphiles, and as such they prefer to aggregate into lamellar phases. This alone already has one remarkable consequence: as classical micellization theory shows, the size distribution of aggregates depends crucially on their morphology [5]. Spherical micelles tend to be very monodisperse, having a size distribution peaked around some particular mean value that accommodates the spontaneous curvature of the surfactants. In contrast, cylindrical micelles have a much wider exponential size distribution with a mean length that scales with the square root of surfactant concentration. Lamellar-forming amphiphiles differ even more strongly: it turns out that they follow a distribution that, even though initially exponentially decaying, again peaks at the size corresponding to a *single aggregate containing all amphiphiles*. Hence, bilayers have an intrinsic tendency to be "infinite." In consequence, it is only too natural to apply continuum theory to them: the mechanism of self-assembly transforms individual entities, only a couple of nanometers big, into bilayer sheets extending over many microns. This extremely large aspect ratio also points toward the proper effective theory that would need to be developed on the micron scale: membranes are essentially two-dimensional surfaces.

Let us begin to develop a model for bilayer elasticity. The simplest deformation we can do to a bilayer is stretch it. Experiments of this type can be done with very high accuracy using micropipette techniques [10,11]. The idea is essentially this: grab a large vesicle (many microns in size) with a micropipette (having a bore on the order of a micrometer) and gently suck the vesicle in, thereby putting it under tension. Since one can measure both the pressure difference between inside and outside and relative area changes and the vesicle size very accurately, one can accurately probe the stress–strain relation. Neglecting fluctuations, which are important at low tension, one finds that the bilayer can essentially be stretched like a harmonic spring, obeying a stretching energy of the form

$$E_{\text{stretch}} = \frac{1}{2} K_{\text{stretch}} \frac{(A - A_0)^2}{A_0}. \tag{2.1}$$

Here, K_{stretch} is the stretching modulus. Typical values valid for phospholipids are on the order of 250 mN/m [11]. Notice that the membrane tension σ is given by the derivative of the energy with respect to the area (at constant number of lipids, N):

$$\sigma = \left(\frac{\partial E_{\text{stretch}}}{\partial A}\right)_N = K_{\text{stretch}} \frac{A - A_0}{A_0} = K_{\text{stretch}} u, \tag{2.2}$$

where $u = (A - A_0)/A_0$ is the dimensionless (relative) *strain*. In other words, what we have here is nothing but Hooke's law: *a linear stress–strain relation*.

What happens now if we bend such a membrane? Since the membrane has a finite thickness, bending implies that the "outside" part of the bilayer will be stretched a little bit while the inside part will be compressed a little bit. The total energy of the bent membrane can then be obtained by a volume integral of the stretching energy of infinitesimal volume elements comprising the bilayer. The full calculation can for instance be found in Landau and Lifshitz [12]. Here we will restrict ourselves to a simplified presentation of the matter.

In order to understand the stretching behavior of volume elements, we will use an energy identical in spirit to Eq. (2.1): a deviation from a given volume will cost quadratically in energy:

$$E_{\text{stretch}} = \frac{1}{2} Y \frac{(V - V_0)^2}{V_0}, \tag{2.3}$$

where Y is called *Young's modulus*. Let us consider a quadratic membrane patch of side length L and thickness h and bend it along one direction into an arc of curvature radius R, measured from the midplane of the membrane, also known as the *neutral surface* (see Fig. 2.1). Assuming that the stretching along the x-direction will only change the shape of an infinitesimal cube along the same direction (and not shrink it, for instance, in a direction perpendicular to x), we see that the length dx is stretched by a factor $(R + z)/R = 1 + z/R$, where z is the perpendicular distance of that cube from the neutral surface. Hence, the stretching energy of this cube is given by

$$\frac{1}{2} Y \frac{(dx'\, dy'\, dz' - dx\, dy\, dz)^2}{dx\, dy\, dz} = \frac{1}{2} Y \frac{(dx(1 + z/R) dy\, dz - dx\, dy\, dz)^2}{dx\, dy\, dz} = \frac{1}{2} Y \frac{z^2}{R^2} dx\, dy\, dz, \tag{2.4}$$

so that the total stretching/compression energy of the membrane patch is given by

$$E_{\text{bend}} = \frac{Y}{2R^2} \int_{-L}^{L} dx \int_{-L}^{L} dy \int_{-h/2}^{h/2} dz\, z^2 = \frac{L^2 Y h^3}{24 R^2}. \tag{2.5}$$

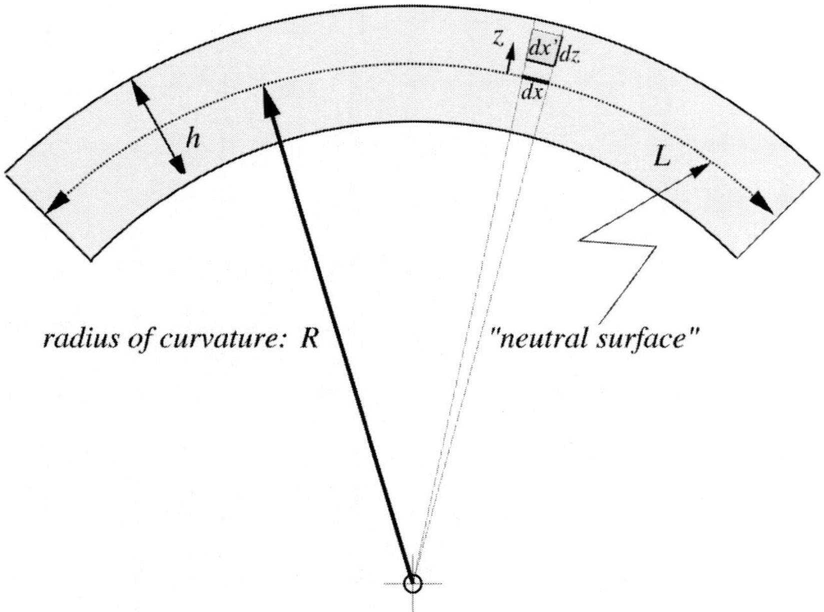

Figure 2.1. How to derive bending energy from stretching. A thin elastic sheet is bent along one direction into an arc of curvature radius *R*, measured from the neutral surface. A small volume element a distance *z* from that neutral surface is stretched along the *x*-direction if it is on the outer side, and compressed if it is on the inner side.

This implies a bending energy density per unit area of

$$e_{\text{bend}} = \frac{E_{\text{bend}}}{L^2} = \frac{1}{24} Yh \left(\frac{h}{R} \right)^2 . \tag{2.6}$$

At this point several comments are in order:

1. For simple dimensional reasons, Young's modulus, stretching modulus, and membrane thickness are related by $K_{\text{stretch}} = Yh$.
2. Since typically $h \ll R$, i.e., the radius of curvature is much larger than the membrane thickness, the bending energy per unit area can be quite small even if the stretching modulus is large. This is why we can make soft springs out of rather stiff metal: by making them thin enough.
3. The bending energy scales as the square of the curvature. Using the abbreviation $\kappa = \frac{1}{12} Yh^3 = \frac{1}{12} K_{\text{stretch}} h^2$, we can then write the bending energy density as

$$e_{\text{bend}} = \frac{1}{2} \kappa \frac{1}{R^2} . \tag{2.7}$$

It should be noted that, even though this looks like a simple quadratic elastic theory ("Hooke's law for curvatures"), curvature is not just another form of strain, and this theory is subtly different from standard elastic theories.[2]

4. Since lipid membranes are composed of two monolayers that can slide past each other, no transversal stress can be transmitted across the neutral surface. In consequence, the stretching and compression of the outer and inner monolayer will be smaller than if both leaflets were rigidly coupled, as Figure 2.2 illustrates. It is in fact more appropriate to describe the overall deformation energy as two times the deformation of a monolayer that has half the thickness of the membrane. These two factors of 2 do not cancel, since the thickness enters not linearly but cubically in the bending energy. We should therefore expect

$$e_{\text{bend,bilayer}}(h) = 2 \times e_{\text{bend.monolayer}}(h/2) = \frac{1}{96} Yh \frac{h^2}{R^2}. \tag{2.8}$$

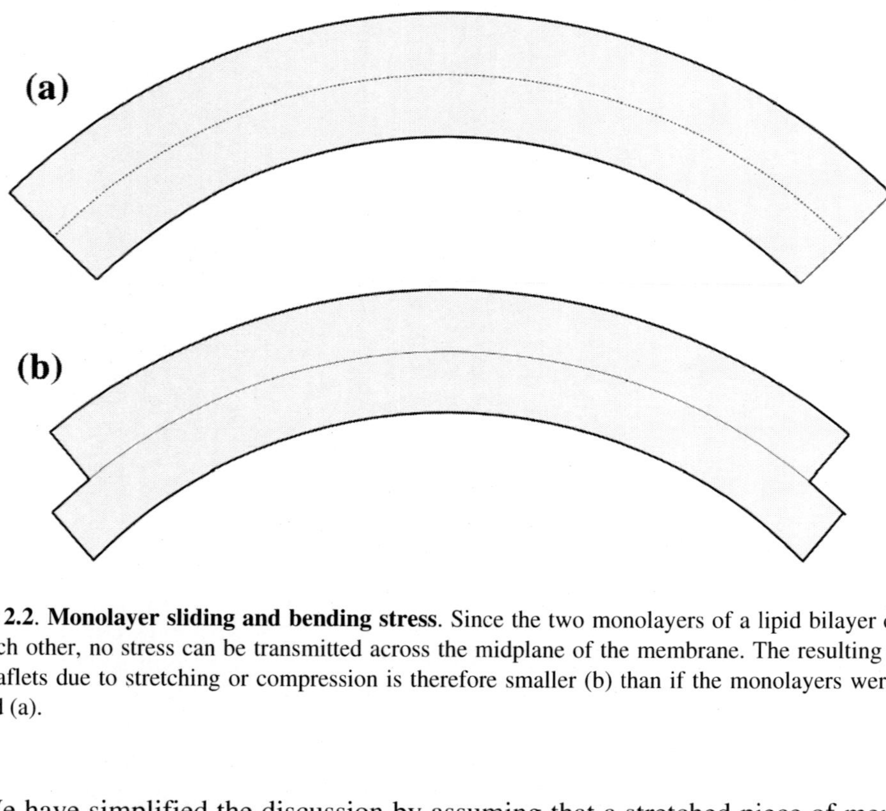

Figure 2.2. Monolayer sliding and bending stress. Since the two monolayers of a lipid bilayer can slide past each other, no stress can be transmitted across the midplane of the membrane. The resulting strain in both leaflets due to stretching or compression is therefore smaller (b) than if the monolayers were rigidly coupled (a).

5. We have simplified the discussion by assuming that a stretched piece of membrane will not shrink in the direction perpendicular to the applied strain, but there is no principal justification for this. In fact, taking a piece of material of length L and width w, a longitudinal strain $u = \Delta L / L$ also implies a transversal relative width change $\Delta w / w = -\nu u$, where $-1 \leq \nu \leq \frac{1}{2}$ is *Poisson's ratio* [12]. If the material shows no transverse response to a longitudinal strain, as we have pretended so far, we have the special case $\nu = 0$. It is also easy to see that in the limit of an incompressible material, where the transverse shrinking has to compensate for the longitudinal extension, we must have $\nu = \frac{1}{2}$.[3] It may be shown that Poisson's ratio

modifies our formulas by effecting the replacement $Y \to Y/(1-\nu^2)$ [12]. Hence, a better formula for the bending modulus than the one we have initially written down would be

$$\kappa = \frac{K_{\text{stretch}} h^2}{48(1-\nu^2)}. \tag{2.9}$$

This includes the "bilayer = 2 monolayers" correction discussed above.

6. Let us put it in numbers: using Eq. (2.9) in the limit of an incompressible material ($\nu = \frac{1}{2}$, such that the prefactor 48 turns into 36) and taking typical values for the stretching modulus ($K_{\text{stretch}} = 250$ mN/m) and the bilayer thickness (h = 4 nm) we arrive at $\kappa = 27 k_B T_r$, where $k_B T_r \approx 4.1 \times 10^{-21} J \approx 0.6$ kcal/mol is the thermal energy at room temperature. This is quite a typical value for phospholipids, maybe a bit on the large side [11]. But beyond that actual number, what really matters is the fact that it is a *fair* amount larger than thermal energy, so fluctuations do not tear the bilayer into bits and pieces, while it is not *tremendously* much larger than thermal energy, so nanoscopic sources of energy, provided, e.g., by adhering proteins or ATP hydrolysis, are capable of deforming the lipid membrane.

2.3. CURVATURE ELASTICITY

Based on what we have learned in the previous section, we suspect that bilayer elasticity will turn out to be a "curvature squared theory," where the elastic constants — e.g., κ — might even be derivable from other known elastic constants (such as Young's modulus and Poisson's ratio). However, we have actually only looked at a rather simple deformation: homogeneous bending into one direction. While it is possible to extend this line of reasoning to more complicated deformations — details can be found in Landau and Lifshitz [12] — it proves more instructive to "derive" a phenomenological curvature-based theory from scratch. Based only on the expected symmetries and the knowledge that the shape of the deformed surface will matter, one can essentially write down the correct Hamiltonian. The most transparent way to do this is within a differential geometric framework. In order to clarify the notation, we will first and very briefly revisit a few basic points. These can by no means replace a good introduction into the field, and the reader is encouraged to consult, for instance, the classical texts by do Carmo [13], Spivak [14], or Kreyszig [15].

2.3.1. Differential Geometry *in nuce*

Let us consider a two-dimensional surface embedded into three-dimensional Euclidean space $\mathbf{R}^3$. We can locally describe it by a parametrization, which for now we will take to be a mapping from a two-dimensional region of $\mathbf{R}^2$ into $\mathbf{R}^3$, given by the function $\mathbf{x} = \mathbf{X}(u^1, u^2)$, where the two parameters u^a ($a \in \{1,2\}$) are the (curvilinear) surface coordinates. Using the embedding function $\mathbf{X}$, we can define tangent and normal vectors on the surface:

$$\mathbf{e}_a = \frac{\partial \mathbf{X}}{\partial u^a} = \partial_a \mathbf{X}, \tag{2.10}$$

$$\mathbf{n} = \frac{\mathbf{e}_1 \times \mathbf{e}_2}{|\mathbf{e}_1 \times \mathbf{e}_2|}. \tag{2.11}$$

Notice that while the normal vector **n** is indeed normalized to unit length, the tangent vectors $\mathbf{e}_a$ are in general neither normalized nor perpendicular to each other.[4]

With these vectors we are now in the position to define two important tensors on the surface: the metric tensor g_{ab} and the (extrinsic) curvature tensor K_{ab}:

$$g_{ab} = \mathbf{e}_a \cdot \mathbf{e}_b, \tag{2.12}$$

$$K_{ab} = \mathbf{e}_a \cdot \partial_b \mathbf{n} = -\mathbf{n} \cdot \partial_b \mathbf{e}_a = -\mathbf{n} \cdot \partial_a \partial_b \mathbf{X}. \tag{2.13}$$

As the definitions show, both tensors are symmetric. The metric will be concerned with scalar products and thus distance information on the surface. Consider for instance two vectors $\mathbf{v} = \sum_{a=1,2} v^a \mathbf{e}_a \equiv v^a \mathbf{e}_a$ and $\mathbf{w} = w^b \mathbf{e}_b$ defined at the *same* point of the surface. Their scalar product is then given by

$$\mathbf{v} \cdot \mathbf{w} = \left(v^a \mathbf{e}_a\right) \cdot \left(w^b \mathbf{e}_b\right) = v^a w^b \;\; \mathbf{e}_a \cdot \mathbf{e}_b = v^a w^b g_{ab} = v^a w_a, \tag{2.14}$$

where we also have shown how indices can be lowered by the metric. Indices can also be raised with the tensor g^{ab}, whose components are defined such that the following holds:

$$g^{ab} g_{bc} = \delta^a_c, \tag{2.15}$$

where δ^a_c is the Kronecker-delta, which is 1 if both indices are identical and zero if they are different. So g^{ab} is essentially the inverse of g_{ab}. Lower indices are referred to as "covariant coordinates," while upper indices are "contravariant coordinates." The determinant of the metric is abbreviated as $g = \det(g_{ab})$ and can also be expressed as $|\mathbf{e}_1 \times \mathbf{e}_2|^2$. From this we see immediately that the proper area element on the surface is given by $dA = \sqrt{g}\; du^1\; du^2$.

The curvature tensor K_{ab} measures local *extrinsic curvature*, i.e., bending of the two-dimensional surface into the third space dimension. This is intuitively reasonable, since it monitors how the normal vector changes as we move along the surface. If it stays constant, the surface is flat; if it is not, the surface must be curved. Raising one index of the curvature tensor gives the matrix $K^b_a = K_{ac} g^{cb}$, whose eigenvalues are the *principal curvatures*, and whose eigenvectors point into the *principal directions*. The *trace* of the curvature tensor is called the total extrinsic curvature, $K = \mathrm{Tr}(K_{ab}) = g^{ab} K_{ab} = K^a_a$. The determinant of the matrix K^b_a is called the Gaussian curvature, $K_{\mathrm{G}} = \det(K^b_a)$. Notice that in the present case of 2 × 2 matrices it can also be written as $K_{\mathrm{G}} = \frac{1}{2}(K^2 - K_{ab}K^{ab})$, which is (half) the difference between the square of the trace and the trace of the square of the extrinsic curvature tensor.

Since the surface coordinates are curvilinear, they depend on position. Hence, the ordinary partial derivative of surface vectors or tensors will, via the product rule, also contribute terms from the positional dependence of the local basis. This ultimately implies that, for instance, $\partial_a v_b$ is *not* a tensor of second rank, since these additional terms spoil the transformation law. It is, however, possible to restore this property by defining a covariant derivative ∇_a that incorporates the additional terms. When acting on scalars, it is identical to the usual partial derivative (since scalars do not involve the position-dependent basis!), while its action on co- or contravariant vectors is given by

$$\nabla_a v_c = \partial_a v_c - \Gamma^b_{ac} v_b \,, \tag{2.16}$$

$$\nabla_a v^c = \partial_a v^c + \Gamma^c_{ab} v^b \,. \tag{2.17}$$

The new object Γ^c_{ab} is called the *Christoffel symbol* (of second kind) and is defined as

$$\Gamma^c_{ab} = \tfrac{1}{2} g^{cd} \left(\partial_a g_{bd} + \partial_b g_{da} - \partial_d g_{ab} \right). \tag{2.18}$$

Observe that Γ^c_{ab} is symmetric in its two lower indices. We also have a Christoffel symbol of the first kind, $\Gamma_{abc} = g_{cd}\Gamma^d_{ab}$. Since it's the "job" of the Christoffel symbols to "undo" the non-tensor character of the partial derivative, they cannot be tensors themselves.[5] Covariant derivatives of objects with more than one co- or contravariant index follow the same pattern as seen in Eqs. (2.16) and (2.17), i.e., each covariant index necessitates the subtraction of a Christoffel symbol term, and each contravariant index necessitates its addition (the placement of the indices on the Christoffel symbol is always obvious).

It turns out that the metric is constant with respect to the covariant derivative, $\nabla_a g_{bc} = 0$, a fact known as the Lemma of Ricci. The practically extremely convenient consequence is that *raising and lowering of indices commutes with the process of covariant differentiation.*

On the downside: covariant derivatives no longer commute! Instead, we have

$$\left(\nabla_a \nabla_b - \nabla_b \nabla_a \right) v_c = R_{abcd} v^d \,, \tag{2.19}$$

a relation which defines the Riemann tensor R_{abcd}, whose components — from Eq. (2.19) — can be worked out to be

$$R_{abcd} = \partial_c \Gamma_{bda} - \partial_d \Gamma_{bca} + \Gamma^e_{bd} \Gamma_{eca} - \Gamma^e_{bc} \Gamma_{eda} \,. \tag{2.20}$$

The contraction of the Riemann tensor with respect to the first and third index is called the *Ricci tensor*:

$$R_{bd} = g^{ac} R_{abcd} \,, \tag{2.21}$$

whose repeated contraction gives the Ricci scalar:

$$R = g^{bd} R_{bd} \,. \tag{2.22}$$

In general, the Riemann tensor has more independent components than the Ricci tensor, which in turn has more independent components than the Ricci scalar.[6] However, for two-dimensional surfaces all of the three tensors have exactly one independent component, and so the Riemann tensor is completely determined by the Ricci scalar, namely via

$$R_{abcd} = \tfrac{1}{2} R (g_{ac} g_{bd} - g_{ad} g_{bc}) \qquad (\textit{only} \text{ in two dimensions!}). \tag{2.23}$$

Finally, it turns out that there are integrability conditions to be satisfied (e.g., $\partial_a \partial_b \mathbf{e}_c = \partial_b \partial_a \mathbf{e}_c$), which link the extrinsic geometry associated with the curvature tensor with the intrinsic geometry associated with the metric.[7] These are called the equations of Gauss-Codazzi-Mainardi, and they state the following:

$$\nabla_a K_{bc} - \nabla_c K_{ab} = 0\,, \tag{2.24}$$

$$K_{ac} K_{bd} - K_{ad} K_{bc} = R_{abcd}\,. \tag{2.25}$$

If we contract Eq. (2.25) with respect to *ac*, we find

$$K K_{bd} - K_{ad} K^a_b = R_{bd}\,. \tag{2.26}$$

Contracting it once more with respect to *bd*, we find the truly remarkable result

$$2K_{\text{G}} = K^2 - K^{ab} K_{ab} = R\,. \tag{2.27}$$

This final equation states that the Gaussian curvature K_{G}, which has been defined using the *extrinsic* curvature tensor (i.e., by making use of a normal vector pointing into the third space dimension!), can in fact be expressed by the Ricci scalar curvature, which originates from the entirely *intrinsic* Riemann tensor (i.e., without any reference to an embedding into 3-space). Hence, the Gaussian curvature can be obtained exclusively from distance measurements *within* the surface. This discovery is due to Gauss, who was so much impressed by it that he called it the *Theorema Egregium*. He then went on to execute a precision triangulation measurement of a triangle formed by three mountaintops near Göttingen (namely Hohenhagen, Brocken, and Inselberg). He wanted to check whether the Euclidean result that the sum of the interior angles must be 180° is satisfied, because he realized that *the Euclidean geometry of space is an empirical question, subject to verification or disproof by measurement* [16].

This should suffice for now as a minute survey of differential geometry. Since none of these formulas have been properly derived here and mathematical subtleties have been nonchalantly swept under the rug, the interested reader is highly encouraged to look up the details in [13–15]. This section should rather be seen as a convenient introduction of the geometric *dramatis personae* as well as a quick reference for some useful formulas that will be needed below. They are typically applied extensively throughout all calculations, but luckily, when the dust has settled, most of the final formulas will express relations between extremely intuitive geometric quantities and not complicated tensors. So even if one is unwilling to delve into differential geometry, one will be able to make use of many of its beautiful results.

2.3.2. Curvature Elasticity from a Formal Point of View

As Eqs. (2.7) and (2.8) suggest, the key physical characteristic of bending elasticity is its proportionality to squared curvature. The prefactor κ measures the corresponding bending rigidity, which can be linked to other (maybe more fundamental) material properties, such as for instance Young's modulus. This step, however, requires both modeling and certain non-obvious assumptions (concerning, for instance, the homogeneity of the material or the lateral stress profile). We might thus alternatively consider κ as a phenomenological modulus that can be determined in experiment. If we take this point of view, we may more generally view the surface energy density as reflecting the first terms of an expansion of the generalized elastic energy in surface scalars, ordered by dimensionality, each term multiplied by a phenomenological modulus. Which terms one can write down depends on the available fields and symmetries. In the simplest case we have only the geometry of the membrane, fully determined by the two tensors

g_{ab} and K_{ab}. The first few terms have been discussed for instance by Capovilla, Guven, and Santiago [17], and we briefly review their results:

- The simplest scalar that can be constructed from them is $g_a^a = 2$, just a number. The corresponding surface integral gives the surface area, and the corresponding "modulus" is the surface tension.
- The next simplest scalar — of dimension $(\text{length})^{-1}$ — is $K_a^a = K$, the total extrinsic curvature. Since K depends on the orientation of the normal vector, it can only appear in a physical Hamiltonian if the surface in question is in some way asymmetric with respect to its two sides. The appearance of a linear term is therefore typically associated with the existence of a spontaneous curvature, but it can even matter for symmetric membranes by virtue of the following observation: the variation of the local area element is proportional to the local value of the curvature. It follows that the *difference in area* between the two close but not identical surfaces representing the two leaflets of the membrane is proportional to the *surface integral over K*. If for some reason the area of these two surfaces is *individually* conserved (for instance, because lipids might not have sufficient time to flip between the two leaflets during the timescale of observation), one can enforce this constraint by an additional term in the Hamiltonian that is proportional to K and whose prefactor is the Lagrange multiplier enforcing this constraint. More details on this, and its relation to so-called "area difference elasticity" can be found in [18].
- For dimension $(\text{length})^{-2}$ we have two independent scalars: the square of the trace and the trace of the square, $K^2 = K_a^a K_b^b$ and $K_a^b K_b^a$, but it proves more convenient to use K^2 and the linear combination $R = K^2 - K_a^b K_b^a = 2K_G$ to work with. The reason is that the Gaussian curvature integrates to a topological invariant (by virtue of the Gauss-Bonnet theorem [13–15]) and can thus often (but by no means always!) be neglected in the description.
- On order $(\text{length})^{-3}$ we have three independent terms, which can for instance be chosen to be K^3, KR, and $K^{ab}R_{ab}$. More terms could be written down, but these would no longer be independent. For instance, by using the Codazzi-Mainardi equations, it is easy to see that $K_a^b K_b^c K_c^a = K^3 - KR - K^{ab}R_{ab}$, and thus no new invariant results.
- On order $(\text{length})^{-4}$ it happens for the first time that the independent terms are not of the same order in surface derivatives. An independent set would for instance be K^4, $K^2 R$, R^2, and $(\nabla_a K)(\nabla^a K)$. While the first three terms are of order curvature in terms of surface derivatives, the fourth term contains the derivative of curvature.

Stopping at this order, the energy density can be written as

$$
\begin{aligned}
e_{\text{surface}} = {} & \sigma \\
& + \kappa_1 K \\
& + \tfrac{1}{2}\kappa K^2 + \tfrac{1}{2}\bar{\kappa} R \\
& + \tfrac{1}{3}\kappa_3 K^3 + \kappa_{3\times} KR + \kappa_{3R} K^{ab} R_{ab} \\
& + \tfrac{1}{4}\kappa_4 K^4 + \kappa_{4\times} K^2 R + \tfrac{1}{2}\bar{\kappa}_2 R^2 + \tfrac{1}{2}\kappa_\nabla (\nabla_a K)(\nabla^a K).
\end{aligned} \tag{2.28}
$$

Typically, the linear term is incorporated in the quadratic one by defining a spontaneous curvature K_0 and writing $\frac{1}{2}\kappa(K-K_0)^2$. Up to a constant (which enters the surface tension) this amounts to the choice $\kappa_1 = \kappa K_0$.

Terms such as the ones featured in Eq. (2.28) — as well as more complicated ones — have been used to model corrugated ("egg-carton") surfaces [19]. However, for most applications going up to quadratic order is fully sufficient. In fact, probably the most important subset of terms from Eq. (2.28) is the quadratic order without linear term (i.e., for up–down symmetric surfaces):

$$e_{\text{Helfrich}} = \sigma + \tfrac{1}{2}\kappa K^2 + \tfrac{1}{2}\bar{\kappa} R \,. \tag{2.29}$$

This Hamiltonian (actually, including the spontaneous curvature) was originally proposed by Wolfgang Helfrich as a suitable model for bilayer elasticity [9], after Canham [8] had three years earlier suggested a Hamiltonian with only one bending term (of the form $K_{ab}K^{ab}$ in our notation).

2.3.3. Additional Fields

Curvature elasticity involves energetic changes of a surface due to deformation of its *shape*. However, other sources of energy are possible if additional fields "live" on the membrane. For instance, a lipid membrane might consist of different types of lipids. If these completely phase separate, then the membrane consists of patches that differ in their elastic properties. However, if these components mix — not necessarily ideally, though — a scalar concentration field ϕ may be defined on the membrane whose local value will generally influence the (free) energy.[8] This can happen in different ways:

- The elastic constants become functions of the concentration; e.g., the bending modulus might depend on the local lipid composition: $\kappa = \kappa(\phi)$.
- New scalars can be constructed by using the geometric variables and non-geometric fields. For instance, a term of the form $\frac{1}{2}\kappa\left(K - \phi c_0\right)^2$, where ϕ is a composition field and c_0 the spontaneous curvature at $\phi = 1$, creates a composition-dependent spontaneous curvature. Of course, one might alternatively look at this as a ϕ-dependent linear K-coupling ($\kappa_1 = \kappa\phi c_0$) combined with a ϕ-dependent extra surface tension ($\sigma \to \sigma + \frac{1}{2}\kappa c_0^2\phi^2$), but the spontaneous curvature interpretation is more natural in this case.
- The field itself contributes to the overall free energy (e.g., in the form of energy/entropy of mixing). The most obvious examples would be a Ginzburg-Landau type functional of the form $V(\phi) + \frac{1}{2}\eta(\nabla_a\phi)(\nabla^a\phi)$, where $V(\phi)$ is a potential and the second term the covariant generalization of a squared gradient.

These examples all involve a single scalar field ϕ. However, one might have more than one field. Or one might have a vector field. A good example for the latter would be *lipid tilt*. This could be described by the average projection of a lipid into the membrane plane. Without tilt, lipids always point into the same direction as the membrane normal; with tilt they acquire a tangential component that can be expanded in the local tangential frame: $\mathbf{m} = m^a\mathbf{e}_a$. The associated field (components) m^a can again give rise to energy contributions. Since these have to be scalar, one either must construct scalars from these vectors (e.g., $m_a m^a$, $\nabla_a m^a$,

$(\nabla_a m^b)(\nabla_b m^a)$), for which one already needs the metric, or one involves the curvature tensor (e.g., $K_{ab} m^a m^b$). Evidently many more possibilities exist, but these are also strongly restricted by symmetry considerations. An insightful discussion of this has been presented by Nelson and Powers [20,21].

Frequently such fields are "excited" by proteins embedded in the membrane. Rather than guessing the phenomenological form or strength of this coupling, one can try to construct a more detailed approach to describe the local physics. The chapter by Frischknecht in this book provides an example of how this can be done.

2.4. MEMBRANE STRESSES AND SHAPE EQUILIBRIA

The total elastic energy of a membrane is the surface integral over its energy density — containing all contributions discussed so far, geometric or field-derived. In mechanical equilibrium[9] the membrane will choose a configuration that minimizes this overall energy. This naturally leads to a functional minimization problem and corresponding Euler-Lagrange differential equations for the membrane shape and all fields defined on it. But there is also a different way to look at this: in equilibrium all stresses have to balance, meaning that there ought to be a local stress tensor that is divergence-free. Indeed, these two ways of looking at things are closely related, and the divergence of the stress tensor indeed turns out to be nothing but the Euler-Lagrange derivative. Yet, looking at the problem from a point of stresses might provide complementary insight that the usual energy approach does not necessarily offer. This is all the more true as it turns out that the stress tensor in question can be expressed *completely* in terms of the local deformations, thus establishing a strong link between the geometry of a deformed membrane and the forces it transmits. This approach has been pioneered in a series of papers by Capovilla and Guven [17,22–26], and we will in the following summarize some of their key findings.

2.4.1. The Membrane Stress Tensor

If we think about surface stresses, we invariably first think of surface tension. Yet, this easiest of stresses is way too special to serve as a generic guidance: the stress in a soap bubble or in the surface of a pending water droplet is constant over the surface, always tangential, and locally isotropic. In general, however, stresses can vary from place to place, also point out of the surface plane, and depend on the direction at which one places a fictitious cut through the surface. Instead of the standard constant scalar surface tension σ, we have a tensor f^{ai}, where $a \in \{1,2\}$ describes the direction in the two-dimensional surface (along $\mathbf{e}_1$ or $\mathbf{e}_2$) and $i \in \{1,2,3\}$ numbers the directions in three-dimensional space. This description in terms of a 2 × 3 matrix is a bit unhandy, though, and it turns out to be more convenient to combine the i-components into a vector that gets one surface index: $\mathbf{f}^a$. The interpretation is that $\mathbf{f}^1$ is the force per unit length along a cut in the 1-direction and $\mathbf{f}^2$ the force per unit length along a cut in the 2-direction. If we want, we can expand the stress tensor into its tangential and normal components:

$$\mathbf{f}^a = f^{ab}\mathbf{e}_b + f^a\mathbf{n} \,. \tag{2.30}$$

The link to ordinary surface tension is quite simple: in this case the stress is purely tangential, isotropic, and constant, hence f^{ab} must be proportional to the metric — and the constant of proportionality is precisely (minus) the surface tension σ — and the normal component f^a vanishes:

$$\mathbf{f}^a = -\sigma g^{ab}\mathbf{e}_b = -\sigma \mathbf{e}^a \text{ (for surface tension)} . \tag{2.31}$$

The minus sign is ultimately a convention. Here it is chosen like this: place a cut along a surface and locally define a unit vector $\mathbf{l}$ that is normal to the curve of the cut but tangential to the surface (hence we can expand it as $\mathbf{l} = l^a \mathbf{e}_a$). The cut locally separates the surface into two regions — A and B, say — and let us assume that $\mathbf{l}$ points from the A to the B-side. Then the contraction $l_a \mathbf{f}^a$ denotes the *force per unit length that the A-side exerts on the B-side*. For the case of ordinary surface tension we have $l_a \mathbf{f}^a = -\sigma l_a \mathbf{e}^a = -\sigma \mathbf{l}$, and the minus sign convention becomes understandable: surface tension is *pulling*.

The concept of the stress tensor is so useful because, as has just been mentioned, one can write down the stress tensor also for more complicated surface Hamiltonians. There are several ways to see how this comes about. One possibility is to recognize its existence as a consequence of Noether's theorem: Every continuous symmetry implies a conservation law "on shell," i.e., a law satisfied by the solutions of the corresponding Euler-Lagrange equations. Since the Hamiltonian of a membrane — or in fact any purely surface bound Hamiltonian — is evidently invariant with respect to translations of the entire surface, this implies a conserved Noether current, which indeed turns out to be the stress tensor. This line of reasoning has been outlined by Capovilla and Guven [22]. Later Guven has presented an alternative derivation, which provides a technically very powerful approach to the variation problem of a geometric functional [25]. We will briefly illustrate the main point of this second approach.

What makes the variation of geometric surface Hamiltonians so tedious is that the observables on which the energy density depends most directly — metric, metric determinant, curvature trace, etc. — are fairly involved and indirect functions of the embedding function $\mathbf{X}(u^1,u^2)$. It would be much easier if one could directly vary the functional with respect to, say, the curvature tensor or the metric. However, these "high-end" observables are not independent: they are related by integrability conditions, as we have seen above. However, it turns out that these complications can be accounted for by enforcing all the interrelations that geometry requires in the form of *constraints* to the functional. This is most directly done by enforcing the very *definitions* of the respective geometric objects and thus look at an amended functional of the following form [25]:

$$\begin{aligned} H_{\text{constraint}} = \int dA \, \{ & \mathcal{H}(g_{ab}, K_{ab}) \\ & + \mathbf{f}^a \cdot (\mathbf{e}_a - \nabla_a \mathbf{X}) + \lambda_\perp^a (\mathbf{e}_a \cdot \mathbf{n}) + \lambda_n (\mathbf{n}^2 - 1) \\ & + \lambda^{ab} (\mathbf{e}_a \cdot \mathbf{e}_b - g_{ab}) + \Lambda^{ab} (\mathbf{e}_a \cdot \nabla_b \mathbf{n} - K_{ab}) \} . \end{aligned} \tag{2.32}$$

Here, $\mathcal{H}(g_{ab}, K_{ab})$ is a Hamiltonian density depending explicitly only on the two surface tensors g_{ab} and K_{ab}. The terms in the second line define the local coordinate system, and the terms in the third line define the two surface tensors themselves. Notice that all constraints are at most quadratic. Moreover, the embedding function $\mathbf{X}(u^1,u^2)$ only appears at *one* place in the entire functional, namely at the constraint defining the tangent vectors. Varying this constraint functional with respect to the embedding function (and partially integrating once) yields the following very simple Euler-Lagrange equation:

$$\frac{\delta H_{\text{constraint}}}{\delta \mathbf{X}} = 0 \Rightarrow \underline{\nabla_a \mathbf{f}^a = 0}. \tag{2.33}$$

This proves instantly that there must exist some quantity that is covariantly conserved: the Lagrange multiplier $\mathbf{f}^a$ that pins the tangent vectors to the surface. This turns out to be the stress tensor, as is carefully explained in [27]. The other variations (with respect to $\mathbf{e}_a$, $\mathbf{n}$, g_{ab}, and K_{ab}) help to identify $\mathbf{f}^a$ in terms of the surface geometry. One easily finds [25]

$$\mathbf{f}^a = \left(T^{ab} - \mathcal{H}^{ac} K_c^b\right)\mathbf{e}_b - \left(\nabla_b \mathcal{H}^{ab}\right)\mathbf{n}, \tag{2.34}$$

where we have also defined the two additional surface tensors

$$T^{ab} = -\frac{2}{\sqrt{g}} \frac{\delta\left(\sqrt{g}\mathcal{H}\right)}{\delta g_{ab}}, \tag{2.35}$$

$$\mathcal{H}^{ab} = \frac{\delta \mathcal{H}}{\delta K_{ab}}. \tag{2.36}$$

Remarkably, the overall *structure* of the stress tensor as given in Eq. (2.34) is completely independent of the surface Hamiltonian. The latter only enters in the specific form of the two tensors $\mathcal{H}^{ab}$ and T^{ab} (the latter is sometimes also termed *metric stress*). If $\mathcal{H}$ does not explicitly depend on derivatives of metric or curvature tensor, the respective functional derivatives reduce to partial ones. In that case the metric stress tensor can also be written as $T^{ab} = -\mathcal{H} g^{ab} - 2\partial\mathcal{H}/\partial g_{ab}$, where we made use of the formula $\partial g / \partial g_{ab} = g g^{ab}$. The fact that the metric stress has two contributions results from the fact that not just the Hamilton density $\mathcal{H}(g_{ab}, K_{ab})$ but also the area element $dA = \sqrt{g}\, du^1 du^2$ depends on the metric.

That the stress tensor is divergence-free — Eq. (2.33) — is *almost* equivalent to the Euler-Lagrange equation. Almost, because $\nabla_a \mathbf{f}^a = 0$ really constitutes *three* equations, while the vanishing of the (scalar!) Euler-Lagrange derivative is only a single equation. Where do two additional equations come from? The answer is that the actual Euler-Lagrange derivative only corresponds to the *normal component* of $\nabla_a \mathbf{f}^a = 0$, since this is the one related to normal variations of the surface [22]. In contrast, tangential variations only correspond (in first order) to reparametrizations of the surface [17] and will thus only result in *consistency conditions* for the stress tensor components. Using the expansion of $\mathbf{f}^a$ into tangential and normal components introduced in Eq. (2.30), this can be written as [22]

$$\begin{aligned} 0 &= \nabla_a f^{ab} + K_a^b f^a \quad \text{(tangential)}, \\ \mathcal{E}(\mathcal{H}) &= \nabla_a f^a - K_{ab} f^{ab} \quad \text{(normal)}, \end{aligned} \tag{2.37}$$

where $\mathcal{E}(\mathcal{H})$ is the scalar Euler-Lagrange derivative that vanishes in equilibrium. Notice that the tangential projection vanishes even if $\mathcal{E}(\mathcal{H})$ does not — it has nothing to do with equilibrium and energy minimization.

If we consider a *closed* surface, there will generally be a pressure difference P between the inside and outside. From a point of view of forces, it is clear that this will act as a source of constant normal stress. In consequence, the Euler-Lagrange derivative is not zero in equilibrium but equal to this pressure difference [22]:

$$\mathcal{E}(\mathcal{H}) = P\,. \tag{2.38}$$

The easiest way to see this formally is to add a term $-PV$ to the energy functional, which may be viewed either as a Lagrange multiplier to fix the internal volume or as a Legendre transform to an ensemble of constant pressure difference between the two sides. Since the volume V of some region $\mathcal{R}$ can be expressed as

$$V = \int_{\mathcal{R}} dV = \frac{1}{3}\int_{\mathcal{R}} dV\ \nabla\cdot\mathbf{X} = \frac{1}{3}\int_{\partial\mathcal{R}} dA\ \mathbf{n}\cdot\mathbf{X}\,, \tag{2.39}$$

it can be written as a surface integral and thus be included in the above formalism.

Now that we know how to express the Euler-Lagrange derivative in terms of the stress tensor, we can work it out in terms of the surface geometry. The only remaining bits of help we need are the equations of Gauss and Weingarten, $\nabla_a \mathbf{e}_b = -K_{ab}\mathbf{n}$ and $\nabla_a \mathbf{n} = K_a^b \mathbf{e}_b$, which tell us how the local coordinate system moves if we move along the surface (they are the surface analogs of the Frenet-Serret formulas for curves [13–15]). Using this, and the normal part of Eq. (2.37), we arrive at

$$(P =)\ \mathbf{n}\cdot\nabla_a \mathbf{f}^a = \mathcal{E}(\mathcal{H}) = -K_{ab}T^{ab} + \left(K_{ac}K_b^c - \nabla_a\nabla_b\right)\mathcal{H}^{ab}\,. \tag{2.40}$$

2.4.2. Three Examples

To illustrate the above abstract formalism, let us now look at three specific examples of increasing complexity (in terms of the Hamiltonian density). We will look at the stresses and field equations they imply and later use them to learn about mediated interactions between objects that couple to these fields.

2.4.2.1. Scalar Field

The first example we will look at is a simple scalar field ϕ defined on the membrane, which is characterized by a Ginzburg-Landau (free) energy [28] of the form

$$\mathcal{H}(\phi) = \frac{1}{2}\eta(\nabla_a\phi)(\nabla^a\phi) + V(\phi)\,. \tag{2.41}$$

The only complication here is that we replaced the conventional gradient term $(\nabla\phi)^2$ by its covariant version (which we will subsequently take the liberty to again abbreviate as $(\nabla\phi)^2$). The field equation for ϕ for such a Hamiltonian can be obtained by a simple variation with respect to ϕ, and we obtain

$$0 = \mathcal{E}(\mathcal{H}) = -\eta\Delta\phi + V'(\phi)\,, \tag{2.42}$$

where $\Delta = \nabla_a \nabla^a$ is the covariant Laplacian and V' is the derivative of V with respect to its argument. Using Eq. (2.34), we equally quickly find the associated stress tensor [24]:

$$\mathbf{f}^a = \left[\eta(\nabla^a \phi)(\nabla^b \phi) - \left(\tfrac{1}{2}\eta(\nabla\phi)^2 + V(\phi)\right) g^{ab}\right] \mathbf{e}_b \,. \tag{2.43}$$

This "Ginzburg-Landau stress" is of course not new. It has been used frequently in the past within the context of statistical field theory, specifically for the treatment of forces induced by critical fluctuations (the Casimir effect) [29–33].

Since we will ultimately need projections of the stress tensor normal to some arbitrary cut through the surface, let us introduce a local coordinate system that is adapted to such a cut. Assume we have a curve with some local normalized tangent vector $\mathbf{t} = t^a \mathbf{e}_a$ and that perpendicular to this curve we have a normal vector $\mathbf{l} = l^a \mathbf{e}_a$ (still tangent to the surface, of course). Hence, $(\mathbf{l}, \mathbf{t}, \mathbf{n})$ is a convenient local orthonormal basis, which we furthermore choose to be right-handed. Notice that for reasons of completeness we also have $l_a l_b + t_a t_b = g_{ab}$. Using this coordinate system, the important normal projection of the stress tensor (2.43) is readily found to be

$$l_a \mathbf{f}^a = \left[\tfrac{1}{2}\eta\left((\nabla_\perp \phi)^2 - (\nabla_\parallel \phi)^2\right) - V(\phi)\right]\mathbf{l} + \eta(\nabla_\perp \phi)(\nabla_\parallel \phi)\mathbf{t} \,, \tag{2.44}$$

where we also defined the two directional derivatives

$$\nabla_\perp = l_a \nabla^a, \qquad \nabla_\parallel = t_a \nabla^a \,. \tag{2.45}$$

Given that the surface Hamiltonian (2.41) is purely intrinsic, we are not surprised that the corresponding stress is purely tangential. However, contrary to the case of a simple surface tension, it is no longer isotropic, since it contains the term $(\nabla^a \phi)(\nabla^b \phi)$ that is not proportional to g^{ab}.

2.4.2.2. Vector Field

As explained above, a good example of a vector field on a lipid membrane is *lipid tilt*. Let us describe this by the tangential surface vector $\mathbf{m} = m^a \mathbf{e}_a$ (which is not necessarily normalized). In this case we already have quite a number of possibilities to create a scalar energy density from this, even at minimal coupling to the geometry (i.e., $\partial_a \to \nabla_a$). A fairly general case is studied in [27]; here we will restrict to the simpler example

$$\mathcal{H}(m^a) = \tfrac{1}{2}\lambda M^2 + V(m^2), \quad \text{with} \quad M = \mathrm{div}(m^a) = \nabla_a m^a \,. \tag{2.46}$$

The Euler-Lagrange equation for the field m^a is found to be

$$0 = \mathcal{E}(\mathcal{H}) = -\lambda \nabla_a M + 2V'(m^2) m_a \,, \tag{2.47}$$

and the stress tensor is given by [27]

$$\mathbf{f}^a = \left[\left(\tfrac{1}{2}\lambda M^2 + \lambda m^c \nabla_c M - V(m^2)\right) g^{ab} - 2V'(m^2) m^a m^b\right] \mathbf{e}_b \,, \tag{2.48}$$

which is again tangential but not isotropic. Notice, however, that while in the scalar case the *gradient* term gave rise to the anisotropic stress, here it is the *potential* term. In $(\mathbf{l}, \mathbf{t})$ coordinates the projection on l_a reads

$$l_a \mathbf{f}^a = \left[\tfrac{1}{2}\lambda M^2 + \lambda\left(m_\perp \nabla_\perp + m_\parallel \nabla_\parallel\right) M - V(m^2) - 2V'(m^2) m_\perp^2\right]\mathbf{l} - 2V'(m^2) m_\perp m_\parallel \mathbf{t} \,. \quad (2.49)$$

Curiously, a vector field leads to a substantially more complicated stress than a scalar field. If we include all possible quadratic gradient terms, it becomes even more daunting [27].

2.4.2.3. Membrane Shape

In some sense, we can think of membrane shape as a tensor field, since the membrane is characterized by the two tensors g_{ab} and K_{ab}. However, as has been mentioned repeatedly, these tensors are not independent, since they both have to correspond to one and the same surface.

Once shape and tensors are involved, a large number of scalars is possible, but from a physical point of view only a few of them will usually contribute to the energy. As we have discussed above when we generalized the naïve bending expressions (2.7) or (2.8), one can systematically construct a shape-based surface density by expanding it in scalar invariants — allowed by symmetry and ordered by dimensionality. In this context the by far most-frequent Hamiltonian in use is the symmetric quadratic energy expression due to Helfrich, Eq. (2.29). We will only look at this example.

Since $K = g^{ab} K_{ab}$, we easily find $\partial K / \partial K_{ab} = g^{ab}$ and $\partial K / \partial g_{ab} = -K^{ab}$. For the latter we need to remember that $\partial g^{ab} / \partial g_{cd} = -\tfrac{1}{2}(g_{ac} g_{bd} + g_{ad} g_{bc})$. Writing $R = K^2 - K_{ab} K^{ab}$, we can also calculate that $\partial R / \partial K_{ab} = 2(K g^{ab} - K^{ab})$ and $\partial R / \partial g_{ab} = -2(K K^{ab} + K^a_c K^{bc}) = -2R^{ab}$, where the second step follows from Eq. (2.26).[10] We now only need to know two more things: (i) the Einstein tensor $G_{ab} = R_{ab} - \tfrac{1}{2} R g_{ab}$ vanishes identically in two dimensions (a simple consequence of Eq. (2.23)); and (ii) the tensor $K g^{ab} - K^{ab}$ is divergence-free, as follows from a contraction of Eq. (2.24). Putting all this together, we arrive at

$$\mathbf{f}^a = \left[\kappa K\left(K^{ab} - \tfrac{1}{2} K g^{ab}\right) - \sigma g^{ab}\right]\mathbf{e}_b - \kappa(\nabla^a K)\mathbf{n} \,. \quad (2.50)$$

Unlike the scalar- and vector-example from above, this tensor is no longer tangential: it has a normal component proportional to the gradient of the curvature. Moreover, the tangential part f^{ab} is again not isotropic — its eigenvectors coincide with those of the curvature tensor; it is thus diagonal in a coordinate system aligned with the principal directions. Notice finally that the Gaussian modulus $\bar{\kappa}$ has dropped out: *the Gaussian curvature term does not contribute to curvature stress.*

Performing the projection on $(\mathbf{l}, \mathbf{t})$ coordinates, we find

$$l_a \mathbf{f}^a = \left[\tfrac{1}{2}\kappa\left(K_\perp^2 - K_\parallel^2\right) - \sigma\right]\mathbf{l} + \kappa K K_{\perp\parallel}\, \mathbf{t} - \kappa(\nabla_\perp K)\mathbf{n} \,, \quad (2.51)$$

where we also introduced the coordinate-adapted components of the curvature tensor

$$K_\perp = l^a l^b K_{ab}, \quad K_\parallel = t^a t^b K_{ab}, \quad K_{\perp\parallel} = l^a t^b K_{ab} \,. \quad (2.52)$$

Observe the reappearance of the "$\perp^2 - \parallel^2$" motif that we've already seen in the scalar case.

From the divergence of the stress we obtain the shape equation. For a closed vesicle it reads

$$P = \mathcal{E}(\mathcal{H}) = \mathbf{n} \cdot \nabla_a \mathbf{f}^a = \kappa\left[K\left(R - \tfrac{1}{2} K^2\right) - \Delta K\right] + \sigma K \,. \quad (2.53)$$

We see that this Euler-Lagrange equation contains (i) a cubic of the curvature and (ii) its (surface) Laplacian. The tension enters linearly in the curvature. If we set the bending rigidity to zero, we arrive at the well-known Young-Laplace law $P=\sigma K$, relating the excess pressure inside a water droplet or a soap bubble to its curvature and surface tension. In fact, for a *spherical* vesicle this equation *also* reduces to $P=\sigma K$, since in this case both $R-\frac{1}{2}K^2=0$ and $\Delta K=0$; as far as bending goes, spherical surfaces do not create a normal force density.[11]

2.5. MEMBRANE-MEDIATED INTERACTIONS

The idea that interactions between membrane-associated proteins can be created by membrane-based fields (provided that the proteins somehow "couple" to these fields) is a very fruitful and well-studied one. In 1984 Mouritsen and Bloom introduced the mattress model, in which elastic stretching and compression of the lipid bilayer due to hydrophobically mismatched proteins creates interactions between these proteins and affects their phase behavior [34]. Their treatment was later refined by several other groups [35–39]. Coupling to a composition field was investigated in the context of wetting [40–43] and charge-demixing [44–46]. Goulian, Bruinsma, and Pincus first studied interactions mediated by curvature [47,48], a subject taken up by many other groups [27,49–63].

After having introduced the concept of a stress tensor, membrane-mediated interactions between bound particles can be described in a remarkably clear way: the forces between these particles are encoded in the stress tensors pertaining to the field(s) to which the particles couple. And just as in classical three-dimensional elasticity theory, the total force on a particle can be identified with the flux of stress through a closed surface (here: loop) around that particle. However, before this can be exploited, one more subtle point must be clarified.

If several such particles indeed exert forces onto each other, this will generally induce them to *move*. In this case the situation is no longer static, and we would for instance also have to consider viscous stresses. In order to avoid this complication we will assume that the relative positions of the particles are fixed. There are still membrane-mediated forces, but they are now counterbalanced by the constraints. Or, conversely, *the constraints act as sources of stress* for the membrane that can be picked up by closed loop integrals, such that the force on particle i can be written as

$$\mathbf{F}_i=-\oint_i ds\ l_a \mathbf{f}^a, \tag{2.54}$$

where the loop encircles exclusively particle i.

Below we will look at examples of mediated interactions for different fields, but always the same overall geometry: two identical particles are separated a distance d from each other, and their membrane perturbation is such that mirror symmetry with respect to the plane between them holds (see Fig. 2.3). The force on, say, the left particle can then most conveniently be obtained via a contour that passes exactly along the symmetry curve between the particles, while the three remaining branches — 2, 3, and 4 — are drawn out to infinity, since there the stress tensor either vanishes or is at least very simple.

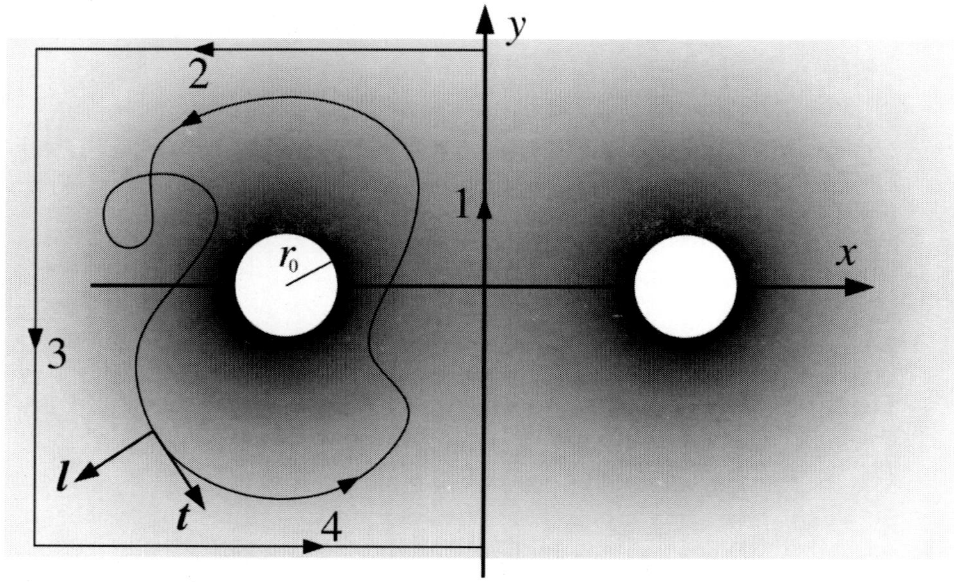

Figure 2.3. Mediated interactions via stress tensor integrals. Two circular particles of radius r_0 and mutual separation d locally disturb a field and thus interact. The force on, say, the left particle can then be obtained by a closed-loop integral of the flux of stress. Let this loop locally have a tangent vector $\mathbf{t} = l^a \mathbf{e}_a$ and an outward-pointing unit normal vector $\mathbf{l} = l^a \mathbf{e}_a$. Since the stress tensor away from external perturbations is divergence-free, the contour can be conveniently deformed such as to conform to the existing mirror symmetry. Furthermore, branches 2, 3, and 4 can then be drawn out to infinity, where the stress tensor typically assumes a very simple form.

The route to interactions à la stresses then proceeds along the following two steps:

1. Simplify the stress integral as far as possible by exploiting all available symmetries. This leads to an exact analytical connection between field and force.
2. Solve the field equation for the two particles and insert into the previously obtained force formula. This, of course, is the hard part. Below we will restrict to linearized situations and superposition approximations. It should be noted, however, that the usefulness of the stress tensor approach does not rest on these approximations. For instance, the exact analytical formula can unveil interesting information about the force even in the absence of a solution to the field equation.

2.5.1. Exact Analytical Interaction Formulas

2.5.1.1. Scalar Field

For the scalar field ϕ we need to integrate the stress flux from Eq. (2.44) along branches $1 \to 2 \to 3 \to 4$. If far away from the two particles the field approaches the value $\phi = 0$, and if furthermore $V(\phi = 0) = 0$, the stress flux vanishes on branches 2, 3, and 4. We then readily find

$$\mathbf{F} = -\int_1 ds \left\{ \tfrac{1}{2}\eta \left[(\nabla_\perp \phi)^2 - (\nabla_\parallel \phi) \right] - V(\phi) \right\} \mathbf{l} \ . \tag{2.55}$$

Moreover, since mirror symmetry holds, $\nabla_{\perp}\phi$ vanishes on branch 1. We hence obtain for the force on the left particle

$$\mathbf{F} = \int_{1} ds \left\{ \tfrac{1}{2}\eta \left(\nabla_{\parallel}\phi \right)^{2} + V(\phi) \right\} \mathbf{l} \ . \tag{2.56}$$

If $V(\phi) \geq 0$, we see that the force is attractive, since $\mathbf{l}$ always points in the positive x-direction (i.e., for the left particle that we're concerned about it points toward the right particle). This result holds true no matter how complicated $V(\phi)$ is otherwise, i.e., no matter how nonlinear and intractable the field equations are.

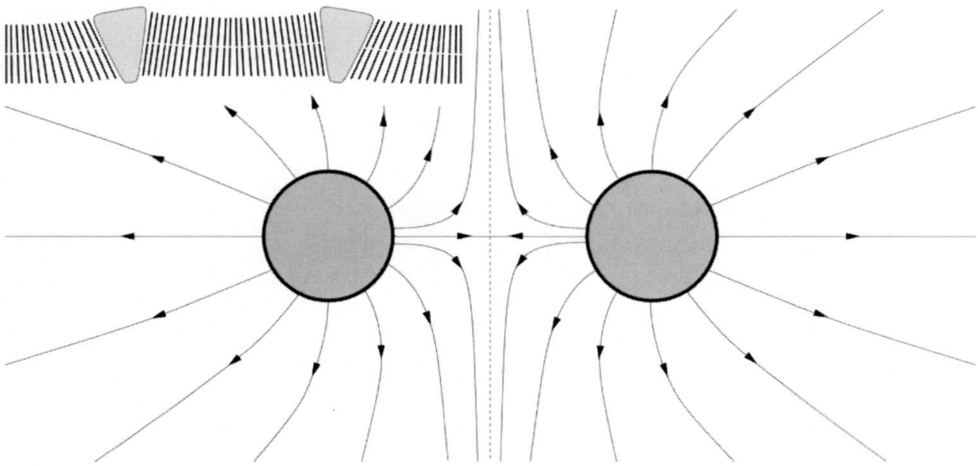

Figure 2.4. Lipid tilt field lines. Two-lipid tilt imposing circular particles symmetrically inserted into a bilayer gives rise to tilt field lines as qualitatively illustrated in this sketch. A visual analogy to electrostatics suggests that in this situation the two particles should repel — a suspicion confirmed by the actual calculation, see Eq. (2.69). Reprinted with permission from [27]. Copyright © 2005, American Physical Society.

2.5.1.2. Vector Field

For the vector field m^a we need to look at the stress (2.49). We will make the same assumption as for the scalar field, namely that m^a and $V(m^2)$ vanish far away from the perturbations, such that we again only need to consider branch 1. The stress tensor is a fair bit more complicated than in the scalar case, but mirror symmetry will save the day: most terms vanish. To understand why this is so, it is helpful to visualize the situation, which is done in Figure 2.4.

To begin with, $m_{\perp} = l^a m_a$ must vanish on the axis, since a vector field that has a mirror symmetry with respect to some plane must on that plane have a vanishing normal component. This alone removes three of the terms in Eq. (2.49). Next, the term $m_{\parallel}\nabla_{\parallel}M$ can be rewritten as $\nabla_{\parallel}(m_{\parallel}M) - M\nabla_{\parallel}m_{\parallel}$. Here, the total derivative is *along* the direction of integration and will thus only contribute two boundary terms — but these vanish if the field vanishes at the boundaries (which we took to be infinitely far away). Since furthermore we have $M = \nabla_\perp m_\perp + \nabla_\parallel m_\parallel$, we readily find

$$\mathbf{F} = -\int_{\mathrm{I}} ds \left\{ \tfrac{1}{2}\lambda \left[\left(\nabla_\perp m_\perp\right)^2 - \left(\nabla_\parallel m_\parallel\right)^2 \right] - V(m^2) \right\} \mathbf{1} \ . \tag{2.57}$$

Notice that despite a substantially different stress tensor the structure of this term is vexingly similar to the general scalar expression in Eq. (2.55). However, in this case the $\perp$-term does *not* drop out: Even though $m_\perp$ vanishes on the mid-curve, $\nabla_\perp m_\perp$ generally does not. And since this term has the opposite sign compared to the two other terms, the *overall* sign of the interaction (i.e., attraction vs. repulsion) is no longer obvious. Notice, again, the appearance of the "$\perp^2 - \parallel^2$" motif.

2.5.1.3. Membrane Shape

The relevant stress tensor is now the one from Eq. (2.51). In this case we have to be a bit careful, because even though the curvatures will approach zero far away from the perturbation (since we have an asymptotically flat membrane in mind), the surface tension σ will not vanish. However, it is easy to see what its contribution is: branches 2 and 4 will cancel against each other. On branches 1 the direction is always in the *positive* x-axis (in particular, it has no vertical component, since the surface on that branch is horizontal in x-direction for symmetry reasons). On branch 3 the direction is always in the *negative* x-axis. The only difference of these two branches is a difference in the *length* of the integration contour: While branch 3 is a straight line, branch 1 is generally curved, since the surface close to the membrane-deforming objects is not flat. There will thus be an *extra length* ΔL by which branch 1 is longer than branch 3,[12] thus implying an extra force $\sigma \Delta L$ in the positive x-direction (i.e., a tension-mediated attraction). As far as curvatures are concerned, both $KK_{\perp\parallel}$ and $\nabla_\perp K$ are zero on branch 1. The first one vanishes because branch 1 is a *line of curvature* on which the off-diagonal element of the curvature tensor, $K_{\perp\parallel}$, is equal to zero; the second vanishes obviously for reasons of mirror symmetry. We therefore arrive at the curvature-mediated force

$$\mathbf{F} = \sigma \Delta L \mathbf{x} - \tfrac{1}{2}\kappa \int_{\mathrm{I}} dx \left\{ K_\perp^2 - K_\parallel^2 \right\} \mathbf{x} \, . \tag{2.58}$$

There it is again: the "$\perp^2 - \parallel^2$" motif in the mediated force, and again the fact that it's the *difference* between squares that matters prevents a simple determination of the sign of the interaction. While the surface tension term and the $\parallel$-derivative drive an attraction, the $\perp$-derivative drives a repulsion.

We can, however, specialize to a case in which the sign becomes clear. Think of two very long particles of length L that lie parallel to each other, e.g., two actin bundles adsorbed on a membrane. In the limit in which we can ignore end-effects (say, $L \gg d$, where d is their mutual separation), this becomes a quasi-one-dimensional problem, because the translationally invariant direction along the particles drops out, and we need to look at a *force per length*, F/L, between these particles. There is still an excess length ΔL, but it does not scale with the particle length, so its contribution to the force per length vanishes for large L. Something similar happens with the $K_\parallel$-contribution: it no longer scales with the length of the particle, because it is only different from zero at the two ends — between the ends branch 1 is essentially a straight line. In consequence, the $K_{\parallel}$ contribution per length also vanishes in the limit of long particles. However, the $K_{\perp}$-contribution does not vanish: In fact, $K_{\perp}$ will have some constant value anywhere between the ends of the long particles, and the deviations at the ends will again vanish in the limit. Hence, in the long-particle limit we arrive at

$$\mathbf{F}/L = -\tfrac{1}{2}\kappa K_{\perp}^{2}\mathbf{x} \quad \text{(long particles, } L \gg d\text{)}, \tag{2.59}$$

and we immediately see that we have a repulsion. It should be pointed out that even though this formula no longer seems to contain the tension σ, this does by no means imply that the force is independent of it: it enters implicitly by codetermining the curvature $K_{\perp}$ on branch 1, since the tension σ has of course *not* dropped out of the shape equation.

2.5.2. Interaction Forces for Simple Linearized Situations

In the previous section we have discussed ways to derive interaction forces in terms of simple integrals over the stress tensor. These furnish tight connections between geometry and forces and often have a remarkably simple structure. They are analytically exact, but they are also formal, since one cannot extract actual numbers out of them without knowing the values for the respective fields. The latter requires one to solve the shape equations, and this is generally difficult, because for curved surfaces they are invariably nonlinear.

However, in order to illustrate the usefulness of the approach, we will now derive explicit force formulas between two particles for situations in which the shape equations can indeed be solved — for instance, by restricting to scalar or vector fields on flat membranes, or to small deviations from flatness such that the Hamiltonian can be linearized.

It should also be remarked that even if the differential equation is linear, solving it in the absence of sufficient symmetry can be quite a task. The reason is that boundary conditions need to be adjusted at the surfaces of two particles. One might want to object that the superposition principle should still hold — and of course it does — but what is sometimes not sufficiently well appreciated is how this principle may (or rather: may not) be used if the sources of fields are *not* point particles. One needs to write the general solution as a superposition of, say, eigenfunctions of the differential operator and *the boundary conditions need to be applied to the superposition*. It'd be so much easier if one could instead look at a single particle — a much more symmetric situation! — adjust the boundary conditions, and then superimpose two such solutions belonging to two different particles. But, alas, after boundary conditions have been fixed, one can no longer superimpose the solutions, because solution 1 of particle 1 added up onto solution 2 of particle 2 will destroy the boundary conditions at particle 2, which only went into the construction of solution 2 (and, of course, vice versa). But be this as it may, this incorrect superposition can still be a good *approximation* to the two-particle solution, and this "cheat" is more politely termed "superposition approximation" or "Nicolson approximation" [64]. For the sake of clarity we will employ this shortcut here, since we are more interested in illustrating the geometry–force relation rather than how to solve partial differential equations with tedious boundary conditions. Under what conditions a superposition approximation gives at least the correct far-field behavior is a nontrivial question that we will not delve into here.

2.5.2.1. Scalar Field

The treatment of the scalar field theory on a membrane we have provided above is valid for arbitrarily curved membranes. Remember that the curvature enters into the covariant gradient. It results in a Laplace operator in the field equation that in truth is the curvilinear Laplace-Beltrami operator, and it ends up having the metric tensor appear in the stress tensor.[13] So, even if the scalar field theory is quadratic, the resulting field equation may be very difficult because the Laplace-Beltrami operator is nonlinear.

To avoid all these complications, let us therefore restrict to a situation in which the surface on which the field lives is in fact flat. Hence, the geometry becomes trivial, but of course the forces do not. The Euler-Lagrange Eq. (2.42) becomes $-\eta\Delta\phi + V'(\phi) = 0$, where Δ now is the "ordinary" Laplacian in the plane. For the simple quadratic case $V(\phi) = \frac{1}{2}t\phi^2$ we therefore get the Helmholtz equation

$$\left(\Delta - \ell^{-2}\right)\phi = 0, \quad \text{with} \quad \ell = \sqrt{\eta / t}\,. \tag{2.60}$$

Let us assume that a circular particle of radius r_0 has adsorbed onto the surface, which locally fixes the value of the scalar field ϕ to the value ϕ_0. The problem then acquires cylindrical symmetry, and Eq. (2.60) can be rewritten as

$$\tilde{r}^2\phi'' + \tilde{r}\phi' - \tilde{r}^2\phi = 0, \quad \text{with} \quad \tilde{r} = r/\ell \quad \text{and} \quad ' = \partial/\partial\tilde{r}\,. \tag{2.61}$$

This is a modified Bessel equation, whose general form is $x^2y'' + xy' - (x^2 - n^2)y = 0$. The solution of this equation are the modified Bessel functions of first and second kind, $\mathrm{I}_n(x)$ and $\mathrm{K}_n(x)$, respectively [65]. At large argument the former asymptotically scale like $\mathrm{I}_n(x) \approx e^x / \sqrt{2\pi x}$ and the latter like $\mathrm{K}_n(x) \approx \pi e^{-x} / \sqrt{2\pi x}$. The solution of the field equation which satisfies the boundary condition $\phi(r_0) = \phi_0$ at the rim of the particle and which vanishes at $r \to \infty$ is therefore given by

$$\phi(r) = \phi_0^* \,\mathrm{K}_0(r/\ell), \quad \text{with} \quad \phi_0^* = \phi_0 / \mathrm{K}_0(r_0/\ell)\,. \tag{2.62}$$

We see that the field decays essentially exponentially with a characteristic length ℓ determined by the (square root of the) ratio between the two coupling constants η and t entering the Hamiltonian.

If we now have two such particles at a center-to-center distance d, and if we use the superposition approximation, the total field ends up being the sum of two such decaying Bessel functions, centered at the respective particles. We can then insert that field into the interaction equation (2.56) and obtain, after a little bit of algebra, the following expression for the force:[14]

$$\begin{aligned} F(d) &= 2t\ell\,\phi_0^{*2} \int_{-\infty}^{\infty} d\xi \left\{ \frac{\xi^2 \mathrm{K}_1^2\left(\sqrt{\xi^2 + (d/2\ell)^2}\right)}{\xi^2 + (d/2\ell)^2} + \mathrm{K}_0^2\left(\sqrt{\xi^2 + (d/2\ell)^2}\right) \right\} \\ &= 2\pi\, t\ell\,\phi_0^{*2}\,\mathrm{K}_1(d/\ell). \end{aligned} \tag{2.63}$$

The pair potential is then the integral over this force:

$$U(d) = -2\pi\eta\phi_0^{*2}\ \mathrm{K}_0(d/\ell)\,. \tag{2.64}$$

Just like the field itself, the interaction decays like the modified Bessel function of second kind and order 0, and thus essentially exponentially with length scale ℓ.

Notice that ℓ diverges as $t \to 0$, and the interaction thus becomes logarithmic and very long ranged. However, two cautionary remarks should be made then:

1. The potential $V(\phi)$ will most likely have higher-order terms (since in the spirit of a Landau theory we think of it as an expansion in the field). Likely this might be a term of quartic order. This will change the naïve solution obtained for $t \to 0$, but it is hard to calculate this because then the differential equation is no longer linear.
2. As is well known, a potential $V(\phi) = \frac{1}{2}t\phi^2 + \frac{1}{4}u\phi^4$ (with $u > 0$) approaches its critical point in the limit $t \to 0$ [28]. This also means that critical fluctuations will become important, and they of course must also contribute to the interaction, in the spirit of a Casimir force [29–33]. Since for a simple scalar field theory the upper critical dimension is $d = 4$, this means that for $d < 4$ the fluctuation part to the free energy at the critical point will be more important than the mean field contribution [28]. Hence, even though our calculation for the interaction force becomes large and significant in this limit, it also becomes questionable, because the critical contribution must be expected to be larger.

2.5.2.2. *Vector Field*

Restricting again to the case of a planar surface, the Euler-Lagrange equation (2.47) for the vector field m^a that belongs to the Hamiltonian density (2.46) simplifies to

$$-\lambda\nabla(\nabla\cdot\mathbf{m}) + 2V'(m^2)\mathbf{m} = 0\,, \tag{2.65}$$

where we now use the terminology $\mathbf{m}$ to describe the two-dimensional vector field in the plane. Notice that the two del-operators in a row are "grad-div" and not "div-grad", i.e., they are *not* the Laplacian of the vector $\mathbf{m}$. Assuming a quadratic potential $V(m^2) = \frac{1}{2}tm^2$, this equation becomes

$$-\nabla(\nabla\cdot\mathbf{m}) + \ell^{-2}\mathbf{m} = 0 \quad \text{with} \quad \ell = \sqrt{\lambda/t}\ , \tag{2.66}$$

where the characteristic length ℓ is again the square root of the ratio of the two moduli that define the Hamiltonian.

Restricting again to axisymmetry and writing $\mathbf{m}(r) = m(r)\mathbf{e}_r$, this equation becomes

$$\tilde{r}^2 m'' + \tilde{r}m' - (\tilde{r}^2 + 1)\phi = 0, \quad \text{with} \quad \tilde{r} = r/\ell \quad \text{and} \quad {}' = \partial/\partial\tilde{r}\,. \tag{2.67}$$

After what we have learned from the scalar case, this is readily recognized again as a modified Bessel equation, this time however of order 1. The radially symmetric solution that imposes the boundary condition $m(r_0) = m_0$ at the rim of a circular tilt-imposing particle of radius r_0 and that vanishes at infinity is given by

$$m(r) = m_0^*\ \mathrm{K}_1(r/\ell), \quad \text{with} \quad m_0^* = m_0/\mathrm{K}_1(r_0/\ell)\,. \tag{2.68}$$

This looks similar to the scalar case, but here it is the modified Bessel function of second kind and order 1 that solves the problem. Still, the tilt again decays essentially exponentially with characteristic length ℓ.

Two such particles at a center-to-center distance d, within superposition approximation, create a total field that can be inserted into interaction equation (2.57). Once more a little straightforward algebra is required and one obtains

$$\begin{aligned} F(d) &= 4t\ell\, m_0^{*2} \int_{d/2\ell}^{\infty} d\xi\, \frac{\left(\xi^2 - 2(d/2\ell)^2\right) \mathrm{K}_0(\xi)\mathrm{K}_2(\xi) + \left(\xi^2 - (d/2\ell)^2\right) \mathrm{K}_1^2(\xi)}{\xi\,\sqrt{\xi^2 - (d/2\ell)^2}} \\ &= -2\pi\, t\ell\, m_0^{*2}\, \mathrm{K}_1(d/\ell). \end{aligned} \tag{2.69}$$

While the integral looks substantially different from the scalar case, the final result is almost identical — *with one absolutely crucial difference*: there is an additional minus sign out front, which means that the interaction is this time *repulsive*! Integrating the force, we find the pair potential for the symmetric tilt-mediated interaction to be

$$U(d) = 2\pi\lambda m_0^{*2}\, \mathrm{K}_0(d/\ell)\ . \tag{2.70}$$

As has been pointed out above, the Hamiltonian (2.46) for the vector field is not the most general possible. Even if we restrict to the quadratic level, there are two more first-derivative terms possible that would consequently enter into the stress tensor and the Euler-Lagrange equation. Remarkably, it turns out that they hardly change the interaction formula (2.70): all that happens is that the modulus λ is replaced by the expression $\lambda + 2\mu$, where μ is a second modulus, multiplying the quadratic term $M_{ab}M^{ab}$ in the extended Hamiltonian, where $M_{ab} = \frac{1}{2}(\nabla_a m_b + \nabla_b m_a)$ is the *symmetrized* derivative of the vector field. The modulus ν of a third term $\frac{1}{4}\nu\, F_{ab}F^{ab}$, where $F_{ab} = \nabla_a m_b - \nabla_b m_a$ is the *antisymmetrized* derivative, does not enter the force at all. This is discussed in detail in [27].

It is indeed curious to see that the form of the interaction potential compares in every detail to the scalar case — except the sign. This shows that any plausibility argument concerning the question of whether membrane-perturbing particles attract or repel should be eyed with some suspicion. Apparently, a good sanity check is this: would the prediction of one's favorite argument depend on whether the particles perturb a scalar or a vector field?

2.5.2.3. Membrane Shape

The last case we would like to look at is curvature elastic membranes. In this case the surface of course cannot be assumed to be flat, and in order to arrive at tractable equations, we need to at least expand the integrand of the Helfrich functional up to quadratic order. The well-known result is

$$E_{\text{curv.lin}} = \tfrac{1}{2}\int dx\, dy\, \left\{\kappa\, \left(\Delta h(x,y)\right)^2 + \sigma\, \left(\nabla h(x,y)\right)^2\right\}, \tag{2.71}$$

where $h(x,y)$ now describes the surface by its deviation from the flat reference state $h = 0$ and Δ is once more the ordinary planar Laplacian. This simplified version of the energy functional is good as long as gradients are small, i.e., as long as we have $|\nabla h| \ll 1$.

The easiest way to obtain the Euler-Lagrange equations for this expanded functional is to actually vary it directly. One quickly obtains the result

$$\Delta\left(\Delta - \ell^{-2}\right)h = 0 \quad \text{with} \quad \ell = \sqrt{\kappa / \sigma}\,, \tag{2.72}$$

showing that the solutions will be the eigenfunctions of the Laplacian with eigenvalues 0 and ℓ^{-2}. Hence, in the axisymmetric situation the general solution can be written as

$$h(r) = c_1 + c_2 \ln(r/\ell) + c_3 \mathrm{K}_0(r/\ell) + c_4 \mathrm{I}_0(r/\ell)\,. \tag{2.73}$$

The solution which does not diverge at infinity and which would be created by a circular particle of radius r_0, which at its rim imposes a nonzero membrane angle α, is then given by

$$h(r) = \alpha^* \ell\left(\mathrm{K}_0(r_0/\ell) - \mathrm{K}_0(r/\ell)\right) \quad \text{with} \quad \alpha^* = \alpha / \mathrm{K}_1(r_0/\ell)\,. \tag{2.74}$$

The two-particle solution is again obtained approximately via superposition of two such one-particle solutions. Knowing the shape, we can then insert it into the interaction formula (2.58). For this we need to know that $K_\perp = \partial^2 h/\partial x^2\big|_{x=0}$ and $K_\parallel = \partial^2 h/\partial y^2\big|_{x=0}$. Also, the excess length is obtained by integrating $\sqrt{1+(\partial h/\partial y)^2} - 1 \approx \frac{1}{2}(\partial h/\partial y)^2$ along the line with $x = 0$. Working out all these derivatives is a bit tedious, but after the necessary algebra we end up with the interaction force

$$\begin{aligned} F(d) = 4\sigma\ell\,\alpha^{*2}\int_0^\infty d\xi\, &\left\{\frac{\left(\xi^2 - (d/2\ell)^2\right)\mathrm{K}_0^2\left(\sqrt{\xi^2 + (d/2\ell)^2}\right) + \xi^2\,\mathrm{K}_1^2\left(\sqrt{\xi^2 + (d/2\ell)^2}\right)}{\xi^2 + (d/2\ell)^2}\right. \\ &\left. + \frac{2\left(\xi^2 - (d/2\ell)^2\right)\mathrm{K}_0\left(\sqrt{\xi^2 + (d/2\ell)^2}\right)\mathrm{K}_1\left(\sqrt{\xi^2 + (d/2\ell)^2}\right)}{\left(\xi^2 + (d/2\ell)^2\right)^{3/2}}\right\} \\ = -2\pi\,\sigma\,\ell\,\alpha^{*2}\,\mathrm{K}_1(d/\ell)\,. \end{aligned} \tag{2.75}$$

After having recovered from the amazement that such complicated Bessel integrals have such simple solutions, we see that the force is again repulsive. The interaction potential is then found to be

$$U(d) = 2\pi\kappa\,\alpha^{*2}\,\mathrm{K}_0(d/\ell)\,, \tag{2.76}$$

and is — again! — of the same structural form as all the other equations for the other fields that we have obtained so far.

However beautiful, the result in Eq. (2.76) should be viewed with a bit of skepticism: this is a situation where the superposition approximation really becomes questionable. The reason is, among other things, that in the presence of two curvature creating particles there is one important "mode" with which the particles can react to their neighbor: They can *tilt* toward or away from each other. This process, if it happens, will clearly lower the overall free energy and thereby have an impact on the interaction. In order to account for this, one needs to do better than a simple superposition approximation, and consequently the treatment becomes significantly more involved. Yet, the calculation has been performed by Weikl, Kozlov, and Helfrich [51], even for the case of different local curvature imprints (i.e., different detachment angles α_1 and α_2 at the two particles). Since satisfying the boundary conditions exactly is extremely cumbersome, they instead expand the solution in powers of r_0/ℓ and determine the correct conditions up to this order. Setting $\alpha_1=\alpha_2=\alpha$, their result for the first two nontrivial orders becomes

$$U_{\text{WKH}}(d)=2\pi\kappa\left(\frac{\alpha\, r_0}{\ell}\right)^2\left\{\text{K}_0(d/\ell)+\left(\frac{r_0}{\ell}\right)^2\text{K}_2^2(d/\ell)\right\}. \tag{2.77}$$

This indeed looks fairly different from Eq. (2.76). Does it at least in some limit approach our formula? Yes, but two conditions are required: first, if $\ell\gg r_0$ we can make use of the small argument asymptotics $K_1(x)=1/x+O(\ln(x))$ to see that $\alpha^*\to\alpha\, r_0/\ell$, the prefactor in (2.77). Second, if $d\gg\ell$, the second term in Eq. (2.77) will vanish compared to the first one. So, both results coincide in the fairly restrictive regime $d\gg\ell\gg r_0$. Unfortunately, this regime does not include the interesting zero-tension limit $\ell\to\infty$ at fixed separation *d*. Our formula in this limit again reduces to a logarithm, while the result of Weikl, Kozlov, and Helfrich becomes

$$\lim_{\ell\to\infty}U_{\text{WKH}}(d)=8\pi\kappa\alpha^2\left(\frac{r_0}{d}\right)^4, \tag{2.78}$$

which is the well-known correct expression. Our simple result in Eq. (2.76) does evidently *not* have this property, thus warning us against a too-confident use of the superposition approximation.

As a final side note, it might be worth pointing out that Eq. (2.78) did have a convoluted history, though. The fact that tensionless membranes mediate interactions was first shown by Goulian, Bruinsma, and Pincus [47]. These authors indeed find a $(r_0/d)^4$ decay, but with a prefactor of $-4\pi(2\kappa+\bar{\kappa})\alpha^2$. Since for reasons of mechanical stability $2\kappa+\bar{\kappa}$ must always be positive,[15] the minus sign out front implied an *attraction*. In an Erratum, which appeared soon after [48], they corrected the minus sign. Later, other researchers pointed out that the dependence on the Gaussian modulus $\bar{\kappa}$ is also erroneous, since by virtue of the Gauss-Bonnet theorem [13–15] the Gaussian curvature contribution must drop out of the problem [50,51]. One then ends up at Eq. (2.78).

2.6. SUMMARY

In this chapter we have seen how membrane-mediated interactions of various types can be studied in a continuum treatment of the lipid bilayer. While the Helfrich Hamiltonian itself has been known and studied for more than three decades, the differential geometric techniques — specifically the stress tensor — are more modern or, in any case, have not been widely used. This is to some degree understandable, since this approach requires a familiarity with nonstandard mathematics. However, the remarkable geometric insight that they provide more often than not seems worth the effort. For instance, the combination of a differential geometric framework, variational calculus, and the stress tensor language proves singularly suitable for the treatment of boundary conditions of adhering or contacting surfaces [23,66]. Notice also that recently a discussion of the stress tensor in the more familiar Monge gauge has been presented [67].

Besides exact but formal expressions, we have also derived several explicit force formulas. In order to do this analytically, we had to make certain approximations (linearization, superposition) to avoid obfuscating the line of thought. It ought to be noted that these formulas are therefore not as fundamental and trustworthy as what can in principle be achieved with the stress tensor approach. Ultimately one needs to use some kind of numerics in order to solve what invariably ends up being partial nonlinear differential equations with difficult boundary conditions. But even then the stress approach will be advantageous, since it might, for instance, be numerically more stable to determine the force by a line integral over the final numerically determined solution, rather than to obtain the energy as a function of distance as a surface integral and then numerically differentiate. Notice also that a stress analysis based on membrane shape is of course not restricted to shapes calculated mathematically. One can just as well attempt to analyze shapes determined *experimentally*, for instance, using fluorescent microscopy or various cryo-EM tomography techniques. The challenge then would be to determine these shapes accurately enough such that stresses can be extracted. In other words, one needs to be able to differentiate the shape function *three times* to get the stresses and *still* have a meaningful signal left. Such a strategy holds the potential to offer unique insights into the stresses acting in living cells, and attempts to realize such a program are currently underway [68].

Everything we have talked about in this chapter was ground state theory. No fluctuations were considered. We have seen that they might in certain cases be important — maybe more important than the ground state itself — and so one continuously needs to keep them in mind. However, even then the approach to forces via stresses is very transparent, as a recent publication by Fournier and Barbetta illustrates, in which the contribution to the lateral mechanical tension of a membrane due to its fluctuations is determined by averaging the fluctuating stress tensor [69].

ACKNOWLEDGMENTS

I am immensely grateful to Jemal Guven and Martin Müller for the productive and highly enjoyable collaboration over the past few years, during which a fair amount of the specific results presented here were obtained. I would also like to thank my colleagues at the Theory Department of the MPI for Polymer Research, Mainz, and at Carnegie Mellon University, Pittsburgh, for the scientifically stimulating atmosphere, and the German Science Foundation for their financial support under grants De775/1–2,3,4.

NOTES

1. In many other cases we at least *know* what needs to be fixed, but we are too clumsy to actually do it. For instance, the "trivial" solution to any cancer is to eliminate all the bad cancer cells, and we can tell quite confidently which are the bad ones when we see them; but sadly, they are so difficult to hunt down among the healthy ones.

2. Some people prefer to restrict the word "elastic" to the context of a "gradient-square theory." Bending turns out to be a "Laplacian-square-theory." To keep the terminology comparable but emphasize the difference, one can speak of "curvature-elasticity."

3. If Poisson's ratio is negative, this means that a stretched material expands in the perpendicular direction. Remarkably, $\nu < 0$ is thermodynamically permissible, but such materials (called *auxetics*) are extremely rare.

4. For obvious reasons we require them to *not* be parallel at any point: they would otherwise not span a surface there.

5. Whatever "transformational mess-up" the expression $\partial_a v_c$ suffers from, the correction term $-\Gamma^b_{ac} v_b$ must show an opposite mess-up in order to cancel the first one. Hence, if $\partial_a v_c$ is no tensor, $-\Gamma^b_{ac} v_b$ cannot be a tensor either, for otherwise their *combination* could not be a tensor. Moral: not everything that has a bunch of indices is a tensor.

6. Our world, as we know it, depends on that! In four dimensions — relevant for space-time! — the Riemann tensor has 20 components, but the Ricci tensor has only 10. Einstein's field equations uniquely determine the Ricci tensor in terms of the energy momentum tensor of matter (e.g., the mass distribution). Hence, inside the Earth the Ricci tensor has some nontrivial value, while outside the Earth it is essentially zero. If the Ricci and Riemann tensor had the same number of components, then Ricci would uniquely determine the Riemann. But then the Riemann tensor would also vanish outside the Earth. Since a vanishing Riemann tensor implies a flat space-time, there would be no gravitational field outside the Earth — or *any* mass distribution for that matter! Curiously enough, in *three* dimensions the Ricci and Riemann tensor do have the same number of components. Gravity in three dimensions is therefore something quite different from what we know from our four-dimensional world.

7. The origin of these integrability conditions is connected with the following question: If we randomly pick a metric g_{ab} and a curvature tensor K_{ab}, will these two tensor fields describe a surface? The answer is generally "no." Unlike for the one-dimensional case, where every choice of a curvature and a torsion function leads to a well-defined curve (even unique up to translations and rotations, if we neglect subtleties with segments where the curvature vanishes), the same does not hold for surfaces. There is just too much freedom of choice with two tensor fields. One thus needs to make sure that the intrinsic geometry, dictated by the metric, is compatible with the extrinsic geometry, encoded by the curvature tensor. That's ultimately what the integrability conditions ensure.

8. Membrane patches resulting from a liquid–liquid coexistence in ternary lipid mixtures have been studied with great intensity in the context of "lipid rafts." The chapter by McConnell in this book provides a historical overview.

9. Notice the important distinction between *mechanical equilibrium* and *thermal equilibrium*: in mechanical equilibrium all stresses balance and the membrane finds the shape that minimizes the Hamiltonian. One might alternatively call this the ground state. In thermal equilibrium the membrane will additionally fluctuate, leading to additional free energy contributions and possibly shifts in ground-state properties. Owing to the softness of bending modes, such fluctuations always have the potential of being relevant, yet in the present chapter we will only consider questions pertaining to mechanical equilibrium.

10. Since $R = R^{ab} g_{ab}$, we might naïvely have expected $\partial R / \partial g_{ab} = R^{ab}$, but this is wrong: R^{ab} itself depends also on the metric and needs to be differentiated. In fact, looking at the definition of the Riemann tensor, Eq. (2.20), we see that it depends in a very complicated way on the

metric, and a purely intrinsic calculation is quite involved. Remarkably, the extrinsic detour is *significantly* less painful.

11. This is also related to the *scale invariance* of the bending energy, i.e., its independence of absolute size. If a spherical vesicle had a normal force density that would like to move the surface, it would have to be the same everywhere (since all points on a sphere are equivalent). But this would correspond to a tendency to uniformly expand or shrink the sphere. But since uniform scaling does not change the energy, the force must indeed vanish.

12. If the reader is concerned by the fact that both branch 1 and branch 3 will have infinite length once the contour has been extended to infinity, it might be advisable to calculate the difference in length *before* that limit is performed. It is then easy to see that since both branch 1 and 3 become straight lines far away from the curvature perturbation; the excess length is indeed a well-defined quantity.

13. This is in fact the only way in which the curved geometry enters. Nowhere does a curvature tensor explicitly show up. Hence, the nontrivial geometry affects the physics in the smallest possible way, and this is sometimes referred to as "minimal coupling to curvature."

14. While it is not hard to derive the integral, solving it is quite a task! The author has to confess ashamedly that he was unable to formally derive the solution. Instead, he essentially *guessed* the answer (after noticing striking similarities in the series expansion of the integral) and then confirmed it numerically up to machine precision. The same is true for the even more formidable Bessel integrals that will occur in Eqs. (2.69) and (2.75).

15. If the two principal curvatures are c_1 and c_2, we have $K = c_1 + c_2$ and $K_G = c_1 \cdot c_2$. The Helfrich Hamiltonian (density) is then $\mathcal{H} = \frac{1}{2}\kappa K^2 + \bar{\kappa} K_G = \frac{1}{2}(c_1; c_2) \cdot \begin{pmatrix} \kappa & \kappa + \bar{\kappa} \\ \kappa + \bar{\kappa} & \kappa \end{pmatrix} \cdot \begin{pmatrix} c_1 \\ c_2 \end{pmatrix}$. Requiring (the matrix of) this quadratic form to be positive definite gives rise to the conditions $2\kappa > -\bar{\kappa} > 0$.

PROBLEMS

2.1. Prove Ricci's lemma, namely, that both $\nabla_a g_{bc} = 0$ and $\nabla_a g = 0$.

2.2. Prove that the Gaussian curvature can be written as $K_G = \frac{1}{2}\left(K^2 - K_{ab}K^{ab}\right)$.

2.3. Show that the metric determinant is indeed given by the square of the modulus of the cross-product between the two tangent vectors, i.e., $g = |\mathbf{e}_1 \times \mathbf{e}_2|^2$.

2.4. Why is it true that $\mathbf{e}_a \cdot \partial_b \mathbf{n} = -\mathbf{n} \cdot \partial_b \mathbf{e}_a$?

2.5. Given that the stretching modulus of typical phospholipid bilayers is $K_{\text{stretch}} \approx 250\ \text{mN/m}$, what value for Young's modulus of the membrane interior would this imply, if we make the simplifying assumptions that such an identification is permissible? Is that value reasonable?

2.6. Show that Poisson's ratio for an incompressible material has the value $\nu = \frac{1}{2}$.

FURTHER STUDY

The topics covered in this chapter spanned from membrane biology up to differential geometry, and naturally there's no single book to study all of this. All biological aspects are nicely summarized in many excellent textbooks on molecular cell biology, such as the ones by Lodish et al. [1] or Karp [2]. Physical and in particular thermodynamic properties of membranes are covered in the book by Heimburg [3]. Much about the physics of membranes viewed as curvature-elastic fluctuating surfaces can be learned in the comprehensive review article by Seifert [18] or in the excellent *Proceedings of the Fifth Jerusalem Winter School on the Statistical Mechanics of Membranes and Surfaces*, which have recently appeared in a substantially extended second edition [70]. Differential geometry — as far as it is needed here — is best learned in the older texts such as the classical ones by do Carmo [13], Spivak [14], or Kreyszig [15]. The reader will find a more modern and abstract introduction, but with physicists in mind, in the book by Frankel [71]. The union of differential geometry and variational calculus is the topic of a beautiful book by Lovelock and Rund [72]. The stress tensor, as it has been discussed in this chapter, has been introduced by Capovilla and Guven, and their series of publications is highly recommended by virtue of being both succinct and clear. In [22] the tensor is first introduced and its existence is connected to Noether's famous theorem. A technically very different but highly efficient method both for doing the variation as well as for identifying the stress tensor has later been proposed by Guven [25] (this is the route I indicated in this chapter). An excellent coverage of the topics discussed in this chapter can also be found in the thesis by Müller, which is available online [73].

REFERENCES

1. Lodish H, Berk A, Zipursky SL, Matsudaira P, Baltimore D, Darnell J. 2000. *Molecular cell biology*, 4th ed. New York: W.H. Freeman & Company.
2. Karp G. 2007. *Cell and molecular biology: concepts and experiments*, 5th ed. Hoboken, NJ: Wiley.
3. Heimburg T. 2007. *Thermal biophysics of membranes*. Weinheim: Wiley VCH.
4. Lipowsky R, Sackmann E, eds. 1995. *Structure and dynamics of membranes, i: from cells to vesicles*. Amsterdam: Elsevier.
5. Jones RAL. 2002. *Soft condensed matter*. Oxford: Oxford UP.
6. Hemmingsen AM. 1960. Energy metabolism as related to body size and respiratory surfaces, and its evolution. *Rep Steno Meml Hosp (Copenhagen)* **9**:1–110.
7. Brown JH, West GB, eds. 2000. *Scaling in biology*. Oxford: Oxford UP.
8. Canham PB. 1970. Minimum energy of bending as a possible explanation of biconcave shape of human red blood cell. *J Theor Biol* **26**(1):61–81.
9. Helfrich W. 1973. Elastic properties of lipid bilayers — theory and possible experiments. *Z Naturforsch C* **28**(11):693–703.
10. Kwok R, Evans E. 1981. Thermoelasticity of large lecithin bilayer vesicles. *Biophy J* **35**(3):637–652.
11. Rawicz W, Olbrich KC, McIntosh T, Needham D, Evans E. 2000. Effect of chain length and unsaturation on elasticity of lipid bilayers. *Biophys J* **79**(1):328–339.
12. Landau LD, Lifshitz EM. 1999. *Theory of elasticity*, 3rd ed. Oxford: Butterworth-Heinemann.
13. Do Carmo M. 1976. *Differential geometry of curves and surfaces*. Englewood Cliffs, NJ: Prentice Hall.
14. Spivak M. 1979. *A comprehensive introduction to differential geometry*, 2nd ed. Boston: Publish or Perish.
15. Kreyszig E. 1991. *Differential geometry*. New York: Dover.
16. Hartle JB. 2003. *Gravity: an introduction to Einstein's general relativity*. San Francisco: Addison-Wesley.
17. Capovilla R, Guven J, Santiago JA. 2003. Deformations of the geometry of lipid vesicles. *J Phys A: Math Gen* **36**(23):6281–6295.

18. Seifert U. 1997. Configurations of fluid membranes and vesicles. *Adv Phys* **46**(1):13–137.
19. Goetz R, Helfrich W. 1996. The egg carton: theory of a periodic superstructure of some lipid membranes. *J Phys II* **6**(2):215–223.
20. Nelson P, Powers T. 1992. Rigid chiral membranes. *Phys Rev Lett* **69**(23):3409–3412.
21. Nelson P, Powers T. 1993. Renormalization of chiral couplings in tilted bilayer-membranes. *J Phys II* **3**(10):1535–1569.
22. Capovilla R, Guven J. 2002. Stresses in lipid membranes. *J Phys A: Math Gen* **35**(30):6233–6247.
23. Capovilla R, Guven J. 2002. Geometry of lipid vesicle adhesion. *Phys Rev E* **66**(4):041604.
24. Capovilla R, Guven J. 2004. Stress and geometry of lipid vesicles. *J Phys Cond Matter* **16**(22):S2187–S2191.
25. Guven J. 2004. Membrane geometry with auxiliary variables and quadratic constraints. *J Phys A: Math Gen* **37**(28):L313–L319.
26. Capovilla R, Guven J. 2005. Helfrich-Canham bending energy as a constrained nonlinear sigma model. *J Phys A: Math Gen* **38**(12):2593–2597.
27. Müller MM, Deserno M, Guven J. 2005. Interface-mediated interactions between particles: a geometrical approach. *Phys Rev E* **72**(6):061407.
28. Binney J-J, Dowrick NJ, Fisher AJ, Newman MEJ. 1995. *The theory of critical phenomena*. Oxford: Clarendon Press.
29. Krech M. 1994. *The Casimir effect in critical systems*. Singapore: World Scientific.
30. Krech M. 1997. Casimir forces in binary liquid mixtures. *Phys Rev E* **56**(2):1642–1659.
31. Krech M. 1999. Fluctuation-induced forces in critical fluids. *J Phys Cond Matter* **11**(37):R391–R412.
32. Brown LS. 1980. Dimensional regularization of composite-operators in scalar field-theory. *Ann Phys* **126**(1): 135–153.
33. Eisenriegler E, Stapper M. 1994. Critical behavior near a symmetry-breaking surface and the stress tensor. *Phys Rev B* **50**(14):10009–10026.
34. Mouritsen OG, Bloom M. 1984. Mattress model of lipid–protein interactions in membranes. *Biophys J* **46**(2):141–153.
35. Fattal DR, Ben Shaul A. 1993. A molecular model for lipid–protein interaction in membranes—the role of hydrophobic mismatch. *Biophys J* **65**(5):1795–1809.
36. Aranda Espinoza H, Berman A, Dan N, Pincus P, Safran S. 1996. Interaction between inclusions embedded in membranes. *Biophys J* **71**(2):648–656.
37. Nielsen C, Goulian M, Andersen OS. 1998. Energetics of inclusion-induced bilayer deformations. *Biophys J* **74**(4):1966–1983.
38. May S, Ben Shaul A. 1999. Molecular theory of lipid–protein interaction and the L-alpha–H-II transition. *Biophys J* **76**(2):751–767.
39. May S, Ben Shaul A. 2000. A molecular model for lipid-mediated interaction between proteins in membranes. *Phys Chem Chem Phys* **2**(20):4494–4502.
40. Gil T, Sabra MC, Ipsen JH, Mouritsen OG. 1997. Wetting and capillary condensation as means of protein organization in membranes. *Biophys J* **73**(4):1728–1741.
41. Gil T, Ipsen JH. 1997. Capillary condensation between disks in two dimensions. *Phys Rev E* **55**(2):1713–1721.
42. Gil T, Ipsen JH, Mouritsen OG, Sabra MC, Sperotto MM, Zuckermann MJ. 1998. Theoretical analysis of protein organization in lipid membranes. *Biochim Biophys Acta* **1376**(3):245–266.
43. Gil T, Ipsen JH, Tejero CF. 1998. Wetting controlled phase transitions in two-dimensional systems of colloids. *Phys Rev E* **57**(3):3123–3133.
44. May S, Harries D, Ben Shaul A. 2000. Lipid demixing and protein–protein interactions in the adsorption of charged proteins on mixed membranes. *Biophys J* **79**(4):1747–1760.
45. May S, Harries D, Ben Shaul A. 2002. Macroion-induced compositional instability of binary fluid membranes. *Phys Rev Lett* **89**(26):268102.
46. Mbamala EC, Ben Shaul A, May S. 2005. Domain formation induced by the adsorption of charged proteins on mixed lipid membranes. *Biophys J* **88**(3):1702–1714.
47. Goulian M, Bruinsma R, Pincus P. 1993. Long-range forces in heterogeneous fluid membranes. *Europhys Lett* **22**(2):145–150.
48. Goulian M, Bruinsma R, Pincus P. 1993. Long-range forces in heterogeneous fluid membranes (Vol. 22, p. 145, 1993). *Europhys Lett* **23**(2):155.

49. Park JM, Lubensky TC. 1996. Interactions between membrane inclusions on fluctuating membranes. *J Phys I* **6**(9):1217–1235.
50. Fournier JB, Dommersnes PG. 1997. Long-range forces in heterogeneous fluid membranes—Comment. *Europhys Lett* **39**(6):681–682.
51. Weikl TR, Kozlov MM, Helfrich W. 1998. Interaction of conical membrane inclusions: effect of lateral tension. *Phys Rev E* **57**(6):6988–6995.
52. Dommersnes PG, Fournier JB, Galatola P. 1998. Long-range elastic forces between membrane inclusions in spherical vesicles. *Europhys Lett* **42**(2):233–238.
53. Dommersnes PG, Fournier JB. 1999. N-body study of anisotropic membrane inclusions: membrane-mediated interactions and ordered aggregation. *Eur Phys J B* **12**(1):9–12.
54. Turner MS, Sens P. 1999. Inclusions on fluid membranes anchored to elastic media. *Biophys J* **76**(1):564–572.
55. Marchenko VI, Misbah C. 2002. Elastic interaction of point defects on biological membranes. *Eur Phys J E* **8**(5):477–484.
56. Fournier JB, Dommersnes PG, Galatola P. 2003. Dynamin recruitment by clathrin coats: a physical step? *CR Biol* **326**(5):467–476.
57. Bartolo D, Fournier JB. 2003. Elastic interaction between "hard" or "soft" pointwise inclusions on biological membranes. *Eur Phys J E* **11**(2):141–146.
58. Evans AR, Turner MS, Sens P. 2003. Interactions between proteins bound to biomembranes. *Phys Rev E* **67**(4):041907.
59. Weikl TR. 2003. Indirect interactions of membrane–adsorbed cylinders. *Eur Phys J E* **12**(2):265–273.
60. Sens P, Turner MS. 2004. Theoretical model for the formation of caveolae and similar membrane invaginations. *Biophys J* **86**(4):2049–2057.
61. Müller MM, Deserno M, Guven J. 2005. Geometry of surface-mediated interactions. *Europhys Lett* **69**(3):482–488.
62. Müller MM, Deserno M, Guven J. 2007. Balancing torques in membrane-mediated interactions: exact results and numerical illustrations. *Phys Rev E* **76**(1):011921.
63. Reynwar BJ, Illya G, Harmandaris VA, Müller MM, Kremer K, Deserno M. 2007. Aggregation and vesiculation of membrane proteins by curvature-mediated interactions. *Nature* **447**(7143):461–464.
64. Nicolson MM. 1949. The Interaction between floating particles. *Proc Cambr Philos Soc* **45**(2):288–295.
65. Abramowitz M, Stegun IA. 1972. *Handbook of mathematical functions*. New York: Dover.
66. Deserno M, Müller MM, Guven J. 2007. Contact lines for fluid surface adhesion. *Phys Rev E* **76**(1):011605.
67. Fournier JB. 2007. On the stress and torque tensors in fluid membranes. *Soft Matter* **3**(7):883–888.
68. Lee HJ, Peterson EL, Phillips R, Klug WS, Wiggins PA. 2005. Membrane shape as a reporter for applied forces *Proc Natl Acad Sci USA* 105(49):19253–19257.
69. Fournier JB, Barbetta C. 2008. Direct calculation from the stress tensor of the lateral surface tension of fluctuating fluid membranes. *Phys Rev Lett* **100**(7):078103.
70. Nelson D, Piran T, Weinberg S, eds. 2004. *Statistical mechanics of membranes and surfaces*, 2nd ed. Singapore: World Scientific.
71. Frankel T. 2003. *The geometry of physics*, 2nd ed. Cambridge: Cambridge UP.
72. Lovelock D, Rund H. 1989. *Tensors, differential forms, and variational principles*. New York: Dover.
73. Müller MM. 2007. *Theoretical studies of fluid membrane mechanics*. Accessed at <http://ubm.opus.hbz-nrw.de/volltexte/2007/1475/>.

3

STRUCTURE AND DYNAMICS OF LIPID MONOLAYERS: THEORY AND APPLICATIONS

Svetlana Baoukina.[1] Siewert J. Marrink,[2] and D. Peter Tieleman[1*]

[1]*Department of Biological Sciences, University of Calgary, Canada*

[2] *Groningen Biomolecular Sciences and Biotechnology Institute, Department of Biophysical Chemistry, University of Groningen, Nijenborgh, The Netherlands*

3.1. INTRODUCTION

Lipid monolayers at the air/water interface offer excellent model systems for various areas in science. They can be used as models to study two-dimensional and surface phenomena in physics and chemistry, such as adsorption, surface activity, wetting, ordering, and phase transitions. In biology, lipid monolayers represent models for biological membranes and biologically important interfaces, such as the gas exchange interface in the lungs and tear film in the eyes. Experimental and theoretical studies on monolayers have been carried out for more than a hundred years, pioneered by the works of Rayleigh, Pockels, and Langmuir. Computational studies on lipid monolayers have been developed during the last two decades to offer a different approach. Computer simulations provide information on monolayer properties at small scales but at a high temporal (picoseconds–microseconds) and spatial (Ångstroms–micrometers) resolution, and thus can complement experimental data and theoretical models.

This chapter summarizes the basic properties of lipid monolayers and gives an overview of experimental and theoretical methods for monolayers. We describe the properties of lung surfactant as an example of biological applications of lipid monolayers. We then proceed to the main focus of this chapter: computer simulations of lipid monolayers at the air/water interface.

Address all correspondence to D. Peter Tieleman, Department of Biological Sciences, University of Calgary, 2500 University Drive NW, Calgary AB T2N1N4 Canada, 403-2202966, 403-2899311 (fax), <tieleman@ucalgary.ca>, <http://moose.bio.ucalgary.ca>

R. Faller et al. (eds.), *Biomembrane Frontiers: Nanostructures, Models, and the Design of Life*, Handbook of Modern Biophysics, DOI: 10.1007/978-1-60761-314-5_3,

3.2. LIPID MONOLAYERS: BASIC PROPERTIES AND APPLICATIONS

3.2.1. Lipid Molecules

Lipids refer to a broad and diverse group of compounds. They include fatty acids, glycerolipids, glycerophospholipids, sphingolipids, sterol and prenol lipids, etc. [1]. Lipid molecules typically consist of a hydrophobic (apolar) tail and a hydrophilic (polar) headgroup (Fig. 3.1).

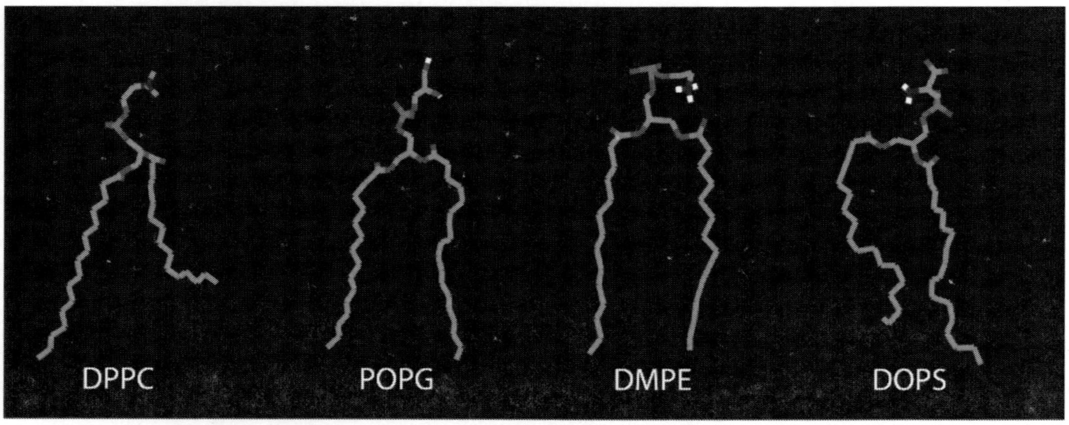

Figure 3.1. Selected glycerophospholipids with different headgroups.

The hydrophobic tail is often represented by hydrocarbon chains. Hydrocarbon chains can contain a varying number of carbons, connected by single bonds (saturated) or double bonds (unsaturated). Saturated hydrocarbon chains have different rotational isomeric states (trans or gauche) separated by small energy barriers, which provides their flexibility and a large number of conformations. In sterol lipids (for example, cholesterol), the hydrophobic part consists mainly of fused carbon rings and is rigid.

In glycerophospholipids, the glycerol backbone connects the hydrocarbon chains to the headgroup. Common headgroup types include phosphatidylcholine (PC), phosphatidylethanolamine (PE), phosphatidylserine (PS), phosphatidylglycerol (PG), and phosphatidylinositol (PI). For cholesterol, the headgroup is simply a hydroxyl group. In sphingolipids, the sphingosine backbone connects the tail to the headgroup to form ceramides, sphingomyelin, cerebrosides, etc.

Lipid molecules are elongated and asymmetric. Their size depends on the type of headgroup and on the length and degree of unsaturation of the hydrocarbon chains. For example, the ability to form intermolecular hydrogen bonds reduces the effective size of the headgroup. An increase in the number of unsaturated bonds in the chains increases the number of chain kinks, making the effective chain cross-section area wider and the molecular length shorter. As an estimate for an average lipid molecule, one can use a length of 2.5 nm (half a typical bilayer membrane) and a cross-sectional area of 0.6 nm^2. Depending on the size of the tail and on the size and the polarity of the headgroup, lipid molecules have different affinities for water and can range from very hydrophobic to amphiphilic [2]. In this chapter, we will focus on the amphiphilic lipids.

Amphiphilic lipids are poorly soluble in both polar and apolar media. Lipid molecules act as surfactants (or surface active agents): they adsorb on the polar–apolar interfaces (e.g., aqueous surfaces, oil/water interface) to form a monomolecular film (or a monolayer). In the monolayer, the headgroups face the polar subphase and the tails face the apolar subphase. The interactions between the polar and apolar media are unfavorable, and the molecular energies close to the surface are higher than in bulk. This results in a surface tension at the interface, given by the excess energy per unit area. Lipid molecules in the monolayer act to lower the surface tension at the interface by reducing the unfavorable polar–apolar contacts. Monolayers at an air–liquid interface formed by molecules insoluble in the subphases are called Langmuir films. In this chapter we will consider lipid monolayers at the air/water interfaces.

3.2.2. Monolayer: Molecular Density and Phase Behavior

Lipid monolayers at the air/water interface can adopt different molecular densities, depending on the amount of lipid and on the available interfacial area. The molecular density can be changed by varying the monolayer area at the interface or by applying an external surface pressure. The monolayer molecular density in turn determines the surface tension at the interface. A bare air/water interface has a surface tension of 72 mN/m at room temperature [3]. The surface tension decreases with increasing number of lipids per unit area and can be reduced to very low values. This reduces significantly the work required to increase the interfacial area. The surface tension at the interface in the presence of lipid molecules, i.e., the surface tension in the monolayer, γ_m, depends on the surface pressure, $\Pi(A_L)$, and the surface tension at the bare interface, γ_0:

$$\gamma_m(A_L) = \gamma_0 - \Pi(A_L), \tag{3.1}$$

where A_L denotes the area per lipid (inversely proportional to monolayer molecular density). The surface pressure equals the external pressure applied to the monolayer that is required to maintain a given area per lipid A_L. In the absence of external pressure and at an unrestricted interfacial area, the monolayer will expand indefinitely ($A_L \to \infty$) to maximize its translational entropy. At a limited interfacial area, the surface pressure in the monolayer is always larger than zero. At a high lipid concentration in the subphase (above the critical micelle concentration), the adsorption of lipid molecules at the interface will continue until the monolayer reaches the so-called equilibrium spreading pressure. For many glycerophospholipids, the equilibrium spreading pressure lies in the range of 30–50 mN/m. Further increase of surface pressure may be possible by lateral compression of the monolayer, i.e., by reducing its available interfacial area.

The surface pressure characterizes the strength of intermolecular interactions in the monolayer. These interactions contain contributions of several opposing forces [4]. Attractive van der Waals interactions between lipids favor close packing of the molecules in the monolayer. The surface tension between the hydrocarbon chains and the air tends to reduce the area per lipid. Its value is 25 ± 4 mN/m for most hydrocarbons. The headgroup–headgroup interactions include repulsive electrostatic contributions from the charges and the dipoles. These repulsive interactions are partially compensated by intermolecular hydrogen bonds and by screening by ions and water molecules that are associated with the headgroups. Interactions between the hydrocarbon tails contain a significant repulsive component, originating from the reduced chain conformational entropy in the presence of neighboring molecules.

The distribution of the intermolecular forces along the normal in lipid monolayers is heterogeneous and gives rise to a lateral pressure profile, similar to lipid bilayers (Fig. 3.2). The lateral pressure distribution in lipid monolayers depends on the molecular density. While the asymmetric pressure distribution in the monolayer gives it a tendency to bend, the surface tension at the air/water interface forces the monolayer to remain flat. Due to the small thickness of monolayers and the low solubility of lipids in water, lipid monolayers at the air/water interface behave essentially as two-dimensional systems.

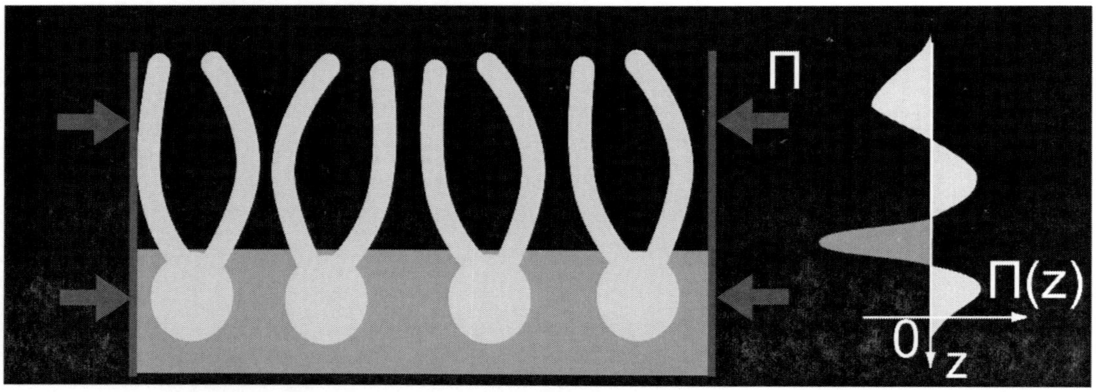

Figure 3.2. Schematic representation of intermolecular interactions and the pressure profile in lipid monolayers.

The properties of lipid monolayers depend dramatically on their molecular density. They can form various phases in two dimensions which are controlled by temperature and by the surface pressure, analogous to the three-dimensional pressure [5]. These phases can be distinguished by several structural and dynamic parameters, which we will briefly describe here.

A typical characteristic of a solid phase is its long-range translational ordering. In the two-dimensional lipid monolayer, translational ordering implies periodicity in the lateral arrangement of the molecules, i.e., the presence of a two-dimensional crystalline lattice. In a liquid phase, the order is short range and normally includes several coordination shells of a molecule. A measure of the structural arrangements of molecules is a radial distribution function, in which periodic oscillations of the density indicate translational ordering.

Hydrocarbon chains of lipid molecules are flexible and have many conformational degrees of freedom. The conformational ordering of the hydrocarbon chains can be characterized by the orientational order parameter. The orientational order parameter of the nth carbon segment in the chain with respect to the monolayer normal is given by

$$S_n^z = \left\langle \tfrac{1}{2}(3 \cdot \cos^2 \theta_n - 1) \right\rangle, \tag{3.2}$$

where θ_n is the angle formed by the vector connecting the $n - 1$ and $n + 1$ segments of the hydrocarbon chain and the monolayer normal z.

Dynamic properties of lipid monolayers can be evaluated using time correlation functions. The (long-range) lateral diffusion coefficient, D, of lipid molecules is given by

$$\langle r^2(t) \rangle = 4 \cdot D \cdot t \quad , \tag{3.3}$$

where $r(t)$ is the lateral displacement of molecules as a function of time. Rotational motions of lipids around their molecular axis can be described by a rotational autocorrelation function for the groups of interest.

The lateral compressibility of the monolayer characterizes the energetic cost to change its molecular density. The area compression modulus, K_A, is inversely proportional to the lateral compressibility and can be found from the pressure–area dependence

$$K_A = -A_L \, \partial\Pi / \partial A_L \quad . \tag{3.4}$$

Structural, dynamic, and elastic properties of a monolayer vary significantly depending on the temperature, surface pressure, and the monolayer molecular density, which in turn determine its phase state.

Monolayer phase behavior can be best described using a surface pressure–area isotherm. A schematic representation of the isotherm for a single component lipid monolayer at a temperature below the main phase transition temperature, T_m , is shown in Figure 3.3a. The main phase transition temperature corresponds to the transition from the liquid-crystalline to the gel phase in the bilayer of the same composition (at the normal atmospheric pressure).

At very low surface densities (or large areas per lipid as compared to molecular dimensions), lipid molecules in the monolayer do not interact. The monolayer is in the gas phase, in which the surface pressure is zero and the hydrocarbon chains make contacts with the water surface.

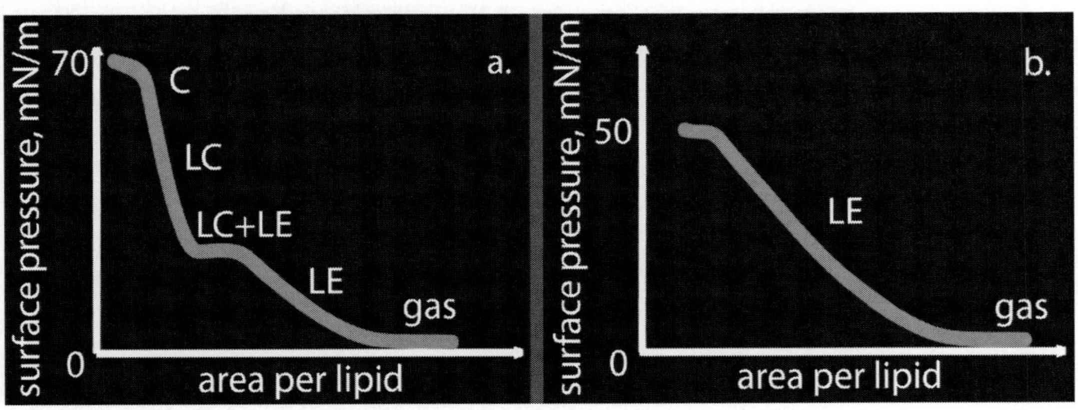

Figure 3.3. Schematic surface pressure–area isotherm of a lipid monolayer at temperatures (**a**) below and (**b**) above the main phase transition temperature.

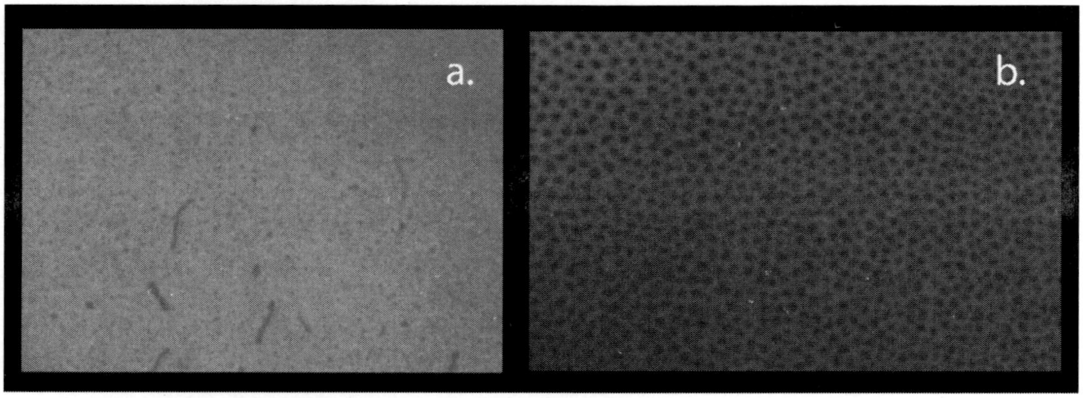

Figure 3.4. Fluorescence microscopy images of a lipid monolayer (DPPC and POPG in ratio 4:1 at 25°C): (**a**) coexistence of LE and gas phases; (**b**) coexistence of LC and LE phases. Please visit http://www.springer.com/series/7845 to view a high-resolution full-color version of this illustration.

At smaller molecular areas, the gas phase coexists with the liquid-expanded (LE) phase. The structure of the coexisting phases is very heterogeneous; the domains of the two phases assume different shapes, with the gas phase forming "bubbles" in the LE phase (see Fig. 3.4a). In this region the surface pressure is very low.

The surface pressure lifts significantly above zero as the LE phase forms an interconnected region, gradually squeezing the gas phase and becoming more homogeneous. In the LE phase, the monolayer is characterized by translational disorder of the lipid molecules and by conformational disorder of the hydrocarbon chains. The translational and rotational motions of the molecules are fast, as the effect of the presence of neighbors is small.

With increasing surface pressure, the interactions between the molecules become stronger. At surface pressures of approximately 35 mN/m, the properties of a lipid monolayer are comparable to those of a bilayer (of the same composition) in the liquid-crystalline phase. At higher surface pressures and smaller molecular areas, the monolayer adopts the liquid-condensed (LC) phase. In the LC phase, the orientational order of the hydrocarbon chains increases with respect to the LE phase. The orientations of molecular axes in this phase also become more ordered. The lateral diffusion coefficient decreases by nearly two orders of magnitude, and the rotational motions of lipids become slower by several orders. The LC phase also shows an increased translational ordering of molecules with respect to the LE phase. A common structural arrangement of molecules in the LC phase is hexagonal packing. The LC phase, however, is not fully ordered. This monolayer phase is similar to the gel phase in lipid bilayers.

In the plateau part of the isotherm the monolayer undergoes a first-order phase transition from the LE into the LC phase. Each point in this region corresponds to the coexistence of the two phases (Fig. 3.4b). The ratio of the two phases can be found by the lever rule. The coexistence region is not strictly horizontal, which is affected by the sizes and the shapes of the domains, determined by the energy and entropy of mixing of the two phases [6]. At very high surface pressures, the monolayer can transform form the LC into the condensed (C) phase (the terminology for the LC and C phases may vary, e.g., the C phase is also sometimes called solid condensed). In contrast to the gel-like LC phase, this phase is solid; the lipids are arranged in a two-dimensional crystalline lattice. Monolayers usually contain many crystalline domains [7]. The transformation from the LC into the solid phase often manifests itself as a kink in the pressure–area isotherm. The solid phase is characterized by even slower dynamics than the LC phase, with the hydrocarbon chains almost fully ordered. On increasing the molecular density in the monolayers from the gas to the solid phase, the slope of the isotherm gradually increases, corresponding to a reduction of the monolayer lateral compressibility.

With increasing temperature, the onset of the LE phase shifts toward larger molecular areas; while the LE–LC phase coexistence region becomes smaller and shifts to higher surface pressures. At temperatures above the main phase transition temperature, T_m, the monolayers may remain in the LE and do not transform into the LC (C) phases (see Fig. 3.3b).

Depending on the nature of the lipid molecules constituting the monolayer and on the number of components, the monolayers can form several LE and LC (C) phases with varying structural characteristics, such as the molecular tilt with respect to the monolayer normal. The LE phase may not fully transform into the LC phase and persist to high surface pressures.

The phases adopted by a monolayer are affected by the rate of change of its molecular density, i.e., on the rate of monolayer lateral compression. Only at very low compression rates can the monolayer achieve equilibrium for every point of the isotherm. At high compression rates, the monolayer is driven out of equilibrium, which may lead to the formation of metastable phases. Whether the compression rate is fast or slow depends on the compression time relative to the relaxation times in the monolayer [8], determined by the translational and rotational motions of lipid molecules, as well as by the cooperative displacements of molecules in response to monolayer deformation.

3.2.3. Monolayer Collapse

The higher the monolayer molecular density, the higher is the surface pressure required to obtain this density and the lower is the resulting surface tension at the interface. At a certain very high molecular density, lipid monolayers become unstable at the interface and collapse [9]. Monolayer collapse is characterized by the loss of lipids from the interface into the subphase(s). The flat two-dimensional monolayer thus explores a third dimension. The maximum surface pressure achieved in the monolayer before collapse occurs is called the collapse pressure.

The collapse pressure depends on the temperature and the lipid composition of the monolayer. At temperatures above T_m, the monolayers usually remain in the LE phase and collapse at relatively low surface pressures of 30–50 mN/m (see Fig. 3.3b). For example, most unsaturated lipids have a low main phase transition temperature below 0°C; monolayers composed of unsaturated lipids at room temperature collapse in the LE phase at low surface pressures. At high compression rates, however, the monolayers can be trapped at the interface and achieve higher pressures, forming a metastable phase [10]. At temperatures below T_m, the monolayers usually transform into the LC and/or solid phase and achieve high surface pressures before collapse. For example, a dipalmitoyl–phosphatidylcholine (DPPC) monolayer at temperatures below T = 41°C transforms into the condensed phase and collapses at a surface pressure of approximately 70 mN/m, allowing a reduction of the surface tension at the interface to values near zero.

Depending on monolayer composition, compression rate, and temperature, monolayer collapse can proceed through different pathways and can result in the formation of various lipid aggregates in the subphases [11]. The lipids ejected from the interface can form, for example, bilayer folds, vesicles, tubes, or micelles. If these aggregates remain associated with the monolayer, their re-spreading at the interface on expansion of the monolayer area is facilitated. In this case, the monolayer collapse is reversible. Otherwise, it leads to irreversible loss of lipid molecules from the monolayer.

The type of subphase for collapse depends on the nature of the lipids constituting the monolayers [12]. The chemical structure of the lipids in turn determines the surface energy of exposure of the headgroups and the tails to the water and the air subphases, respectively. For example, formation of a lipid bilayer in water reduces the energy of the unfavorable contacts of the hydrocarbon chains (in the monolayer) with air, and thus can decrease the energy of the system.

3.2.4. Lung Surfactant

Lung surfactant is a mixture of lipids and proteins forming a monomolecular film at the gas exchange interface in the lung alveoli [8]. Lung surfactant facilitates breathing. Its deficiency or inhibition causes failure of lung function that leads to severe respiratory disorders.

Lung surfactant consists mainly of PC lipids (80% by weight), with DPPC as major component (40% w). Anionic PG and PI lipids account for 8–15% by weight. Lung surfactant also contains cholesterol (5–10% w), PE lipids, fatty acids, and other minor components. Overall, lung surfactant is characterized by a rather complex lipid composition, with an almost equal content of saturated and unsaturated lipids. There are two hydrophobic proteins associated with the surface activity of lung surfactant: surfactant-associated proteins B (SP-B) and C (SP-C). Their total amount does not exceed 3% by weight; however, they play an important role in lung surfactant function.

The main physiological function of lung surfactant is to reduce the surface tension at the aqueous surface of the alveoli. Alveoli represent small globular structures that terminate the airways. Human lungs contain hundreds of millions of alveoli with a radius R, of the order of 100 μm and a total area of gas exchange interface of about 70 m^2. In the absence of surfactant, the air/water interface of the alveoli would have a high surface tension of about $\gamma \sim 70$ mN/m. Combined with the small radius of the alveoli, this interface would be unstable against the so-called Laplace pressure:

$$P = 2\gamma / R \quad (3.5)$$

Lung surfactant stabilizes the gas exchange interface by reducing the surface tension to values near zero. Moreover, it maintains the surface tension at low values at the interface upon dynamic compression and expansion of the lung surface during the breathing cycle [13]. This property is closely associated with surfactant monolayer collapse and coexistence with lipid aggregates in the subphase, called reservoirs [14].

The classical theory for lung surfactant function proposes that excess lipid material is ejected from the monolayer into the reservoirs upon interface compression at exhalation, and then re-incorporated into the interface upon interface expansion at inhalation. Earlier hypotheses proposed that upon interface compression the unsaturated lipid components are "squeezed out" from the monolayer, which would allow the saturated lipids in the monolayer to achieve low surface tensions [15,16]. Recent experimental studies have shown that the redistribution of saturated and unsaturated lipids is not necessary to obtain low surface tensions [17–19]. The protein component is not required to achieve low surface tension either; they are believed to play an important role in the connectivity of the surfactant reservoirs with the monolayer [20]. Atomic force microscopy measurements suggest that lipid reservoirs are stacks of bilayer patches adjacent and parallel to the monolayer [21].

3.2.5. Model Lipid Mixtures

Lipid monolayers can be used as models to study complex phenomena in biological systems. For example, lipid monolayer constitutes the outer layer of tear film in the eyes [22]. Lipid monolayer is the main structural element of lung surfactant, which has very low protein content. While biologically important monolayers are multicomponent systems, model lipid mixtures of simplified composition that reproduce essential properties of real systems can be used for experimental studies. In monolayer experiments it is possible, for example, to monitor the interactions of different compounds with lipid monolayers. Binding to and penetrating into the monolayer can be followed through changes in pressure–area isotherms.

Lipid monolayers at the air/water interface are systems with properties that can be precisely controlled. They can be used to produce lipid bilayers or multilayers by pulling a hydrophobic or a hydrophilic substrate through the monolayer. This procedure is called the Langmuir-Blodgett deposition technique.

Different experimental methods can be used to measure various properties of lipid monolayers at different resolutions. In the next section we will briefly describe several techniques to study model lipid mixtures.

3.3. EXPERIMENTAL STUDIES

3.3.1. Langmuir Trough

Surface pressure–area isotherms of lipid monolayers can be measured experimentally using a Langmuir trough with a Wilhelmy surface balance [23]. The setup includes a rectangular Teflon trough that is filled with water or an aqueous solution. It contains one or two movable (along one direction) Teflon barriers that allow variation of the air/water interfacial area available to the monolayer. A Wilhelmy balance consists of a thin plate composed of filter paper or other material suspended vertically at the interface. The plate is partially immersed in liquid, which forms a meniscus. The surface pressure is measured from the force pulling down the plate as a sum of contributions from buoyancy, gravity, and surface tension. Lipids dissolved in an organic solvent (for example, chloroform) can be spread at the interface with a syringe or injected into the water subphase and then adsorbed at the interface. As the solvent evaporates, the surface pressure can be monitored as a function of the monolayer area. The Langmuir trough is easy to use and allows direct control of the amount of lipids at the interface and thus measurement of the average area per molecule in the monolayer. The monolayer phase behavior can be studied by measuring surface pressure as a function of the area per lipid. The main disadvantages of the technique include monolayer leakage (escape of lipids from the interface) at high surface pressures and slow compression and expansion rates as compared, for example, to the rates in the alveoli during breathing. One of the main advantages of the method is the possibility to combine the measurement with microscopy imaging. This allows monitoring the monolayer structure and phase coexistence directly during its compression or expansion.

3.3.2. Fluorescence Microscopy

Monolayer compression–expansion on the Langmuir trough can be combined with fluorescence microscopy (FM) imaging [24,25]. A small amount of a dye — lipid analogues labeled with a fluorescent group — is introduced to the lipid mixture. The fluorescent groups of the dye molecules are usually bulky compared to other lipids and have a preference for more expanded and disordered lipid phases. Lateral organization of lipid monolayers can thus be visualized with a light microscope. LC or solid domains appear as darker regions, the LE domains correspond to brighter patterns, while the gas phase shows as dark regions due to low density. Examples of various monolayer phases are shown in Figure 3.4. The structure of the coexisting phases can be detected with micrometer resolution.

3.3.3. Brewster Angle Microscopy

Brewster angle microscopy (BAM) also provides real-time imaging of the lipid monolayer in a Langmuir trough setting [26]. The air/water interface is illuminated by a parallel laser beam polarized in the plane of incidence. The incidence angle is set to the Brewster's angle, at which no reflection occurs. The Brewster's angle depends on the refractive indices of the media. In the presence of a monolayer at the interface, a small amount of light is reflected with a different polarization due to a change in the optical properties of the system. Differences in the structure and molecular densities between various phases in the lipid monolayer manifest themselves as a contrast in the images. An advantage of the BAM method over FM is the absence of fluorescent "impurities" in the monolayer. This technique allows monolayer visualization with a micrometer resolution. To obtain a higher resolution, one can use atomic force microscopy.

3.3.4. AFM

Atomic force microscopy (AFM), also called scanning force microscopy, provides information on monolayer topography with Ångstrom resolution. In order to be manipulated with AFM, lipid monolayers have to be transferred from the air/water interface to a solid substrate (e.g., mica) using the Langmuir-Blodgett deposition technique. The microscope includes a sharp tip of nanometer end radius as a probe, which is positioned on a flexible cantilever, and scans across the sample surface. The AFM can be operated in a number of modes, with a static or oscillating cantilever. The interaction forces between the tip and sample are registered through cantilever deflection or through modulation of the oscillation frequencies and amplitudes. The force between the tip and sample and the height of the sample above the substrate can thus be measured. The vertical resolution of the AFM is very high (less than an Ångstrom). The horizontal resolution can reach nanometers for solid-like monolayers, and is lower for more liquid monolayers. The AFM can image lateral organization in lipid monolayers originating from differences in thickness. For example, it is possible to detect more ordered solid-like domains or domains containing longer lipids, as well as the presence of lipid aggregates associated with the monolayer. The main limitation of the method is a possible distortion of the monolayer structure during its deposition on a substrate.

3.3.5. Captive Bubble Surfactometer

A captive bubble surfactometer measures surface tension — area isotherms [27]. The setup consists of a glass chamber that is closed on top by a piston connected to the plunger. Below the piston a slab of agar gel is introduced to form a concave ceiling. The chamber is filled with water and an air bubble is formed in the chamber to float below the ceiling. The chamber is mounted on the microscope stand; a motor drives the piston, which in turn regulates the hydrostatic pressure in the chamber and the volume of the air bubble. A lipid monolayer is spread with a syringe at the air/water interface of the bubble. The surface tension and the area of the bubble interface are calculated from shape analysis (height-to-diameter ratio) of the bubble using the Laplace equation. For the bare air/water interface, the surface tension is high and the bubble will assume an almost spherical shape. In the presence of a monolayer, the surface tension becomes lower and the bubble shape flattens. In contrast to the Langmuir trough, the captive bubble surfactometer allows monolayer area to adjust in response to changes in surface tension induced by hydrostatic pressure. Advantages of the technique over Langmuir troughs include the absence of monolayer leakage, easier regulation of the temperature, and faster compression/expansion rates, close to the physiological rates in the lungs. One of the limitations of the method is the relative difficulty of controlling the exact amount of lipids spread at the air/water interface.

3.4. THEORETICAL MODELS

Theoretical models of lipid monolayers can be useful for interpretation of experimental data. Theoretical models generally aim to describe the equation of state of a lipid monolayer, relating temperature, T, surface pressure, Π, and area per lipid, A_L. The most common models apply a form of the van der Waals equation in two dimensions:

$$\left(\Pi + \frac{a}{A_L^2}\right)\left(A_L - A_0\right) = kT, \quad (3.6)$$

where A_0 is the excluded area per molecule, and a is the interaction term [4]. Another common approach is to use a virial expansion of the pressure as a power law series of molecular density:

$$\Pi = kT\left(\frac{1}{A_L} + \frac{B_2}{A_L^2} + \frac{B_3}{A_L^3} + \ldots\right), \quad (3.7)$$

where B_2 and B_3 are the second and the third virial coefficients, respectively, approximating the intermolecular interactions. The first term on the right-hand side corresponds to an "ideal gas," the second term contains two-body interactions, the third term contains three-body interactions, and so on. These equations describe a smooth variation of surface pressure with area per lipid indicative of a one-phase system.

Phase transitions in monolayers have drawn considerable attention, in particular the non-horizontal character of the phase coexistence plateau in the pressure–area isotherms. To describe phase coexistence, changes in intermolecular interactions between the coexisting phases and nonideal entropy of mixing of the two phases can be included in the model [6]. As a result, the pressure–area dependencies (e.g., in the form (3.6) or (3.7)) are modified and show non-horizontal lines in the two-phase region, which originate from the existence of "surface micelles" [28], or domains of finite size of the coexisting phases. Only in the limit of infinite domain size (i.e., in the case of complete phase separation) the coexistence plateau is horizontal.

An alternative approach to describe monolayer phase behavior is Landau theory, in which the free energy of the system is expanded in a Taylor series of order parameters [7]. The order parameters can represent molecular tilt, chain orientations, translational periodicity, etc. Phase transitions are then characterized by changes in one or several order parameters.

Theoretical models usually disregard or substantially simplify the molecular structure of molecules constituting a monolayer. To include atomic-level information on the nature of lipid molecules in the monolayer one can use computer simulations.

3.5. COMPUTER SIMULATIONS

Computer simulations of lipid monolayers include molecular dynamics (MD) and Monte Carlo (MC) simulations. For a many-body system, MD follows a system in time by solving classical equations of motion, while MC relies on random sampling schemes to calculate ensemble averages of the properties of interest. The purpose of the present chapter is to describe a general approach to MD simulations of lipid monolayers at the air/water interfaces. Details of the simulation parameters and algorithms are available in the literature [29] and manuals of the simulation software (http://www.gromacs.org).

3.5.1. Molecular Dynamics

In MD simulations, classical equations of motions are integrated numerically for a system of many interacting particles. A "particle" can represent an atom, a group of atoms, or even a group of molecules. Particles of different chemical composition are distinguished by different types. The interactions between particles are determined by a set of parameters for bonded and non-bonded interaction potentials, which, together with the equations that describe the interactions, are called a force field. The bonded interactions usually include bonds, bond angles, and proper and improper dihedrals, approximated by empirical functions such as harmonic potentials for bonds and angles and cosine expansions for dihedrals. The non-bonded van der Waals and electrostatic interactions are usually described by Lennard-Jones and Coulomb potentials. The forces on each particle are calculated as derivatives of the potential assuming pairwise additive interactions. A microscopic system of a limited number of particles, corresponding to the macroscopic system of interest, fills the simulation box of a given size with periodic conditions on its boundaries. The volume of the simulation box and the pressure can be controlled through various pressure-coupling algorithms. As an input, initial coordinates are assigned to all particles. The temperature is introduced through the velocity distribution. The equations of motion are integrated for a short time step and an updated configuration of the system is obtained; this procedure is repeated millions of times. The result of the simulation is a trajectory — the coordinates of all particles as a function of time. From the trajectory, various structural, dynamic, and thermodynamic properties of the system can be calculated.

3.5.2. Force Fields

In all force fields, the air (gas) phase is modeled by vacuum, because the density in the gas phase is low and intermolecular interactions are negligible. The force fields can be divided into two groups: atomistic and coarse-grained (CG) models.

Atomistic models describe all or nearly all atoms in a molecule as separate particles. The models may differ in the representations of nonpolar hydrogen atoms. For example, in the GROMOS force field, the carbon segments of the hydrocarbon chains in lipid molecules are treated as united atoms [30]. A DPPC lipid molecule in this force field comprises 50 particles. Simulations with atomistic models provide a higher level of detail at a significant computational expense. With current standard computational resources, they are suitable for studying a system containing about 100 lipid molecules on a timescale of 100 ns.

CG models are based on the grouping of atoms into interaction sites. For example, the MARTINI force field uses an approximate four-to-one mapping: four water molecules are grouped into a water particle and a DPPC molecule contains only 12 particles [31]. With this model, a system of about 10,000 lipid molecules can be studied on a microsecond timescale. The CG models are computationally faster due to a reduced number of degrees of freedom and smoother interaction potentials that allow larger time steps for integration of the equations of motion (ca. 50 fs versus ca. 2 fs for atomistic models). Thus they provide substantial computational speed-up at the expense of loss of atomic detail. The choice of a force field for simulating a lipid monolayer thus should depend on the phenomena of interest.

3.5.3. How to Simulate a Monolayer

3.5.3.1. Statistical Ensembles for Interfaces

Lipid lamellar phases have been the subject of extensive computational studies for more than a decade. The main difference of monolayer simulations is the presence of an interface with air. While the interface provides a well-defined geometry for a monolayer, its phase behavior depends on the choice of force field [32]. Lipid bilayers in equilibrium are tensionless and are characterized by a fixed area per lipid at a given temperature, and can be compressed or expanded laterally to a very low extent (less than 5%). In contrast to bilayers, lipid monolayers can assume significantly different molecular densities or areas per lipid, which in turn modulates the surface tension at the interface.

The surface tension is the conjugate variable to the interfacial area; monolayer simulations thus require selected statistical ensembles, characterized in the early works [33,34]. Valid combinations for pressures and system dimensions include: constant normal pressure and interface area, constant lateral pressure and length normal to the interface, constant volume and surface tension, and constant normal pressure and surface tension. System dimensions and pressure in simulations can be controlled with different pressure-coupling schemes. The presence of explicit vacuum in the box (substituting for air) requires the normal box size to be fixed in all simulations to prevent box shrinkage. Simulation at constant volume implies that all box dimensions are fixed, which provides constant monolayer area. A semi-isotropic pressure coupling scheme or surface tension coupling implies equal pressures in the lateral directions and independent pressure in the normal direction. These schemes can be used to set the surface tension at the interface (if its value is known), and to perform monolayer compression or expansion to the required surface tension. In this case, the monolayer area adjusts to the given surface tension.

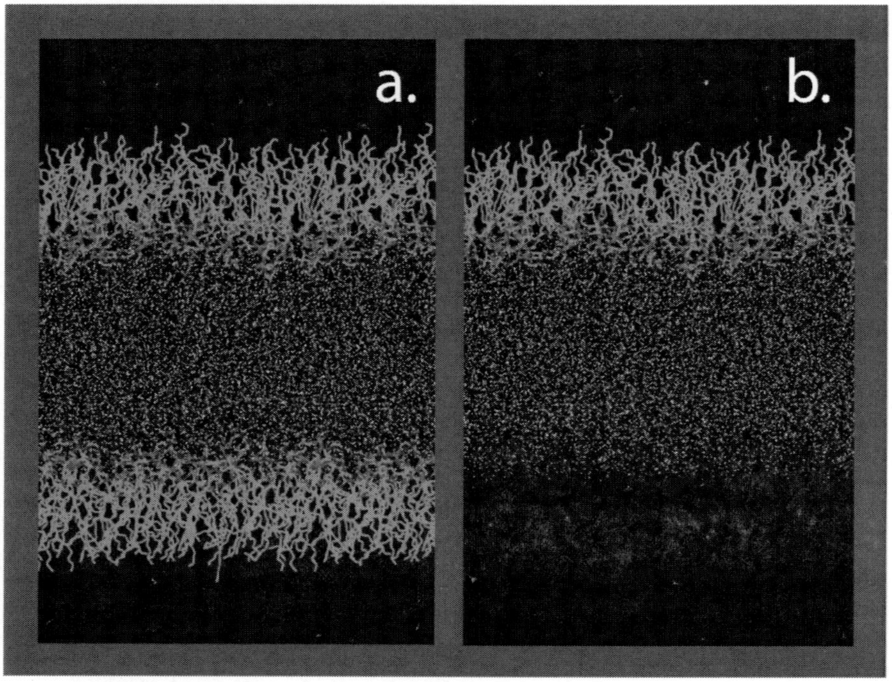

Figure 3.5. System setup for simulations of lipid monolayers consists of a water slab in vacuum and (**a**) two symmetric monolayers at the two water surfaces; (**b**) a single monolayer at one of the surfaces of the water slab.

3.5.3.2. Monolayer at the Interface: System Setup

For simulations of lipid monolayers at the air/water interface, the system setup includes a water slab in a vacuum with two symmetric monolayers at the two vacuum/water interfaces (Fig. 3.5a). The monolayers lie in the *x*-*y* plane and the monolayer normal is parallel to the *z*-axis. The presence of symmetric monolayers provides equal surface tensions at both interfaces. The surface tension in the monolayer γ_{m} is calculated from the average total surface tension in the system during the simulation. The surface tension is given by the difference of the normal, P_{N} and lateral, P_{L} pressures in the box:

$$\gamma_{\mathrm{m}} = \langle (P_{\mathrm{N}} - P_{\mathrm{L}}) \cdot L_z \rangle / 2 \tag{3.8}$$

where L_z is the box normal size, $P_{\mathrm{L}} = (P_{xx} + P_{yy})/2$, and the brackets denote averaging. An alternative system setup includes one single monolayer at the surface of the water slab (Fig. 3.5b). A wall potential can be applied to the second water surface to prevent diffusion of water molecules into vacuum. The resulting total surface tension of the system, γ, is a sum of the monolayer and the wall/water, γ_{ww}, surface tensions:

$$\gamma = \gamma_m + \gamma_{ww} . \tag{3.9}$$

The contribution of the wall potential to the surface tension can be evaluated using a control simulation of the vacuum/water system. While the setup with a single monolayer resembles more the experimental conditions and may somewhat reduce the total number of particles in the box, the setup with two monolayers provides more sampling of the monolayer phase space.

3.5.3.3. How to Compress a Monolayer

In simulations, the monolayer molecular density and surface tension can be changed and set to a selected value. This is achieved by lateral compression or expansion of the monolayer, in the same manner as in experimental methods (e.g., Langmuir trough and captive bubble surfactometer). There are two different simulation approaches: (1) to apply a surface tension; (2) to change the interfacial area available to the monolayer.

In the first method, a required surface tension is applied to the monolayer using the surface-tension coupling scheme. Alternatively, one can apply a lateral pressure using the semi-isotropic pressure coupling scheme. The resulting surface tension in the monolayer is found using Eq. (3.8). Depending on the surface tension in the starting configuration, a negative lateral pressure can produce monolayer compression or expansion. A positive lateral pressure larger than the normal pressure in the system induces a negative surface tension at the interface and thus destabilizes the monolayer. Positive lateral pressures can be used, for example, to simulate monolayer collapse.

The second method allows variation of monolayer area (or area per lipid) at a selected rate to obtain a required value. To this end, the simulation box size should be changed at a certain rate during the simulation. The box x and y dimensions control the monolayer area at the interface, while the z dimension controls the thickness of the vacuum region.

Monolayer compression and expansion can be performed in one or two directions to mimic experimental conditions. To perform a one-directional compression using the first method, an anisotropic pressure coupling scheme has to be used. Due to relatively short simulation times (up to microseconds) as compared to experiments (seconds–minutes), compression/expansion rates obtained in simulations are several orders of magnitude faster than the experimental rates.

3.5.3.4. What Parameters Affect Monolayer Properties?

In simulations, the surface tension at the bare vacuum/water interface is an important parameter (see Eq. (3.1)) that defines the correspondence of the monolayer properties to real systems at the air/water interface. It is worth mentioning here that the surface tension in simulations is a system-size dependent property [35]. It increases with decreasing simulation box size due to the suppressed undulations of the interface in the imposed periodicity at the box boundaries. The surface tension at the vacuum/water interface, γ_{vw}, depends on the water representation in the given force field. For atomistic models, it is also affected by the treatment of the long-range Lennard-Jones and electrostatic interactions. For example, in the TIP3P water model used with the atomistic CHARMM [36] and AMBER [37] force fields, the surface tension at the vacuum/water interface equals $\gamma_{vw} \approx 50–55$ mN/m at 300 K. In the SPC water model used with the

GROMOS force field, the surface tension is somewhat similar: $\gamma_{vw} \approx 55$ mN/m at 300 K. In the CG MARTINI force field, the value is lower: $\gamma_{vw} \approx 32$ mN/m at 300 K. This is a result of the Lennard-Jones representation of water particles in the model, which does not fully include the entropy of water molecules related to the dipole orientations and hydrogen bonding.

For a monolayer at the vacuum/water interface, the hydrocarbon chains of lipid molecules are exposed to vacuum and partially interact with water. The surface tensions at the hydrocarbon chain/vacuum and hydrocarbon chain/water interfaces enter the balance of surface forces in a monolayer. Most force fields closely reproduce the experimental values of these surface tensions (52 and 22 mN/m for the chain/water and chain/air interfaces in bulk alkanes [3], respectively). The underestimated surface tension of the vacuum/water interface in simulations has only a marginal effect on monolayer properties at high molecular densities (or small areas per lipid), when the number of vacuum/water contacts is negligible. However, at low molecular densities, for example, in the LE phase and LE–gas coexistence region, it has a much stronger effect. The underestimated surface tension at the vacuum/water interface artificially stabilizes pores in lipid monolayers, which appear at much higher molecular densities than in real systems. This will prevent actual expansion of the monolayer and can result in an increase in monolayer tension with increasing area.

The ability of the force field to reproduce the structural and dynamic properties and phase transitions in lipid lamellar phases is also an important factor to consider. Depending on the parameterization procedure of the "building blocks" in lipid molecules, the main phase transition temperature in lipid bilayers in simulations may be different from the real (experimental) temperature. This will also affect the phase behavior of lipid monolayers. As another example, in the CHARMM force field, lipid bilayers at the correct (comparable to experimental values) areas per lipid have nonzero tension. Although other bilayer properties are well reproduced, the interpretation of the surface tension in lipid monolayers in this case requires attention. The temperature dependence of lipid properties may be underestimated in simulations. For example, in the CG MARTINI model, due to the reduced number of degrees of freedom, the entropic contributions are compensated by enthalpic contributions to provide the correct free energies. This leads to weaker temperature dependence of lipid monolayer properties, including the surface tension–area isotherm. Most lipid force fields reproduce lipid properties reasonably well at room and physiological temperatures.

3.5.3.5. Comparing Results to Experiments

Due to the factors mentioned above, at any selected surface tension or area per lipid, the monolayer properties in simulations can be somewhat different from real systems. To find the correspondence to real systems, one can compare the monolayer surface pressure (or tension)–area dependence calculated from simulations to an experimentally measured isotherm.

From simulations one can calculate the monolayer surface tension, γ_m, and area per lipid molecule at the interface, A_L. These parameters can be found in two similar ways: (1) by setting the area per lipid to a selected value (simulation at constant monolayer area) and calculating the average surface tension during simulation; or (2) by setting the surface tension to a selected value (surface–tension coupling) and calculating the average area per molecule during simulation. In principle, the monolayer tension–area isotherm can be obtained from a single compression or expansion run. However, a set of simulations at different monolayer areas and tensions is preferable, as it allows equilibration at each point.

To find the surface pressure–area isotherm from the calculated tension–area isotherm, one can simply use Eq. (3.1). In this formula, the experimental surface tension at the air/water interface, γ_0, may be substituted by the surface tension at the vacuum/water interface, γ_{VW}, calculated from simulations. The closer the values of γ_0 and γ_{VW} are, the more straightforward the comparison. If the values differ noticeably, relation (3.1), strictly speaking, does not hold for all points of the isotherm. In this case, a correction can be introduced to compensate for the difference in surface tensions:

$$\gamma_m(A_L) = \gamma^* - \Pi(A_L)\,. \tag{3.10}$$

Here, an effective parameter, γ^*, can be found by fitting the calculated pressure–area isotherm to an experimental isotherm. This parameter may be chosen to vary for different regions of the isotherm, depending on the exposure of the water molecules to vacuum.

Fitting of simulated to experimental isotherms is, however, not straightforward for the following reasons. The experimentally measured isotherms depend on the technique and the experimental details (e.g., monolayer compression rate). Isotherms measured under identical conditions using the same technique may still differ, in particular in the high-pressure region. The discrepancies originate from differences in monolayer structure and phase behavior (e.g., the presence of metastable domains or rippled surface of the monolayer) and also from the loss of material from the interface (e.g., partial monolayer collapse or leakage) [8]. These factors are difficult to control in experiments, but easy to monitor in simulations.

On the other hand, in simulations, the isotherm slope, proportional to the area compression modulus of the monolayer (see Eq. (3.4)), depends on system size. The larger the lateral box size, the better the simulations results for a microscopic system in approximating the isotherms of macroscopic systems measured experimentally.

The most complex part of the isotherm is a phase coexistence region. To simulate phase coexistence, the system size has to be chosen larger than the critical diameter for nucleation of the second phase in a monophase monolayer. Otherwise, the monolayer will assume a metastable state with properties intermediate between the two phases. If the monolayer size is large enough, phase transformations can be induced by changing the temperature or surface pressure (tension) in the monolayer.

3.5.4. Applications

3.5.4.1. Lipid Monolayers in Atomic Detail

A DPPC monolayer is one of the most extensively studied systems. In simulations, the properties and phase behavior for a DPPC monolayer have been investigated using various all-atom models [38–41]. An atomistic representation of the monolayer is shown in Figure 3.6a. It has been possible to reproduce in simulations the gas, LE, and LC phases and their coexistence, and to visualize the monolayer structure and dynamics in these phases. For example, changes in the electron density profiles and radial distribution functions with molecular density for various groups in the monolayer and surrounding water have been calculated. Simulations at 293 K reproduce the coexistence of the LC and LE phases (for a monolayer containing ~200 lipids) for areas per lipid in the range $A_L \sim 0.5$–$0.8\ \text{nm}^2$. The simulations show a significant decrease in the

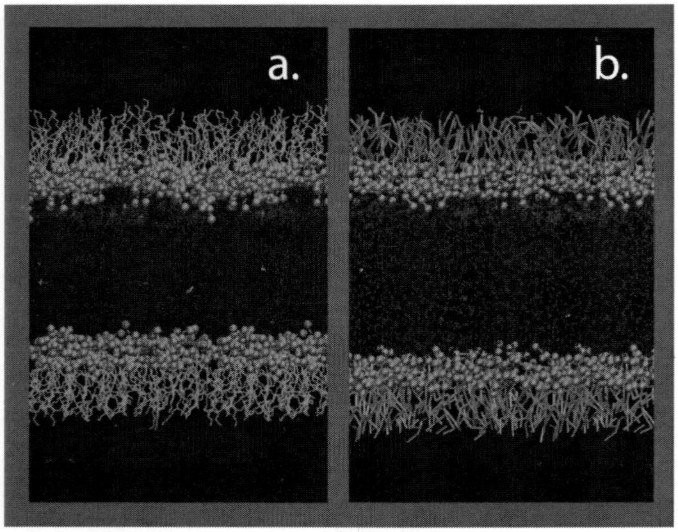

Figure 3.6. DPPC monolayers at the vacuum/water interfaces in (**a**) atomistic and (**b**) coarse-grained force fields.

orientational order parameter of the hydrocarbon chains in the gas phase (at A_L ~ 1.37 nm^2) with respect to the LE phase. Interestingly, there is a sudden increase in the order parameter at A_L ~ 1 nm^2 that has been associated with monolayer rupture and formation of stable pores, corresponding to the onset of an LE–gas coexistence region. Several selected points on the pressure–area isotherm were calculated using Eq. (3.1) at 323 and at 294 K, in qualitative agreement with experimental data.

3.5.4.2. Surface Pressure — Area Isotherms

The entire surface pressure–area isotherm of a DPPC monolayer can be calculated [42] in simulations using the MARTINI CG model (Fig. 3.7). A CG representation of the monolayer is shown in Figure 3.6b. The LC and LE phases and the LE phase with pores were reproduced for small (64 lipids) and large (~4000 lipids) monolayers. In the large monolayers, coexistence of the LC and LE phases was observed for lipid areas in the range A_L ~ 0.48–0.56 nm^2 (Fig. 3.8). The monolayer tension–area isotherms obtained as an output from simulations were fitted to the experimental surface pressure–area isotherms using Eq. (3.9) with γ^* = 55 mN/m. The fitting was performed at the LC–LE phase coexistence plateau. The width of the coexistence plateau in the isotherms depends on the relative main phase transition temperatures in bilayers. The CG model gives the bilayer transition temperature at 295 K, about 20 degrees lower than the experimental value. The best match of the isotherm calculated at 300 K was obtained for an experimental isotherm measured at 310 K [10]. The evolution of the LC and LE domains is shown in Figure 3.9. Phase separation in the monolayers is complete and results in two infinite periodic domains of the two phases spanning the simulation box. Further domain growth is limited by the box size, and the system is too small (~50 × 50 nm^2) to obtain macroscopic phase coexistence.

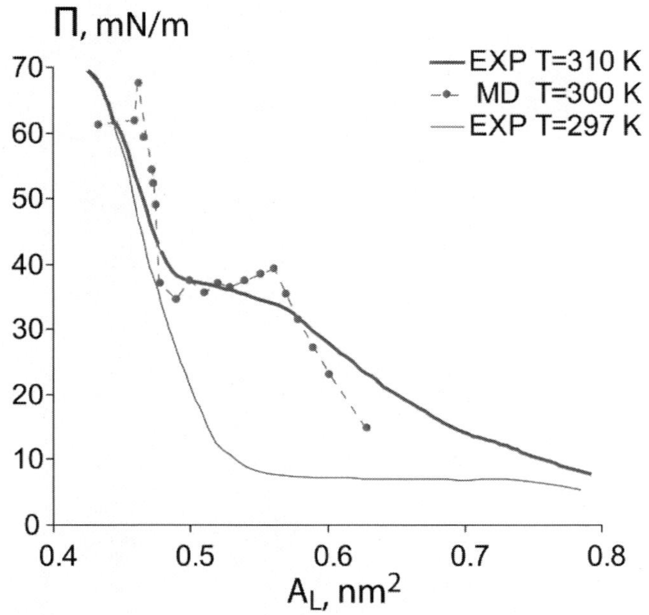

Figure 3.7. Surface pressure–area isotherm of a DPPC monolayer from MD simulations and experiments. Reprinted with permission from [42]. Copyright © 2007, American Chemical Society.

The structural and dynamic properties of the monolayer are distinct between the phases. In the LE phase, there is no long-range translational order. The orientational order parameter for the bonds of the hydrocarbon chains is about half in the LE phase compared to that in the LC phase. The lateral diffusion coefficient (see Eq. (3.3)) in the LE phase is 2×10^{-7} cm^2/s (at A_L = 0.57 nm^2), two orders of magnitude higher than in the LC phase: $D = 1 \times 10^{-9}$ cm^2/s (at A_L = 0.475 nm^2). In the LC phase, lipids assume hexagonal packing with longer-range translational ordering characterized by the oscillations in the radial distribution function (Fig. 3.7). The area compression modulus (see Eq. (3.4)) in the LC phase — $K_A \sim 1400$ mN/m — is about an order of magnitude higher than in the LE phase: $K_A \sim 150$ mN/m (at $A_L = 0.620$ nm^2).

In the high-pressure region, the slope of the isotherm becomes less steep as the monolayer surface becomes rippled, i.e., deviates from flat geometry. This is because the apparent (projected) monolayer area is smaller than the true area of the curved monolayer. On further reduction of the monolayer area, its undulations grow in amplitude and the monolayer collapses by ejecting lipids into the water subphase. This corresponds to the collapse plateau in the pressure-area isotherm.

Simulations of compression/expansion cycling were performed for a DPPC monolayer in the temperature range 293–323 K using the MARTINI CG model [43]. The pressure–area dependencies were calculated using Eq. (3.1) and are in good qualitative agreement with experiments. However, due to an underestimated surface tension of the vacuum/water interface in the CG model, the simulated dependencies are shifted toward higher surface pressures as compared to experimental data. While the presence of compression/expansion hysteresis loops is consistent with the experimental cycles, the hysteresis loops in simulations are of larger width.

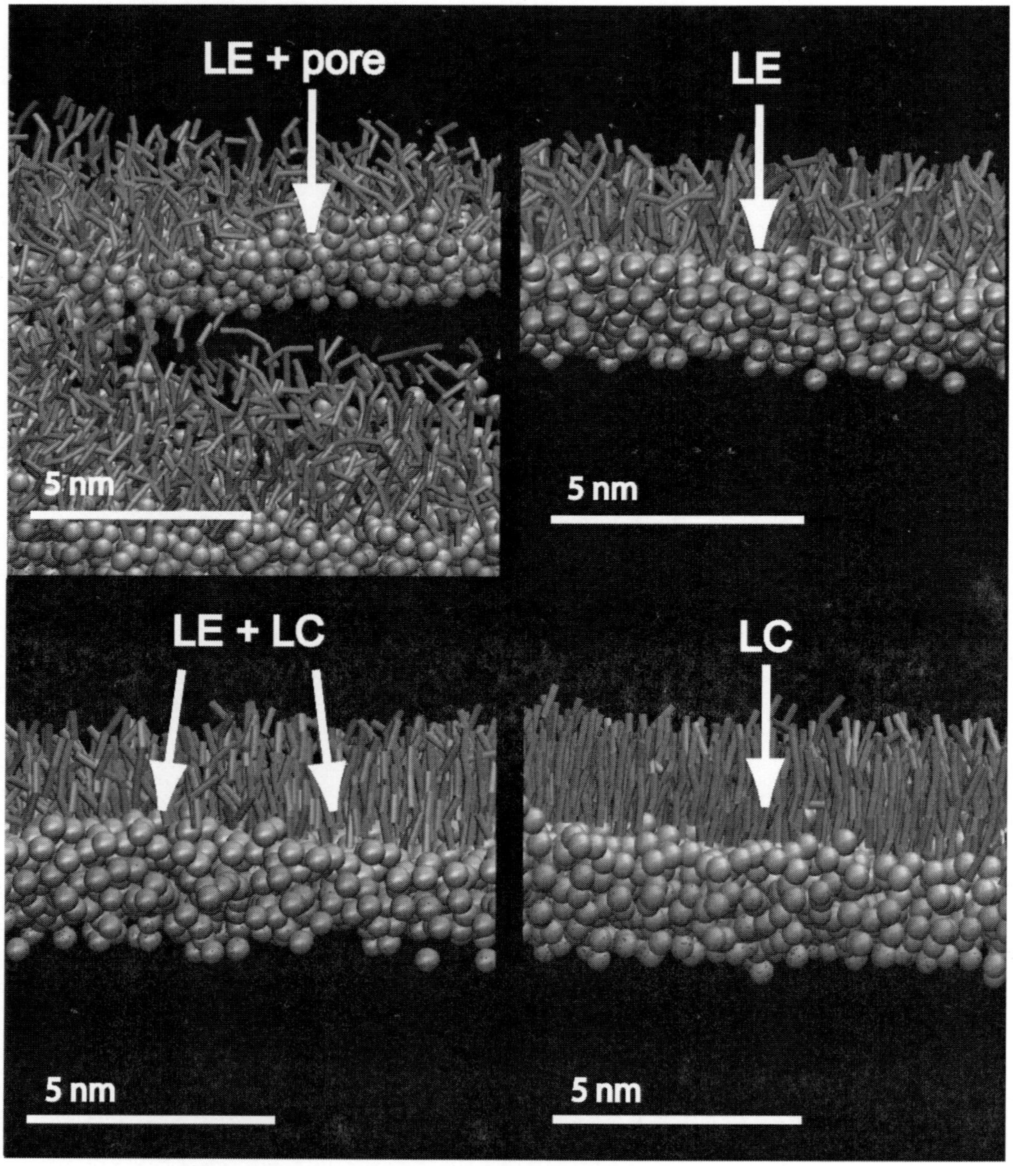

Figure 3.8. Snapshots from simulations of a DPPC monolayer: LE and LC phases, their coexistence region, and LE phase with pores. Reprinted with permission from [42]. Copyright © 2007, American Chemical Society.

3.5.4.3. Collapse Transformations of Lung Surfactant Model

In simulations, it is possible to visualize collapse transformations of lipid monolayers at sub-Ångstrom and picosecond resolution. To simulate monolayer collapse, the system size has to be chosen large enough to allow collective out-of-plane displacements of molecules. In simulations collapse usually starts at negative surface tensions, as the monolayers are artificially stabilized

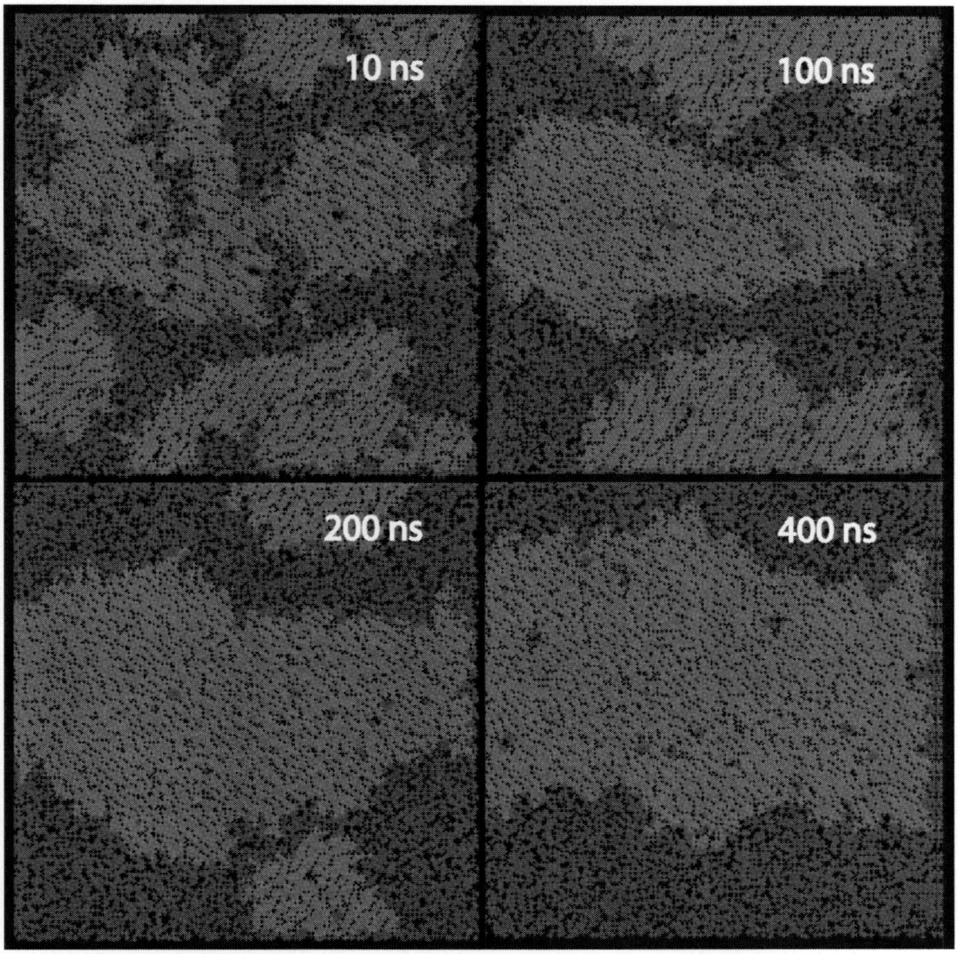

Figure 3.9. Coexistence of LE (dark gray) and LC (light gray) phases in a DPPC monolayer. Distribution of lipids between domains is obtained using Voronoi analysis. Reprinted with permission from [42]. Copyright © 2007, American Chemical Society.

in flat geometry at the interface, due to relatively small system size and the imposed periodicity at box boundaries. Monolayer collapse on lateral compression is shown in Figure 3.10 for a model lung surfactant mixture, containing DPPC and POPG lipids and cholesterol in ratio 8:2:1, and SP-C protein at 310 K [44]. The monolayer collapse is initiated by monolayer undulations, which grow in amplitude and lead to monolayer buckling, followed by folding of the monolayer into a bilayer in the water subphase.

3.6. CONCLUSIONS AND OUTLOOK

In the past decades, great progress has been made in understanding the structure and phase behavior of lipid monolayers at the air/water interface. This progress owes to considerable advancement of experimental techniques complemented by theoretical methods. However, there

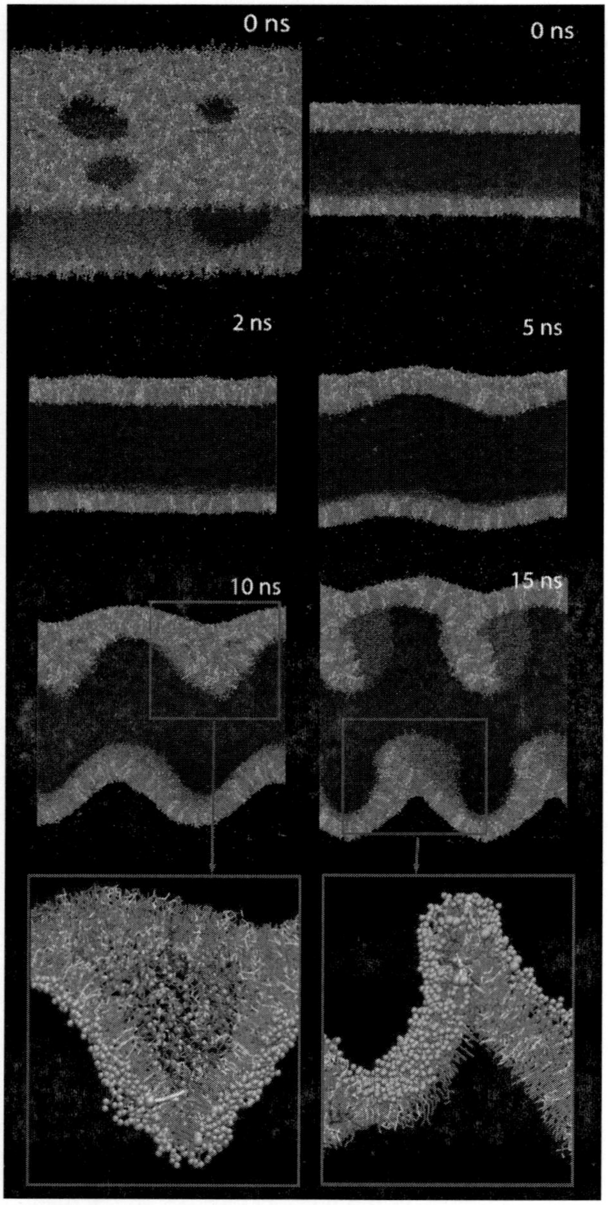

Figure 10. Monolayer transformations on collapse for a lung surfactant model: DPPC (green), POPG (yellow), and cholesterol (cyan) in ratio 8:2:1 and SP-C protein (red) (from ref. 44). Please visit http://www.springer.com/series/7845 to view a high-resolution full-color version of this illustration.

remains significant room for improvement of our understanding of lipid monolayer properties, in particular in application to complex biological systems. Experimental data can be misinterpreted or be inconsistent due to various limitations of the techniques, while theoretical models consider highly idealized systems. Molecular dynamics simulations bring new insights by providing information on the properties of multicomponent lipid monolayers at high spatial and temporal resolution. While simulations are still limited to microscopic systems on relatively short timescales, it has already been possible to visualize domain dynamics in two-phase monolayers and large-scale collective transformations during monolayer collapse. With continuing rapid advancement of computational resources and further development of simulation methods, we believe simulations will become increasingly important to study monolayer phenomena.

ACKNOWLEDGMENTS

SB is an Alberta Ingenuity postdoctoral fellow, and DPT is an Alberta Heritage Foundation for Medical Research Senior Scholar and Canadian Institutes of Health Research New Investigator. This work was supported by the Natural Science and Engineering Research Council (Canada). We thank Jelger Risselada for the implementation of the simulation box rescaling algorithm and his contributions to data analysis.

PROBLEMS

3.1. Which lipid constitutes a better surfactant: triglyceride or DPPC?

3.2. Depending on the temperature and surface pressure, lipid monolayers can form a variety of phases — e.g., gas, liquid-expanded, liquid-condensed, and condensed. At a fixed temperature and surface pressure, which monolayer has a higher molecular density: DPPC or POPC?

3.3. The slope of the pressure–area isotherm gradually increases with increasing surface pressure. What can explain the decrease in isotherm slope at high pressures often observed experimentally?

3.4. In simulations, molecular force fields often underestimate the surface tension at the water/vacuum interface. What is the possible result of a simulation, if the surface tension applied to the monolayer is larger than the surface tension at the water/vacuum interface (and smaller than the surface tension at the air/water interface in real systems).

FURTHER READING

1. Mouritsen, O.G. 2005. *Life — as a matter of fat*. Springer: Heidelberg, 2005.
2. Cevc, G. *Phospholipids handbook.* Marcel Dekker: New York. 1993
3. Gibbs, J.W. *Collected works.* Longmans Green: New York. 1928.
4. Safran, S.A. Statistical Thermodynamics of Surfaces, Interfaces and Membranes. Addison-Wesley: New York, 1994.
5. Israelachvili, J. *Intermolecular and Surface Forces.* Academic press: London.1992.
6. Lee, K. Y. C. Collapse mechanisms of Langmuir monolayers Annu. Rev. Phys. Chem. 2008. 59, 771-791.
7. Zuo, Y. Y., Veldhuizen, R. A. W., Neumann, A. W., Petersen, N. O. & Possmayer, F. *Current Perspectives in Pulmonary Surfactant -- Inhibition, Enhancement and Evaluation.* BBA – Biomembranes. doi:10.1016/j.bbamem.2008.03.021.

REFRENCES

1. Fahy E, Subramaniam S, Brown HA, Glass CK, Merrill AH, Murphy RC, Raetz CRH, Russell DW, Seyama Y, Shaw W, Shimizu T, Spener F, van Meer G, Van Nieuwenhze MS, White SH, Witztum JL, Dennis EA. 2005. A comprehensive classification system for lipids. *Eur J Lipid Sci Technol* **107**(5):337–364.
2. Mouritsen OG. 2005. *Life — as a matter of fat*. Heidelberg: Springer.
3. Aveyard R, Haydon DA. 1973. *An introduction to the principles of surface chemistry*. New York: Cambridge UP.
4. Marsh D. 1996. Lateral pressure in membranes. *Biochim Biophys Acta* **1286**(3):183–223.
5. Knobler CM, Desai RC. 1992. Phase transitions in monolayers. *Annu Rev Phys Chem* **43**:207–236.
6. Ruckenstein E Li, BQ. 1998. Surface equation of state for insoluble surfactant monolayers at the air/water interface. *J Phys Chem B* **102**(6):981–989.
7. Kaganer VM, Mohwald H, Dutta P. 1999. Structure and phase transitions in Langmuir monolayers. *Rev Mod Phys* **71**(3):779–819.
8. Wustneck R, Perez-Gil J, Wustneck N, Cruz A, Fainerman VB, Pison U. 2005. Interfacial properties of pulmonary surfactant layers. *Adv Colloid Interface Sci* **117**(1–3):33–58.
9. Gaines GL. 1966. *Insoluble monolayers at liquid–gas interfaces*. New York: Wiley Interscience.
10. Crane JM, Putz G, Hall SB. 1999. Persistence of phase coexistence in disaturated phosphatidylcholine monolayers at high surface pressures. *Biophys J* **77**(6):3134–3143.
11. Rugonyi S, Smith EC, Hall SB. 2004. Transformation diagrams for the collapse of a phospholipid monolayer. *Langmuir* **20**(23):10100–10106.
12. Lu WX, Knobler CM, Bruinsma RF, Twardos M, Dennin M. 2002. Folding Langmuir monolayers. *Phys Rev Lett* **89**(14):146107.
13. Bachofen H, Schurch S. 2001. Alveolar surface forces and lung architecture. *Comp Biochem Physiol A: Mol Integr Physiol* **129**(1):183–93.
14. Lipp MM, Lee KYC, Takamoto DY, Zasadzinski JA, Waring AJ. 1998. Coexistence of buckled and flat monolayers. *Phys Rev Lett* **81**(8):1650–1653.
15. Bangham AD, Morley CJ, Phillips MC. 1979. The physical properties of an effective lung surfactant. *Biochim Biophys Acta* **573**(3):552–556.
16. Clements JA. 1977. Functions of alveolar lining. *Am Rev Respir Dis* **115**(6):67–71.
17. Piknova B, Schief WR, Vogel V, Discher BM, Hall SB. 2001. Discrepancy between phase behavior of lung surfactant phospholipids and the classical model of surfactant function. *Biophys J* **81**(4)2172–2180.
18. Takamoto DY, Lipp MM, von Nahmen A, Lee KYC, Waring AJ, Zasadzinski JA. 2001. Interaction of lung surfactant proteins with anionic phospholipids. *Biophys J* **81**(1):153–169.
19. Yu SH, Possmayer F. 2003. Lipid compositional analysis of pulmonary surfactant monolayers and monolayer-associated reservoirs. *J Lipid Res* **44**(3):621–629.
20. Serrano AG, Perez-Gil J. 2006. Protein–lipid interactions and surface activity in the pulmonary surfactant system. *Chem Phys Lipids* **141**(1–2):105–118.
21. von Nahmen A, Schenk M, Sieber M, Amrein M. 1997. The structure of a model pulmonary surfactant as revealed by scanning force microscopy. *Biophys J* **72**(1):463–469.
22. Sullivan DA, Stern ME, Tsubota K, Dartt DA, Sullivan RM, Bromberg BB. 2002. *Lacrimal gland, tear film and dry eye syndromes 3*. Boston: Springer.
23. Clements JA. 1957. Surface tension of lung extracts. *Proc Soc Exp Biol Med* **95**(1):170–172.
24. Weis RM. 1991. Fluorescence microscopy of phospholipid monolayer phase transitions. *Chem Phys Lipids* **57**(2–3):227–239.
25. Knobler CM. 1990. Seeing phenomena in flatland—studies of monolayers by fluorescence microscopy. *Science* **249**(4971):870–874.
26. Meunier J. 2000. Why a Brewster angle microscope? *Colloids Surf* **171**(1–3):33–40.
27. Schurch S, Bachofen H, Goerke J, Possmayer F. 1989. A captive bubble method reproduces the in situ behavior of lung surfactant monolayers. *J Appl Physiol* **67**(6):2389–2396.
28. Israelachvili J. 1994. Self-assembly in 2 dimensions—surface micelles and domain formation in monolayers. *Langmuir* **10**(10):3774–3781.
29. Lindahl E, Hess B, van der Spoel D. 2001. GROMACS 3.0: a package for molecular simulation and trajectory analysis. *J Mol Model* **7**(8):306–317.

30. van Gunsteren WF. 1987. *GROMOS: Groningen molecular simulation program package*. Groningen: University of Groningen.
31. Marrink SJ, Risselada HJ, Yefimov S, Tieleman DP, Vries AH. 2007. The MARTINI force field: coarse-grained model for biomolecular simulations. *J Phys Chem B* **111**(27):7812–7824.
32. Ahlstrom P, Berendsen HJC. 1993. A molecular dynamics study of lecithin monolayers. *J Phys Chem* **97**:13691–13702.
33. Feller SE, Zhang YH, Pastor RW. 1995. Computer-simulation of liquid/liquid interfaces, 2: surface-tension area dependence of a bilayer and monolayer. *J Chem Phys* **103**(23):10267–10276.
34. Zhang YH, Feller SE, Brooks BR, Pastor RW. 1995. Computer-simulation of liquid/liquid interfaces, 1: theory and application to octane/water. *J Chem Phys* **103**(23):10252–10266.
35. Lindahl E, Edholm O. 2000. Mesoscopic undulations and thickness fluctuations in lipid bilayers from molecular dynamics simulations. *Biophys J* **79**(1):426–433.
36. MacKerell AD, Bashford D, Bellott M, Dunbrack RL, Evanseck JD, Field MJ, Fischer S, Gao J, Guo H, Ha S, Joseph-McCarthy D, Kuchnir L, Kuczera K, Lau FTK, Mattos C, Michnick S, Ngo T, Nguyen DT, Prodhom B, Reiher WE, Roux B, Schlenkrich M, Smith JC, Stote R, Straub J, Watanabe M, Wiorkiewicz-Kuczera J, Yin D, Karplus M. 1998. All-atom empirical potential for molecular modeling and dynamics studies of proteins. *J Phys Chem B* **102**(18):3586–3616.
37. Case DA, Cheatham TE, Darden T, Gohlke H, Luo R, Merz KM, Onufriev A, Simmerling C, Wang B, Woods RJ. 2005. The Amber biomolecular simulation programs. *J Comput Chem* **26**(16):1668–1688.
38. Dominguez H, Smondyrev AM, Berkowitz ML. 1999. Computer simulations of phosphatidylcholine monolayers at air/water and CCl4/water interfaces. *J Physic Chem B* **103**(44):9582–9588.
39. Knecht V, Muller M, Bonn M, Marrink SJ, Mark AE. 2005. Simulation studies of pore and domain formation in a phospholipid monolayer. *J Chem Phys* **122**(2):024704.
40. Mauk AW, Chaikof EL, Ludovice PJ. 1998. Structural characterization of self-assembled lipid monolayers by N pi T simulation. *Langmuir* **14**(18):5255–5266.
41. Skibinsky A, Venable RM, Pastor RW. 2005. A molecular dynamics study of the response of lipid bilayers and monolayers to trehalose. *Biophys J* **89**(6):4111–4121.
42. Baoukina S, Monticelli L, Marrink SJ, Tieleman DP. 2007. Pressure-area isotherm of a lipid monolayer from molecular dynamics simulations. *Langmuir* **23**(25):12617–12623.
43. Duncan SL, Larson RG. 2008. Comparing experimental and simulated pressure-area isotherms for DPPC. *Biophys J* **93**(8):2965–2986.
44. Baoukina S, Monticelli L, Amrein M, Tieleman DP. 2007. The molecular mechanism of monolayer–bilayer transformations of lung surfactant from molecular dynamics simulations. *Biophys J* **93**(11):3775–3782.

4

MULTISCALE MODELING OF SUPPORTED LIPID BILAYERS

Matthew I. Hoopes,[1] Chenyue Xing,[2] and Roland Faller[1,2]
[1]Graduate Group in Biophysics and [2]Department of Chemical Engineering and Materials Science, University of California Davis

4.1. INTRODUCTION

Cell membranes consist of a multitude of lipid molecules that serve as a framework for the even greater variety of membrane associated proteins [1–4]. As this highly complex (nonequilibrium) system cannot easily be understood and studied in a controlled way, a wide variety of model systems have been devised to understand the dynamics, structure, and thermodynamics in biological membranes. One such model system is a supported lipid bilayer (SLB), a two-dimensional membrane suspended on a surface. SLBs have been realized to be manageable experimentally while reproducing many of the key features of real biological membranes [5,6]. One of the main advantages of supported bilayers is the physical stability due to the solid support that enables a wide range of surface characterization techniques not available to free or unsupported membranes. As SLBs maintain some of the crucial structural and dynamic properties of biological membranes, they provide an important bridge to natural systems. In order to mimic cell membranes reliably, certain structural and dynamic features have to be reliably reproduced in the artificially constructed lipid bilayers. SLBs should display lateral mobility as in living cells, because many membrane activities involve transport, recruitment, or assembly of specific components. It is also critical for membranes to exhibit the correct thermodynamic phase, namely, a fluid lipid bilayer, to respond to environmental stress such as temperature and pressure changes [7].

Please address all correspondence to Roland Faller, Department of Chemical Engineering, Bainer Hall, One Shields Avenue, University of California Davis, Davis, CA 95616 USA, (530) 752-5839, (530) 752-1031 (fax), <rfaller@ucdavis.edu>.

R. Faller et al. (eds.), *Biomembrane Frontiers: Nanostructures, Models, and the Design of Life*, Handbook of Modern Biophysics, DOI: 10.1007/978-1-60761-314-5_4,

There are several ways to fabricate supported lipid bilayers (SLBs) on planar substrates. One can use vesicle fusion on solid substrates [5,8–10] as well as Langmuir-Blodgett deposition [11,12]. Proteoliposome adsorption and subsequent membrane formation on a mica surface was first demonstrated by Brian and McConnell [13]. Because of its simplicity and reproducibility, this is one of the most common approaches to prepare supported membranes. A diverse range of different solid substrates has been used as support material below the bilayer [14,15]. Silicon oxide is the material of choice for vesicle fusion [16]. Polymer cushions dampen the effect of hard surfaces and therefore have been actively investigated [17–20]. However, it is not fully understood which changes the introduction of a solid support introduces into such a biomimetic system. Experimentally it is almost impossible to address such changes, as extremely high-resolution data would be required.

There are several important driving forces in the formation of supported membranes from vesicles that have to be considered in any modeling attempt. The interactions of vesicles and resulting bilayers with surfaces rely on a subtle balance between van der Waals, electrostatic, hydration, and steric forces [21]. A widely held belief is that the interaction of vesicles and the surface plays the determining role in successful fusion and spreading of lipid bilayers [22]. Hydrophilic, or more correctly lipophilic (for the headgroup), surfaces that have strong attractive interactions with lipid headgroups lead often to bilayer formation. The formation process has experimentally been studied by atomic force microscopy (AFM) [15,23] and single-vesicle fluorescence assays [24], which revealed some mechanistic understanding in vesicle-to-bilayer transformation. The details of this process are, however, far from well understood.

An open question in the interaction between membranes and a support is how much water remains between the support and the lipid membrane, as well as the structure and dynamics of this water cushion. It is generally accepted that a thin water layer exists beneath a supported membrane, and there are a few experimental ways to address its thickness, but the structure of this highly confined water is rather unclear. The supporting lubricating layer of water has a thickness of approximately 0.5 to 2 nm, as shown by the results of proton NMR, neutron and X-ray reflectivity, and fluorescence interference-contrast microscopy [25–30].

4.2. MULTISCALE MODELING OF MEMBRANES

Molecular simulations have in recent decades proved to be a powerful tool to reveal the structure, dynamics, and physical interaction mechanisms of simplified biomembranes, mainly unsupported lipid bilayers [31–35]. Both molecular simulations and SLBs provide insight into complex biological systems, elucidate principles, and can test ideas and models for developing biotechnical products [36]. One of the major strengths of simulations is direct access to all particle positions and momenta at all times. Thus, not only averages or local averages can be accessed, but the full distribution of any property can be measured with spatial and temporal resolution. Naturally, the ensuing data are only valid within the approximations of the model under which they have been obtained. But molecular models of lipid assemblies have greatly improved since the early days of modeling. These days the confidence range and limitations of most models are well characterized and understood.

Molecular simulations, actually typically Molecular Dynamics Simulations, of unsupported lipid bilayers have been performed for a diverse spectrum of different lipids, mainly phospholipids and sterols [37–43]. Many key phenomena in membranes have been investigated, such as the lateral mobility of the constituents, the self-assembly process of different structures, and phase transitions or phase separation within the membrane [2,36,44–48]. It is striking that, although supported lipid bilayers have been extensively studied in experiments, they have been largely neglected in simulations and only very recently a few simulations of supported lipid bilayers have appeared in the literature [49–52]. With their unparalleled access to high-resolution atomic information, simulations are ideally suited to address the changes a support inflicts on a membrane. As these changes are expected on a wide variety of length scales, there is an urgent need for computational supported bilayer models on all length and timescales. In this chapter, we present a summary of recent molecular simulations on a variety of length scales to study supported lipid bilayer systems. We investigate various properties of the lipids supported on different model surfaces.

Although computers have become increasingly fast, it is still true that more efficient and better-adapted models are crucial for simulations of systems of increasing complexity. Many different computational techniques and models have been developed and dedicated to study a wide range of systems and applied to many different questions of interest. Atomistic models, i.e., models that represent at least every non-hydrogen atom as its own interaction center in the simulation, are accurate in describing not only the molecular structure, but also the intermolecular interactions such as chemical bonds, electrostatic, and van der Waals interactions. Because of the nature of the interactions, atomistic models have to employ a time step as short as about one tenth of the period of the fastest oscillation in the system [53]. This fastest mode is normally the stretching of a covalent bond or bending of an angle. These models are able to study membranes of a few tens of nanometers over about a hundred nanoseconds using currently available computing facilities. This is probably the most widely used modeling technique in the study of lipid bilayers, usually focusing on the local structure and dynamics of a membrane [35,54,55] or the effect of interactions with other molecules that may be present as inclusion in the membrane or as solute in the water phase [56,57]. Direct comparison with experiments (e.g., NMR [58–60]) have shown the effectiveness and accuracy of such models.

As properties like self-assembly, phase transitions, and phase separations occur on length and timescales beyond the capabilities of atomistic modeling, coarse-grained (CG) models have to be applied. In these models the number of degrees of freedom is reduced by grouping several atoms into one effective particle, thus eliminating short-range dynamics. CG models speed up simulations and can therefore access the length and timescales of collective phenomena [36.46.61–65]. Simulations of such more simplified models of amphiphilic molecules have therefore received increasing attention [44,62,66–77]. The interaction potential between centers representing groups of atoms are not always based on the chemical identity of the system but chosen for computational efficiency. Still, many interesting generic properties of membranes have been elucidated — e.g., the general pathway of lipid bilayer self-assembly [62,65,69,70]. Also, in order to calculate mechanical properties of the membrane [68] or to study membrane fusion [72], coarse-graining techniques are very useful and in some cases even necessary.

In the area of biomembrane modeling, mainly three levels of detail of molecular description have been used. The most detailed is the above-described atomistic modeling. In many cases a first step of coarse-graining is the application of a united atom model in which hydrogen atoms are subsumed into their heavier neighbors. Starting to further remove degrees of freedom, we arrive at a family of models where 4–8 heavy atoms are collected into one interaction site [78–81]. On this level it becomes feasible to describe larger time and length-scale properties, including some collective phenomena of membranes [36,44,46,64,82]. If these models are still too slow or complex for the task at hand we have to resort to water-free models [83–86[. These models still contain individual lipids as molecular particles; however, the number of interaction sites per lipid is often in the low single digits. Water is not explicitly modeled but represented by an interaction potential that is carefully chosen to lead to self-assembly of lipids into a fluid bilayer without explicit solvent. Using this degree of detail we can, for example, begin to study large-scale fluctuations of membranes [34,87,88].

As we expect that supporting surfaces have an influence on the structure and dynamics of the bilayer at all scales, we present simulations using three different models, each representative of one of the above-described model families discussed here. We also suggest the chapters by Gurtovenko and Vattulainen and by Baoukina et al. to the interested reader for examples of different degrees of modeling.

4.3. ATOMISTIC MODELING

Let us now start with the most detailed example of a model — a united atom description of the abundant phospholipid dipalmitoylphosphatidylcholine (DPPC). We are here using the very well-characterized membrane model originally presented by Berger et al. [32] and use the SPC model for water [89]. We use this combination of models as it has been shown to be very efficient and accurate for many detailed membrane studies [38,90,91]. Many different lipids and their interactions have been described by this model or variants of it. So we refer the reader to this vast literature and concentrate here on the implementation of the interaction with the support. For consistency with the lipids we are modeling the support in atomistic detail as well. We are using a regular defect-free lattice. As mentioned above, there is a wide variety of solid supports that have been studied experimentally [5,10,92–96]; so we have to restrict ourselves to a few specific examples. For simplicity we start with homoatomic support structures, specifically we present studies on carbon and on silicon. These are admittedly experimentally not the most popular supports but the simplest ones being actually used [5,97]. Mica or glass — popular support materials — would be structurally and chemically more complex. We expect graphite (i.e., pure carbon) to be hydrophobic (and lipophilic for the tails) and silicon to be intermediately hydrophilic (lipophobic for the tails). We are modeling the system using a hexagonal lattice with an interatomic distance of 1.4 Å for graphite and the same lattice with a lattice constant of 2.3 Å for silicon. The interaction parameters are shown in Table 4.1. The simulations contain the canonical number of 128 lipids (64 in each leaflet). The modeling was performed using the GROMACS simulation suite [98] Version 3.3 with a time step of 1 fs and the Berendsen weak coupling technique for maintaining transversal pressure (perpendicular to the membrane) and temperature constant [99]. Lipophobic supports tend to destabilize the membranes, whereas we are able to study lipophilic supports without significant problems. The proximal leaflet, i.e., the one closer to the support, develops a very short wavelength undulation on the lipophobic support that leads to membrane breakage in some tens of nanoseconds to a few hundred nanoseconds. The distal leaflet remains almost unchanged in comparison to a normal free-standing system. This behavior might be induced by the strong inward tension caused by the lipophobic support. We see that the water between the membrane and the support is at least partially ordered and that the membrane becomes significantly asymmetric.

Table 4.1. Lennard-Jones Interaction Parameters of Carbon and Silicon Support Atoms in Atomistic Simulations

Atom	σ/nm	ε/kJ/mole
C	0.336	1.623
Si	0.339	9.788

In the only other atomistic study we are aware of Heine et al. [49] investigated DPPC with the CHARMM model [100] atomistically on different quartz surfaces. They found an equilibrium distance between bilayer and support of about 3.5 nm. This is independent of the detailed nature of the surface, that is, hydroxylated and bare quartz behave similarly.

4.4. MESOSCALE MODELING

The second model we employ is an intermediately coarse-grained system. The interaction parameters for the lipids are based on the Martini model developed by Marrink et al. [79,80]. We are using the older version of the Martini model developed in 2004 together with an adapted model for water interaction [50]. Again we refer the reader to the original literature for the exact interaction parameters between the lipid parts and water. The Martini model has proved efficient for properties such as phase behavior and phase separation [101,102]. It has been noted that the dynamics is generally sped up by about a factor of 4 [79] and transition temperatures are reproduced semiquantitatively, i.e., within 20–30 K [103]. As we introduce a support, we realize that with the regular interaction strength of water, we find ice developing at the surface instead of fluid water at temperatures around 325 K [50]. In order to avoid water freezing at the surface, which is a known problem with the Martini force-field, we decided to use water with an interaction strength at 76% of the original strength proposed by Marrink in 2004. This reduction in interaction strength is a compromise between the freezing tendency at high interaction and a too low density of liquid water at lower interaction strengths [104]. We are aware that in more recent implementations of the MARTINI model this problem has been alleviated somewhat by modeling water as a binary mixture [75]; however, with a surface the adapted model tends to demix [51].

The model system we are presenting here is again a bilayer of DPPC lipids; GROMACS was used once more to perform simulations. The initial configuration was derived from a study of a pure unsupported DPPC bilayer [64]. We used a screened Coulomb potential with a cutoff at $r_c = 1.2$ nm. The force was smoothly shifted to zero at the cutoff. We used a neighbor-list cutoff between $r_{NL} = 1.2$ and 1.4 nm, and the interaction cutoff was $r_c = 1.2$ nm, shifting the Lennard-Jones potential smoothly to zero starting from $r_s = 0.9$ nm [98]. The reference temperature was 323 K, above the experimental main phase transition temperature of DPPC, which is $T_c =$ 314 K [105]. The CG model allows a larger time step of 40 fs. The model systems discussed contain either 128 lipids and 2593 CG waters or 512 lipids and 10372 CG waters; one CG water represents 4 real waters. The area per molecule is fixed at 62 Å. A free bilayer with our changed water interaction leads to about 76 Å area per molecule, which is clearly larger than desired [51]. The amount of water on either side of the lipid membrane equilibrates by itself within a few microseconds by permeation of water through the bilayer [51]. We set up the simulations with a water cushion of about 1 nm in thickness.

In order to examine supported lipid bilayers, we need a model for the interactions between the support and the water as well as between the support and the lipids. The support consists (for 512 lipids) of 1760 particles arranged at 0.3-nm spacing in a simple square lattice. The support particles have the same 0.47-nm radius as all other CG sites; therefore, adjacent particles overlap. The specific topology of a flat surface is not very crucial [52]. The surface particles interact with water at an interaction strength of level III in the original Martini model [79]. The surface particles interact at the same level with lipid headgroups (both PO_4 and NC_3), as in the original Martini model; they do not interact among themselves. Table 4.2 lists the non-bonded interaction strength between the support and all other particles. Due to periodic boundary conditions, the system was effectively a slab of water and lipids between two identical surfaces; we did not exclude water–water or water–lipid interactions through the support as long as they were within the cutoff. Recent work by us showed that this does not significantly change the behavior [51]. In order to keep the interaction density of the support constant, the lateral area of the simulation box and consequently the bilayer was kept constant. In the x and y directions, the box size was commensurate with the lattice spacing. The direction normal to the bilayer was coupled to a pressure of 1 bar.

Table 4.2. Interaction Parameters of the Mesoscale Support Model Atoms

Atom	σ/nm	ε/kJ/mole
C	0.47	1.8
Si	0.47	3.4

4.5. WATER-FREE LARGE-SCALE MODELING

Let us now present a third, even more coarse-grained, and correspondingly less detailed, model for an SLB. The motivation for such a water-free model derives from the fact that in most biomembrane simulations a lot of computer time is "wasted" on simulating "uninteresting" water. In detailed models where the hydrogen bonding network is of importance this is necessary, but on large scales the water behavior is not of interest. Instead, the focus shifts to the collective phenomena exhibited by large numbers of amphiphilic molecules. By reducing the degrees of freedom in the system on the atomic level, large systems create a new level of complexity and introduce the ability to look at questions of shape and energy. The problem we are facing here is that, in order to obtain self-assembly into a fluid — not solid — bilayer, we need to deviate from the normal Lennard-Jones interaction. We used here a model proposed by Cooke et al. and studied for several membrane applications [85,87,106]. For details the reader is again referred to the original literature [85]. We only mention here that the model uses a tunable long-range attractive potential that reproduces a fluid bilayer with properties commensurate with experimentally measured values.

We now need to define the interaction between the surface and the lipids, which consist of three particles each: one head bead and two tail beads. In contrast to the models discussed above, only one tail — not two — is included. The surface is made of the same type of particles as the bilayer itself [52]. They have been laid out with hexagonal packing, since this provides the best packing density for spheres with their positions fixed. The head beads interact with surface particles using a standard Lennard-Jones interaction while the two tail particles and the surface interact only repulsively using a Weeks-Chandler-Anderson potential [107].

Employing this model, we studied the effect of surface granularity. In order to realize (almost) the same surface energy density with different granularities, we compensate increased packing density, which leads due to particle overlap to a smoother surface, with correspondingly decreased interaction strength per particle. So if we halve the distance between particles, we need to divide the interaction energy by four in order to obtain the same energy per unit area. For a detailed discussion of implementation and results, the reader is referred to [52]. It turns out that the surface energy density per unit area is the crucial parameter and that bilayers behave very similarly if granularity is changed modestly. This is different if the surface particles are also distributed in different distance to the bilayer, that is, if the surface becomes really rough [51]. For the examples to be discussed below, we use an interaction energy of $\varepsilon = 1$, the unit energy for the Lennard-Jones potential, and we use a spacing of unit length $\sigma = 1$, the size of particles. They are laid out so that they are just touching. We are using Langevin dynamics under constant volume conditions. The cutoff for all interactions is 2.5σ and the system temperature is $1.1\,\varepsilon/k_B$.

4.6. VISUALIZATIONS

One of the most revealing analyses is always just to look at the membrane, as visualization allows our own intuition to work best. Figures 4.1–4.3 compare visualizations of the different degrees of coarse graining discussed herein. Figure 4.1 presents the atomistic simulation, Figure 4.2 the MARTINI model, and Figure 4.3 the water-free model. We can see that in all models we are able to obtain homogeneous, continuous membranes. The asymmetry between the two leaflets is clearly seen as well. It appears that the asymmetry becomes more pronounced for the coarser models than for the atomistic model. It may be the case that larger bead sizes lead to longer-range surface effects, as it appears that the surface effects are transmitted a roughly fixed number of bead sizes into the membrane. This would suggest that we may need to adapt the interaction ranges for the intermediate and the coarse models. As the qualitative behavior in all three cases is similar, we can conclude that we are able to obtain a consistent set of multiscale simulations of supported lipid bilayers.

4.7. DENSITY PROFILES

While visualizations allow us to qualitatively describe the system and connect to it in an intuitive fashion, they do not provide any quantitative data. One of the most easily accessible and experimentally relevant quantitative descriptors of an SLB is the density profile. It can be connected to electron density profiles experimentally obtainable by X-rays or scattering length densities in the case of neutrons [108,109]. In simulations we just measure the density in slabs of equal thickness and plot this as a function of slab position. Figures 4.4–4.6 compare the density profiles in all discussed models between supported lipid bilayers and their free counterparts.

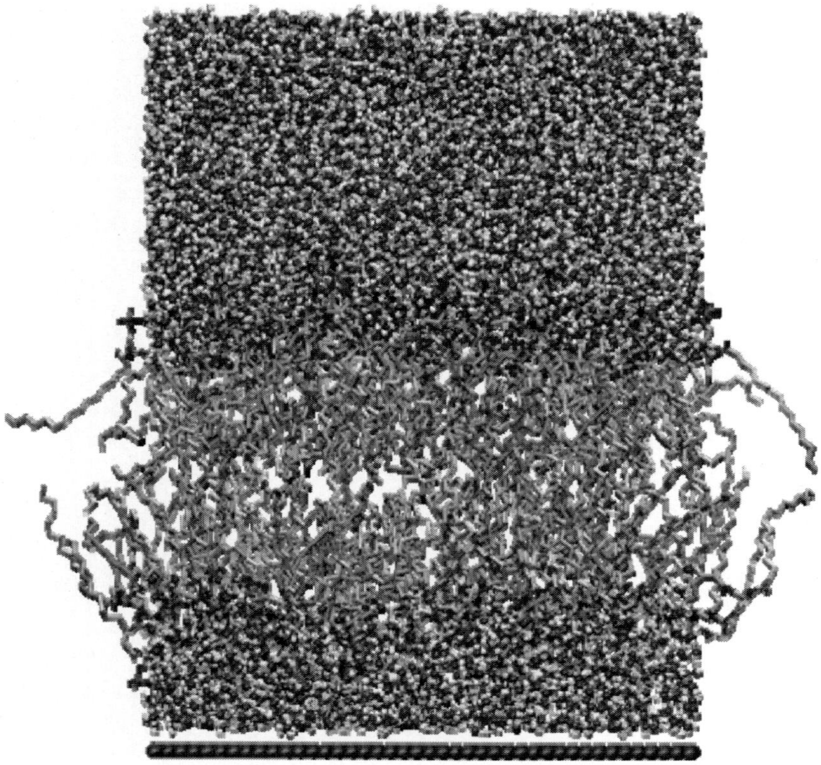

Figure 4.1. Visualization of an atomistic (united-atom) simulation of a supported bilayer. A DPPC bilayer is modeled with the Berger force-field in SPC water. The support is pure silicon. We see an ordering effect of the support on the water and to some extent on the membrane itself. Please visit http://www.springer.com/series/7845 to view a high-resolution full-color version of this illustration.

In the atomistic simulation of a supported bilayer (Fig. 4.4), we see a slight heterogeneity in the density profiles of the two leaflets. The supported bilayer is no longer symmetric around the center of the bilayer, different from that of the free-standing bilayer. The proximal leaflet becomes thicker than the distal leaflet with the presence of a carbon-based support. Minor layering can be observed in the proximal leaflet, especially in the headgroup part. We can see a secondary little peak in the headgroup region of the proximal leaflet for the first layer of the headgroups, also a little step at $Z = 1$ nm, which suggests another layer. Such layering effect of the support on the density profile is more obvious in the larger-scale models, indicating an increasing overestimation of the effect with increasing grain size. The distal leaflet remains essentially the same as the leaflets in the free-standing bilayer system.

In our intermediate scale model (Fig. 4.5) we can see a strong effect of the support on the density profile. For the free bilayer we see a clear minimum density area in the middle and an overall symmetry of the membrane around that lowest density plane. There is not much structure otherwise, showing a lack of order. In the supported bilayer all these features change. There is no symmetry plane in a supported system, as the symmetry is broken, and we see a substantial difference in the structure of the two leaflets. There is significant localization of both

Figure 4.2. Visualization of a mesoscale simulation of a supported bilayer. A DPPC bilayer is modeled with a variant of the MARTINI model. The support is pure silicon. We see stronger ordering of the water and the membrane compared to the atomistic model in Figure 4.1. Please visit http://www.springer.com/series/7845 to view a high-resolution full-color version of this illustration.

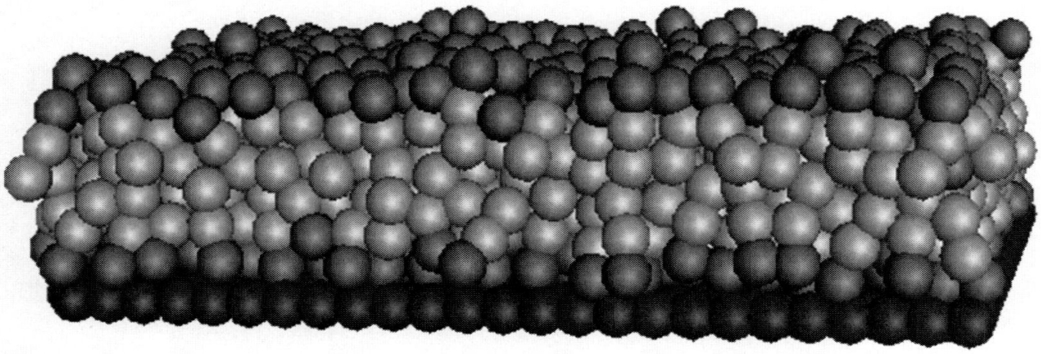

Figure 4.3. Visualization of a water-free large-scale simulation of a supported bilayer. The identity of the lipids cannot easily be identified. Headgroups are dark gray, tails are light gray, and support particles are black. Please visit http://www.springer.com/series/7845 to view a high-resolution full-color version of this illustration.

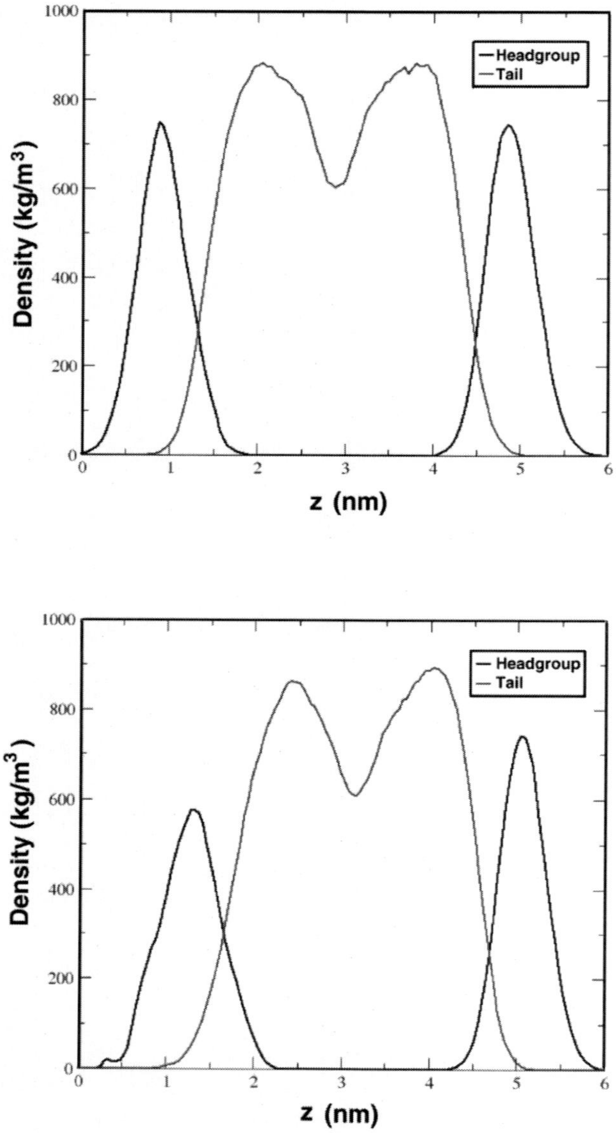

Figure 4.4. Comparison of the density profiles of a carbon-supported (bottom) and an unsupported DPPC (top) system in atomistic detail. We show the headgroups in black and the tails in red. There is a layering effect showing as a pre-peak close to the surface. The difference in the distal leaflet is comparably small. Please visit http://www.springer.com/series/7845 to view a high-resolution full-color version of this illustration.

headgroups and tails. We see increased ordering in the proximal leaflet and much less in the distal leaflet. The proximal headgroup is especially highly localized, and we see a strong interaction with the support. We find an almost crystalline behavior. It also appears that the plane of lowest density is not necessarily the middle of the bilayer any longer. The proximal leaflet is thicker than the distal as the tails are more elongated. It probably is much closer to or even actu-

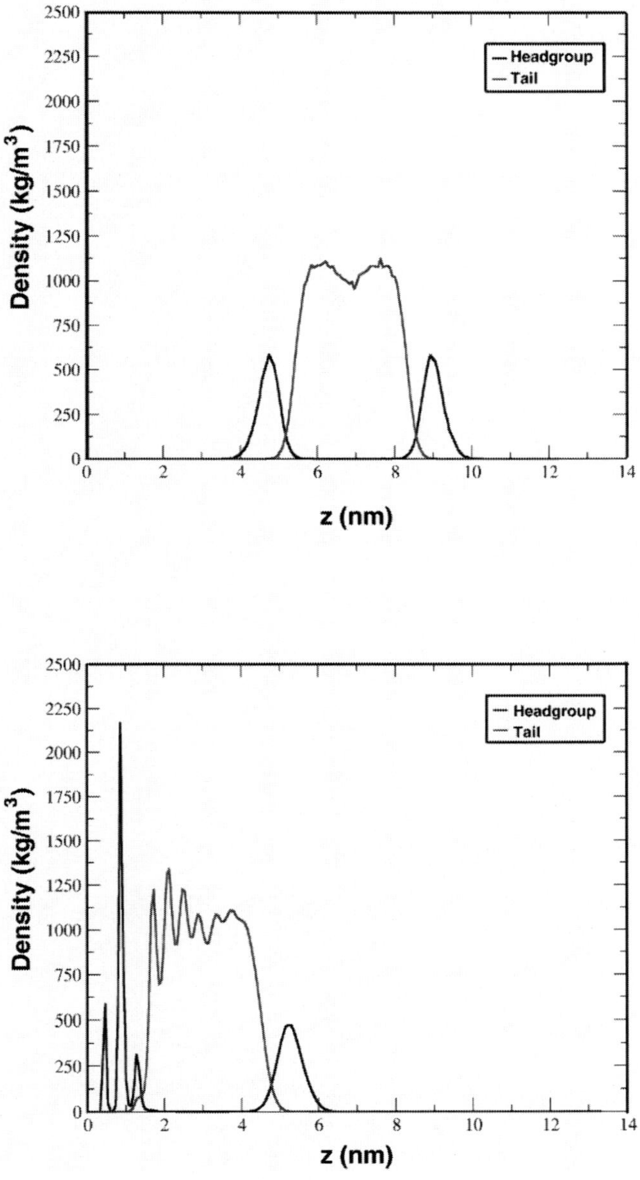

Figure 4.5. Comparison of the density profiles of a supported (bottom) and an unsupported DPPC (top) system using the MARTINI model. We show the headgroups in black and the tails in red. There is a clear layering effect close to the surface. Please visit http://www.springer.com/series/7845 to view a high-resolution full-color version of this illustration.

ally in the gel state. An exact determination of phase boundaries becomes complicated in supported membranes as the equilibration of flip-flops between the two leaflets cannot be achieved in reasonable simulation time. In comparison to the well-characterized density profiles of free bilayers, it appears that the proximal leaflet is significantly altered, whereas the distal leaflet is essentially unaffected. This clearly agrees qualitatively with the atomistic data presented above.

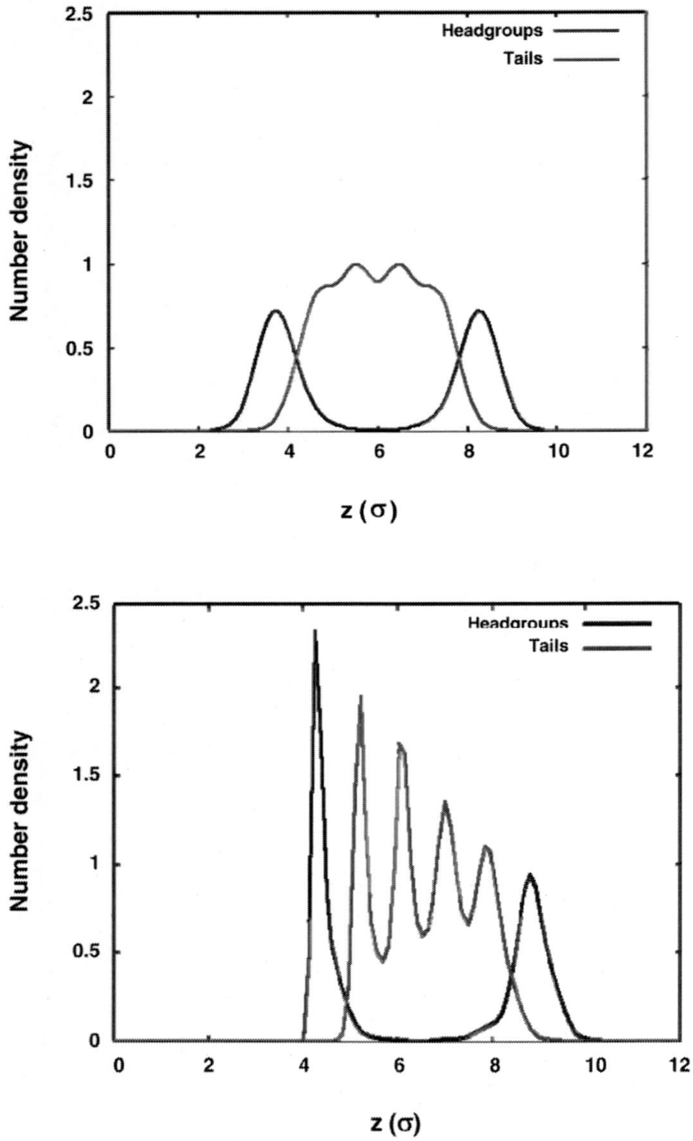

Figure 4.6. Comparison of the density profiles of a supported (bottom) and an unsupported (top) lipid system using the water-free model. We show the headgroups in continuous lines and the tails in broken lines. Layering is visible across the membrane. Please visit http://www.springer.com/series/7845 to view a high-resolution full-color version of this illustration.

If we now focus on the water-free model (Fig. 4.6), we find that the leaflet symmetry is severely disrupted by the solid surface. As this model allows lipid flip-flop within the timescale of the simulation, we do not even expect numerical density symmetry between the leaflets, but the proximal leaflet now contains more lipids than the distal one, which again suggests that the phase transition temperature in the two leaflets is not identical. The overall thickness of the bilayer does not change substantially when considering the headgroup distance across the membrane. We see, however, two main effects: first, the density peaks in the supported system are generally sharper and become sharper with closer approach to the support; second, the peak shape is now asymmetric, especially the headgroup peak in the proximal leaflet, which looks as if it is cut in half, with a very sharp flank facing the support and a much softer slope toward the membrane center. We also find that lipid interdigitation (the degree to which lipids cross over in the opposing leaflet) is slightly reduced in the supported bilayer. The manner in which the two leaflets interact is therefore also changed by the ordering caused by the support.

4.8. PRESSURE AND LATERAL TENSION

Simulations of free phospholipid membranes are traditionally performed under vanishing surface tension by simply performing simulations under constant pressure with independently fluctuating simulation box dimensions. In a supported lipid bilayer there are a variety of reasons why this is in general not possible. There is a nonzero surface tension between water and support in the case of models where water is explicitly present. In any model where our support is made of individual interaction sites rather than an infinitely flat surface, we have to keep the area in the membrane plane constant in order to avoid fluctuating interaction densities or problems with periodic boundary conditions. Essentially, one could measure the lateral pressure in the membrane at a fixed area and then couple the system in the bilayer normal to that pressure. To our knowledge, this has not yet been implemented and also would relate to a rather unnatural environment, as experimentally the normal pressure is ambient pressure.

In order to measure the tension of the system, we are required to measure the lateral and normal pressure separately, as the integral through the corresponding surfaces along the bilayer normal of this difference equals the surface tension of that corresponding surface [110]. So if we just measure the average pressure tensor throughout the system we obtain the sum of surface tensions over all surfaces. For example, in our atomistic system on a carbon surface we obtain an overall surface tension of 147.71 mN/m and in the intermediate-scale Martini model the value is 112.04 mN/m. These are clearly nicely in agreement with each other; however, we have to point out that these values depend on the area per molecule, the system size, and the boundary conditions. Pressure measurements in molecular dynamics simulations are chronically suffering from inaccuracies due to very strong fluctuations that can easily be one or two orders of magnitude larger than the pressure itself at ambient conditions.

If we want to separate out the different surfaces and quantify the effects of the different subsystems (tail, head) on the overall surface tension, we have to measure pressure profiles in the lateral and normal directions. Figure 4.7 (top) shows a sample of a pressure profile in the supported bilayer system at the atomistic level. $Z = 0$ is the center of the bilayer. We can clearly see the asymmetry of the lateral pressure around the center. The proximal leaflet (i.e., the region $Z < 0$) exhibits a larger negative peak at $Z = -2.75$ nm, compared to that in the distal leaflet at $Z = 2.1$ nm. Negative peaks correspond to the water–headgroup interface. A larger negative peak implies a stronger surface tension in the proximal leaflet. The negative peak in the proximal leaflet locates more distant from the center than that in the distal leaflet. This agrees with our observation in the density profiles that the proximal leaflet is thicker. The two extremely sharp peaks at $Z = -3.5 \sim -3$ nm indicate the strong tension at the support–water interface, also implying ordering of the water at the support.

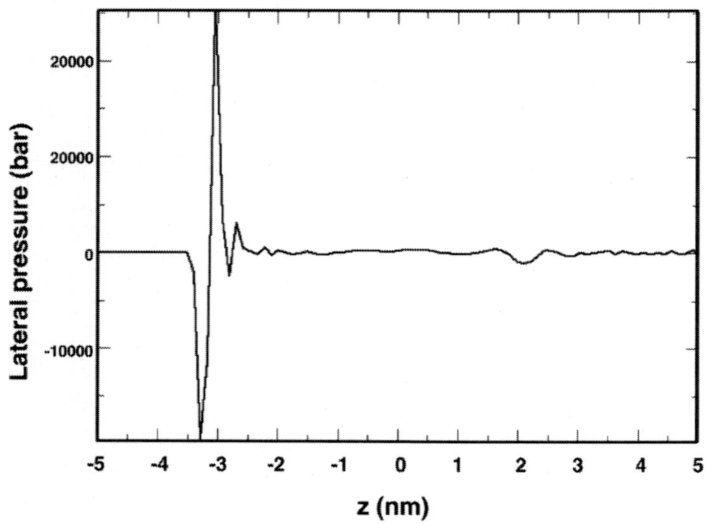

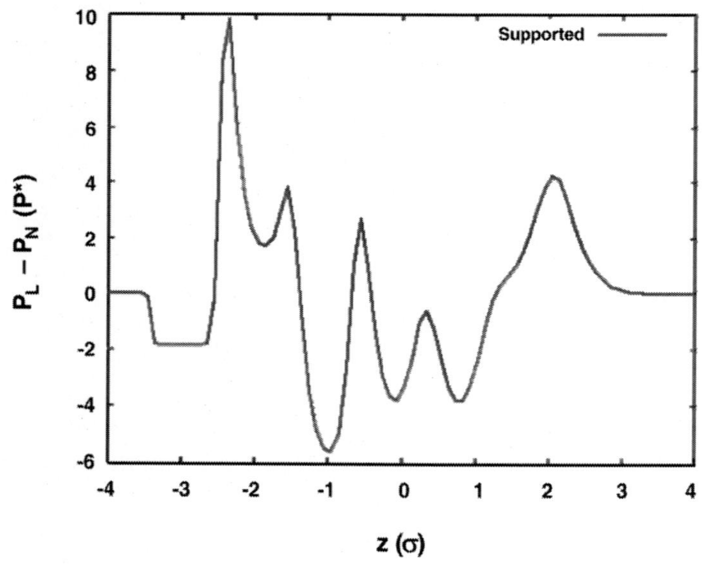

Figure 4.7. Examples of pressure profile in supported systems. Top: lateral pressure, DPPC on carbon. Bottom: water-free model tension profile. We clearly see the different influences of the separate surfaces. $Z = 0$ is the center of the bilayer. Please visit http://www.springer.com/series/7845 to view a high-resolution full-color version of this illustration.

The bottom part of Figure 4.7 is essentially the surface tension profile (difference of lateral and normal pressure) across a supported membrane in the water-free model. Several features are striking here. Due to the missing water and long-range tail interaction, there is a series of large negative peaks in the center of the membrane. The missing symmetry is again clearly due to the presence of the surface, which leads to distinct structures and strong peaks from the surface–head interaction.

4.9. SUMMARY AND OUTLOOK

Modeling of supported membranes is clearly not a solved problem, as it is a very new area of research. We have shown here that simulations are able to reliably model supported lipid bilayers and bring a new perspective to the field, which is different from the many existing experiments.

The central message is that a supported bilayer is significantly different from a free bilayer. This includes differences in the order, area per molecule, thickness, and general structure of the membrane. These differences have to be kept in mind when supported lipid bilayers are used as model membranes, which is often the case. We find that the distal leaflet, that is, the one farther from the support, is typically not strongly influenced and may actually be a good model for a leaflet in a free bilayer.

As simulations are now at the point at which they can contribute to an understanding of such systems, it is clear that they will be able to study problems that are not easily addressable in experiments. One example may be the local influences of support roughness and curvature. Experimentally, the resolution is often too limited to locally resolve the influences. Also, if there are heterogeneities in the interactions, simulations have a better chance to correlate the heterogeneities in the support with heterogeneities in the ensuing assembly.

Another area where simulations will be of important use is the possibility to discern the different influences of supporting structures. It is clear that there are generic and specific interactions between supports and membranes. A generic effect is the suppression of fluctuations due to the support; a specific effect might be direct hydrogen bonding or any hydrophilic and hydrophobic effects. Simulations can now distinguish between different supports and will be able to assess the relative importance of different effects.

Finally, direct access to free energies using advanced simulation techniques will open up a more systematic avenue of phase behavior of systems in confined geometries, of which supported lipid bilayers are a prime example.

ACKNOWLEDGMENTS

We would like to acknowledge the invaluable scientific discussions and suggestions from Marjorie Longo, Ilpo Vattulainen, Samuli Ollila and Markus Deserno. We also acknowledge the hospitality of Tampere University of Technology (CX and RF) and the Max-Planck Institute of Polymer Research (MIH and RF), where part of this work was performed. This research was partially supported by an industry/campus-supported fellowship under the Training Program in Biomolecular Technology (T32-GM08799) at the University of California Davis, as well as the NSF NIRT program under grant CBET 0506662.

PROBLEMS

4.1. What is the main driving force behind the concept of "Multiscale Modeling"?

4.2. Which characteristics of supported bilayers are different or similar to free bilayers?

4.3. Why are supported bilayers such an widely used model system?

FURTHER READING

These books and reviews go into more detail on areas relevant to the current chapter but could not be discussed in all detail. They are sorted according to general topic. There are many more available, so this list is by no means exhaustive.

Molecular Simulation in General

Allen MP, Tildesley DJ. 1987. *Computer simulation of liquids*. New York: Oxford UP.
Frenkel D, Smit B. 2001. *Understanding molecular simulations*, 2nd ed. San Diego: Academic Press.
Leach AR. 2001. *Molecular modeling: principles and applications*, 2nd ed. Reading, MA: Addison Wesley.

Multiscale Modeling of Soft Matter

Voth GA. 2008. Coarse-graining of condensed phase and biomolecular systems. Boca Raton, FL: CRC Press.
Faller R. 2007. Coarse-grain modeling of polymers. In *Reviews in computational chemistry*, Vol. 23, pp. 233–262. Ed KB Lipkowitz, TR Cundari. New York: Wiley-VCH.
Müller-Plathe F. 2003. Scale-hopping in computer simulations of polymers. *Soft Mater* **1**:1–31.
Kremer K, Müller-Plathe, F. 2001. Multiscale problems in polymer science: simulation approaches. *Mater Res Soc Bull* **26**: 205–210.

Modeling of Biomembranes

Tieleman, DP, Marrink, SJ, Berendsen, HJC. 1997. A computer perspective of membranes: molecular dynamics studies of lipid bilayer systems. *Biochim Biophys Acta* **1331**:235–270 (1997)
Forrest LR, Sansom, MSP. 2000. Membrane simulations: bigger and better? *Curr Opin Struct Biol* **10**:174–181.
Feller SE. 2000. Molecular dynamics simulation of lipid bilayers. *Curr Opin Colloid Interface Sci* **5**:217–223.

REFERENCES

1. Kusumi A, Tsuda M, Akino T, Ohnishi S, Terayama Y. 1983. Protein–phospholipid–cholesterol interaction in the photolysis of invertebrate rhodopsin. *Biochemistry* **22**:1165–1170.
2. Falck E, Patra M, Karttunen M, Hyvonen MT, Vattulainen I. 2004. Lessons of slicing membranes: interplay of packing, free area, and lateral diffusion in phospholipid/cholesterol bilayers. *Biophy J* **87**:1076–1091.
3. Pebay-Peyroula E, Rosenbusch JP. 2001. High-resolution structures and dynamics of membrane protein–lipid complexes: a critique. *Curr Opin Struct Biol* **11**:427–432.
4. London E. 2005. How principles of domain formation in model membranes may explain ambiguities concerning lipid raft formation in cells. *Biochim Biophys Acta* **1746**:203–220.
5. Sackmann E. 1996. Supported membranes: Scientific and practical applications. *Science* **271**:43–48.
6. Hianik T. 2006. Structure and physical properties of biomembranes and model membranes. *Acta Phys Slov* **56**:687–806.
7. Shinitzky M. 1984. *Physiology of membrane fluidity*. Boca Raton, FL: CRC Press.
8. Groves JT, Ulman N, Boxer SG. 1997. Micropatterning fluid lipid bilayers on solid supports. *Science* **275**:651–253.
9. Richter RP, Brisson AR. 2005. Following the formation of supported lipid bilayers on mica: a study combining AFM, QCM-D, and ellipsometry. *Biophys J* **88**:3422–3433.
10. Weng KC, Stalgren JJR, Risbud SH, Frank CW. 2004. Planar bilayer lipid membranes supported on mesoporous aerogels, xerogels, and Vycor((R)) glass: an epifluorescence microscopy study. *J Non-Cryst Solids* **350**:46–53.
11. Ulman A. 1991. *An introduction to ultrathin organic films: from Langmuir-Blodgett to self-assembly*. San Diego: Academic Press.

12. Gaines GL. 1996. *Insoluble monolayers at liquid–gas interfaces*. New York: John Wiley & Sons.
13. Brian AA, McConnell HM. 1984. Allogeneic stimulation of cytotoxic t cells by supported planar membranes. *Proc Natl Acad Sci USA* **81**:6159–6163.
14. Groves JT, Ulman N, Cremer PS, Boxer SG. 1998. Substrate–membrane interactions: mechanisms for imposing patterns on a fluid bilayer membrane. *Langmuir* **14**:3347–3350.
15. Reviakine I, Brisson A. 2000. Formation of supported phospholipid bilayers from unilamellar vesicles investigated by atomic force microscopy. *Langmuir* **16**:1806–1815.
16. Groves JT, Ulman N, Boxer SG. 1997. Micropatterning fluid lipid bilayers on solid supports. *Science* **275**:651–653.
17. Knoll W, Frank CW, Heibel C, Naumann R, Offenhusser A, Rhe J, Schmidt EK, Shen WW, Sinner A. 2000. Functional tethered lipid bilayers. *J Biotechnol* **74**:137–158.
18. Sackmann E, Tanaka M. 2000. Supported membranes on soft polymer cushions: fabrication, characterization and applications. *Trends Biotechnol* **18**:58–64.
19. Wagner ML, Tamm LK. 2000. Tethered polymer-supported planar bilayers for reconstitution of integral membrane proteins: silane–polyethyleneglycol–lipid as a cushion and covalent linker. *Biophys J* **79**:1400–1414.
20. Shen WW, Boxer SG, Knoll W, Frank CW. 2001. Polymer-supported lipid bilayers on benzophenone-modified substrates. *Biomacromolecules* **2**:70–79.
21. Israelachvili JN. 1992. Intermolecular and surface forces, 2nd ed. San Diego: Academic Press.
22. Keller CA, Kasemo B. 1998. Surface specific kinetics of lipid vesicles adsorption measured with a quartz crystal microbalance. *Biophys J* **75**:1397–1402.
23. Leonenko ZV, Carnini A, Cramb DT. 2000. Supported planar bilayer formation by vesicle fusion: the interaction of phospholipid vesicles with surfaces and the effect of gramicidin on bilayer properties using atomic force microscopy. *Biochim Biophys Acta* **1509**:131–147.
24. Johnson JM, Ha T, Chu S, Boxer SG. 2002. Early steps of supported bilayer formation probed by single vesicle fluorescence assays. *Biophys J* **83**:3371–3379.
25. Bayerl TM, Bloom M. 1990. Physical properties of single phospholipid bilayers adsorbed to micro glass beads. *Biophys J* **58**:357–362.
26. Johnson SJ, Bayerl TM, McDermott DC, Adam GW, Rennie AR, Thomas RK, Sackmann E. 1991. Structure of an adsorbed dimyristoylphosphatidylcholine bilayer measured with specular reflection of neutrons. *Biophys J* **59**:289–294.
27. Krueger S, Koenig BW, Orts WJ, Berk NF, Majkrzak CF, Gawrisch K. 1996. Neutron reflectivity studies of single lipid bilayers supported on planar substrates. *Basic Life Sci* **64**:205–213.
28. Fromherz P, Kiessling V, Kottig K, Zeck G. 1999. Membrane transistor with giant lipid vesicle touching a silicon chip. *Appl Phys A: Mater Sci Process* **69**:571–576.
29. Miller CE, Majewski J, Gog T, Kuhl TL. 2005. Characterization of biological thin films at the solid–liquid interface by X-ray reflectivity. *Phys Rev Lett* **94**:238104.
30. Miller CE, Majewski J, Watkins EB, Mulder DJ, Gog T, Kuhl TL. 2008. Probing the local order of single phospholipid membranes using grazing incidence X-ray diffraction. *Phys Rev Lett* **100**:058103–4.
31. Hogberg CJ, Lyubartsev AP. 2006. A molecular dynamics investigation of the influence of hydration and temperature on structural and dynamical properties of a dimyristoylphosphatidylcholine bilayer. *J Phys Chem B* **110**:14326–14336.
32. Berger O, Edholm O, Jahnig F. 1997. Molecular dynamics simulations of a fluid bilayer of dipalmitoylphosphatidylcholine at full hydration, constant pressure, and constant temperature. *Biophys J* **72**:2002–2013.
33. Feller SE, Yin DX, Pastor RW, MacKerell AD. 1997. Molecular dynamics simulation of unsaturated lipid bilayers at low hydration: parameterization and comparison with diffraction studies. *Biophys J* **73**:2269–2279.
34. Marrink SJ, Mark AE. 2001. Effect of undulations on surface tension in simulated bilayers. *J Phys Chem B* **105**:6122–6127.
35. Tieleman DP, Berendsen HJC. 1996. Molecular dynamics simulations of a fully hydrated dipalmitoyl phosphatidylcholine bilayer with different macroscopic boundary conditions and parameters. *J Chem Phys* **105**:4871–4880.
36. Muller M, Katsov K, Schick M. 2006. Biological and synthetic membranes: what can be learned from a coarse-grained description? *Phys Rep,* **434**:113–176.

37. Tieleman DP, Berendsen HJC. 1996. Molecular dynamics simulations of fully hydrated dipalmitoylphosphatidylcholine bilayer with different macroscopic boundary conditions and parameters. *J Chem Phys* **105**:4871–4880.
38. Niemela P, Hyvonen MT, Vattulainen I. 2004. Structure and dynamics of sphingomyelin bilayer: insight gained through systematic comparison to phosphatidylcholine. *Biophys J* **87**:2976–2989.
39. Leekumjorn S, Sum AK. 2006. Molecular simulation study of structural and dynamic properties of mixed DPPC/DPPE bilayers. *Biophys J* **90**:3951–3965.
40. Gurtovenko A, Patra M, Karttunen M, Vattulainen I. 2004. Cationic DMPC/DMTAP lipid bilayers: molecular dynamics study. *Biophys J* **86**:3461–3472.
41. Aittoniemi J, Róg T, Niemela P, Pasenkiewicz-Gierula M, Vattulainen I, Karttunen M. 2006. Sterol tilt—major determinant of sterol ordering capability in lipid membranes. *J Phys Chem B (Lett)* **110**:25562–25564.
42. Pandit SA, Bostick D, Berkowitz ML. 2004. Complexation of phosphatidylcholine lipids with cholesterol. *Biophys J* **86**:1345–1356.
43. Pandit SA, Jakobsson E, Scott HL. 2004. Simulation of the early stages of nano-domain formation in mixed bilayers of sphingomyelin, cholesterol, and dioleylphosphatidylcholine. *Biophys J* **87**:3312–3322.
44. Faller R, Marrink SJ. 2004. Simulation of domain formation in DLPC–DSPC mixed bilayers. *Langmuir* **20**:7686–7693.
45. Goetz R, Lipowsky R. 1998. Computer simulations of bilayer membranes: self-assembly and interfacial tension. *J Chem Phys* **108**:7397–7409.
46. Marrink SJ, Risselada J, Mark AE. 2005. Simulation of gel phase formation and melting in lipid bilayers using a coarse grained model. *Chem Phys Lipids* **135**:223–244.
47. Mouritsen OG. 1991. Theoretical models of phospholipid phase transitions. *Chem Phys Lipids* **57**:179–194.
48. Kandt C, Ash WL, Tieleman DP. 2007. Setting up and running molecular dynamics simulations of membrane proteins. *Methods* **41**:475–488.
49. Heine DR, Rammohan AR, Balakrishnan J. 2007. Atomistic simulations of the interaction between lipid bilayers and substrates. *Mol Simul* **33**:391–397.
50. Bennun SV, Dickey AN, Xing CY, Faller R. 2007. Simulations of biomembranes and water: important technical aspects. *Fluid Phase Equilib* **261**:18–25.
51. Xing C, Faller R. 2008. Interactions of lipid bilayers with supports: a coarse-grained molecular simulation study. *J Phys Chem B* **112**:7086–7094.
52. Hoopes MI, Deserno M, Longo ML, Faller R. 2008. Coarse-grained modeling of interactions of lipid bilayers with supports. *J Phys Chem* **129**:175102.
53. Mackerell AD. 2004. Empirical force fields for biological macromolecules: overview and issues. *J Comput Chem* **25**:1584–1604.
54. Aman K, Lindahl E, Edholm O, Hakansson P, Westlund PO. 2003. Structure and dynamics of interfacial water in an L-alpha phase lipid bilayer from molecular dynamics simulations. *Biophys J* **84**:102–115.
55. Pasenkiewicz-Gierula M, Takaoka Y, Miyagawa H, Kitamura K, Kusumi A. 1999. Charge pairing of headgroups in phosphatidylcholine membranes: a molecular dynamics simulation study. *Biophys J* **76**:1228–1240.
56. Hofsass C, Lindahl E, Edholm O. 2003. Molecular dynamics simulations of phospholipid bilayers with cholesterol. *Biophys J* **84**:2192–2206.
57. Tu KC, Klein ML, Tobias DJ. 1998. Constant-pressure molecular dynamics investigation of cholesterol effects in a dipalmitoylphosphatidylcholine bilayer. *Biophys J* **75**:2147–2156.
58. Tieleman DP, Marrink SJ, Berendsen HJC. 1997. A computer perspective of membranes: molecular dynamics studies of lipid bilayer systems. *Biochim Biophys Acta* **1331**:235–270.
59. Feller SE, Brown CA, Nizza DT, Gawrisch K. 2002. Nuclear Overhauser enhancement spectroscopy cross-relaxation rates and ethanol distribution across membranes. *Biophys J* **82**:1396–1404.
60. Pastor RW, Venable RM, Feller SE. 2002. Lipid bilayers, NMR relaxation, and computer simulations. *Acc Chem Res* **35**:438–446.
61. Shelley JC, Shelley MY, Reeder RC, Bandyopadhyay S, Moore PB, Klein ML. 2001. Simulations of phospholipids using a coarse grain model. *J Phys Chem B* **105**:9785–9792.
62. Shelley JC, Shelley MY, Reeder RC, Bandyopadhyay S, Klein ML. 2001. A coarse-grain model for phospholipid simulations. *J Phys Chem B* **105**:4464–4470.
63. Nielsen SO, Lopez CF, Srinivas G, Klein ML. 2004. Coarse grain models and the computer simulation of soft materials. *J Phys Cond Matter* **16**:R481–R512.

64. Marrink SJ, Mark AE. 2004. Molecular view of hexagonal phase formation in phospholipid membranes. *Biophys J* **87**:3894–3900.
65. Lopez CF, Moore PB, Shelley JC, Shelley MY, Klein ML. 2002. Computer simulation studies of biomembranes using a coarse grain model. *Comput Phys Commun* **147**:1–6.
66. Ayton G, Voth GA. 2002. Bridging microscopic and mesoscopic simulations of lipid bilayers. *Biophys J* **83**:3357–3370.
67. Smit B, Hilbers PAJ, Esselink K, Rupert LAM, van Os NM, Schlijper AG. 1990. Computer simulations of a water/oil interface in the presence of micelles. *Nature* **348**:624–625.
68. Goetz R, Gompper G, Lipowsky R. 1999. Mobility and elasticity of self-assembled membranes. *Phys Rev Lett* **81**:221–224.
69. Mouritsen OG. 2000. Computer simulation of lyotropic liquid crystals as models of biological membranes. In *Advances in the computer simulation of liquid crystals*, pp. 139–188. Ed C Zannoni. Dordrecht: Kluwer.
70. Soddemann T, Dünweg B, Kremer K. 2001. A generic computer model for amphiphilic systems. *Eur Phys J E* **6**:409–419.
71. Guo H, Kremer K. 2003. Amphiphilic lamellar model systems under dilation and compression: molecular dynamics study. *J Chem Phys* **118**:7714–7723.
72. Müller M, Katsov K, Schick M. 2003. Coarse grained models and collective phenomena in membranes: computer simulation of membrane fusion. *J Polym Sci B* **41**:1441–1451.
73. Kranenburg M, Smit B. 2005. Phase behavior of model lipid bilayers. *J Phys Chem B* **109**:6553–6563.
74. Marrink SJ, de Vries AH, Mark AE. 2004. Coarse grained model for semiquantitative lipid simulations. *J Phys Chem B* **108**:750–760.
75. Marrink SJ, Risselada HJ, Yefimov S, Tieleman DP, de Vries AH. 2007. The MARTINI force field: coarse grained model for biomolecular simulations. *J Phys Chem B* **111**:7812–7824.
76. Dickey AN, Faller R. 2005. Investigating interactions of biomembranes and alcohols: a multiscale approach. *J Polym Sci B* **43**:1025–1032.
77. Wong BY, Faller R. 2007. Phase behavior and dynamic heterogeneities in lipids: a coarse-grained simulation study of DPPC–DPPE mixtures. *Biochim Biophys Acta* **1768**:620–627.
78. Nielsen SO, Lopez CF, Srinavas G, Klein ML. 2004. Coarse grain models and the computer simulation of soft materials. *J Phys Cond Matter* **16**:R481–R512.
79. Marrink SJ, de Vries AH, Mark AE. 2004. Coarse grained model for semi-quantitative lipid simulation. *J Phys Chem B* **108**:750–760.
80. Marrink SJ, Risselada HJ, Yefimov S, Tieleman DP, de Vries AH. 2007. The MARTINI force field: coarse grained model for biomolecular simulations. *J Phys Chem B* **111**:7812–7824.
81. Shelley JC, Shelley MY, Reeder RC, Bandyopadhyay S, Klein ML. 2001. A coarse grain model for phospholipid simulations. *J Phys Chem B* **105**:4464–4470.
82. Sum AK, Faller R, de Pablo JJ. 2003. Molecular simulation study of phospholipid bilayers and insights of the interactions with disaccharides. *Biophys J* **85**:2830–2844.
83. Brannigan G, Brown FLH. 2004. Solvent-free simulations of fluid membrane bilayers. *J Chem Phys* **120**:1059–1071.
84. Farago O. 2003. "Water-free" computer model for fluid bilayer membranes. *J Chem Phys* **119**:596–605.
85. Cooke IR, Kremer K, Deserno M. 2005. Tunable generic model for fluid bilayer membranes. *Phys Rev E* **72**:011506.
86. Noguchi H, Takasu M. 2001. Self-assembly of amphiphiles into vesicles: a Brownian dynamics simulation. *Physl Rev E* **64**:041913.
87. Harmandaris VA, Deserno M. 2006. A novel method for measuring the bending rigidity of model lipid membranes by simulating tethers. *J Chem Phys* **125**:204905.
88. Farago O. 2008. Mode excitation Monte Carlo simulations of mesoscopically large membranes. *J Chem Phys* **128**:184105-4.
89. Berendsen HJC, Postma JPM, van Gunsteren WF, Hermans J. 1981. Interaction models for water in relation to protein hydration. In *Intermolecular forces*, pp. 331–342. Ed B Pullman. Dordrecht: Reidel.
90. Patra M, Karttunen M, Hyvönen M, Falck E, Lindqvist P, Vattulainen I. 2003. Molecular dynamics simulations: major artifacts due to truncation of electrostatic interactions. *Biophys J* **84**:3636–3645.
91. Dickey AN, Faller R. 2007. How alcohol chain-length and concentration modulate hydrogen bond formation in a lipid bilayer. *Biophys J* **92**:2366–2376.

92. Sackmann E, Tanaka M. 2000. Supported membranes on soft polymer cushions: fabrication, characterization and applications. *Trends Biotechnol* **18**:58–64.
93. Groves JT, Ulman N, Cremer PS, Boxer SG. 1998. Substrate membrane interactions: mechanisms for imposing patterns on a fluid bilayer membrane. *Langmuir* **14**:3347–3350.
94. Johnson JM, Ha T, Chu S, Boxer SG. 2002. Early steps of supported bilayer formation probed by single vesicle fluorescence assays. *Biophys J* **83**:3371–3379.
95. Lenz P, Ajo-Franklin CM, Boxer SG. 2004. Patterned supported lipid bilayers and monolayers on poly(dimethylsiloxane). *Langmuir* **20**:11092–11099.
96. Schönherr H, Johnson JM, Lenz P, Frank CW, Boxer SG. 2004. Vesicle adsorption and lipid bilayer formation on glass studied by atomic force microscopy. *Langmuir* **20**:11600–11606.
97. Meuse CW, Krueger S, Majkrzak CF, Dura JA, Fu J, Connor JT, Plant AL. 1998. Hybrid bilayer membranes in air and water: infrared spectroscopy and neutron reflectivity studies. *Biophys J* **74**:1388–1398.
98. Lindahl E, Hess B, van der Spoel D. 2001. GROMACS 3.0: a package for molecular simulation and trajectory analysis. *J Mol Model* **7**:306–317.
99. Berendsen HJC, Postma JPM, van Gunsteren WF, DiNola A, Haak JR. 1984. Molecular dynamics with coupling to an external heat bath. *J Chem Phys* **81**:3684–3690.
100. Brooks BR, Bruccoleri RE, Olafson BD, States DJ, Swaminathan S, Karplus M. 1983. A program for macromolecular energy, minimization, and dynamics calculations. *J Comp Chem* **4**:187–217.
101. Bennun SV, Longo ML, Faller R. 2007. Phase and mixing behavior in two-component lipid bilayers: a molecular dynamics study in DLPC/DSPC mixtures. *J Phys Chem B* **111**:9504–9512.
102. Faller R, Marrink S-J. 2004. Simulation of domain formation in mixed DLPC–DSPC lipid bilayers. *Langmuir* **20**:7686–7693.
103. Bennun S, Longo ML, Faller R. 2007. The molecular scale structure in fluid–gel patterned bilayers: stability of interfaces and transmembrane distribution. *Langmuir* **23**:12465–12468.
104. Bennun S, Dickey AN, Xing C, Faller R. 2007. Simulations of biomembranes and water: important technical aspects. *Fluid Phase Equilib* **261**:18–25.
105. Silver B. 1985. Physical chemistry of membranes: an introduction to the structure and dynamics of biological membranes. Dordrecht: Kluwer Academic.
106. Reynwar BJ, Illya G, Harmandaris VA, Müller MM, Kremer K, Deserno M. 2007. Aggregation and vesiculation of membrane proteins by curvature-mediated interactions. *Nature* **447**:461–467.
107. Weeks JD, Chandler D, Andersen HC. 1971. Role of repulsive forces in determining the equilibrium structure of simple liquids. *J Chem Phys* **54**:5237–5247.
108. Winter R. 2002. Synchrotron X-ray and neutron small-angle scattering of lyotropic lipid mesophases, model biomembranes and proteins in solution at high pressure. *Biochim Biophys Acta* **1595**:160–184.
109. Sachs JN, Petrache HI, Woolf TB. 2003. Interpretation of small angle X-ray measurements guided by molecular dynamics simulations of lipid bilayers. *Chem Phys Lipids* **126**:211–223.
110. Rowlinson JS, Widom B. 2002. *Molecular theory of capillarity*. New York: Dover.

5

Collective Dynamics in Lipid Membranes: From Pore Formation to Flip-Flops

Andrey A. Gurtovenko[1,2] and Ilpo Vattulainen[3,4,5]

[1] *Institute of Macromolecular Compounds, Russian Academy of Sciences, St. Petersburg, Russia;* [2]*Computational Biophysics Laboratory, Institute of Pharmaceutical Innovation, University of Bradford, West Yorkshire, United Kingdom;* [3]*Department of Physics, Tampere University of Technology, Tampere, Finland;* [4]*Helsinki University of Technology, Helsinki, Finland;* [5]*MEMPHYS — Center for Biomembrane Physics, University of Southern Denmark, Odense, Denmark*

5.1. INTRODUCTION

Biological membranes are excellent examples of biologically relevant soft interfaces. They mediate or even govern a large variety of cellular functions [1–3]. Membranes serve as a host for membrane proteins to carry out their functions, and numerous signaling processes are either conducted inside membranes or mediated by them. Additionally, cellular membranes act as a permeability barrier, allowing only desired particles to permeate through the membrane into and out of the cell, besides which membranes are also involved in a variety of large-scale functions such as in maintaining the osmotic pressure and ion density gradients across the plasma membrane.

The biological relevance of membranes is emphasized by the rather recently proposed lipid raft model [4–7], which essentially stresses the importance of understanding the interplay between lipids and proteins: membrane proteins function together with lipids. Consequently, lipid membrane structures, lipid domain coexistence, and especially the role of cholesterol in the structural properties of membranes have been paid a considerable amount of attention recently. Meanwhile, the dynamics of membranes [3,8,9] has received much less attention despite its substantial importance in, e.g., signaling, domain formation, and diffusion of lipids and proteins in the plane of the membrane.

Address all correspondence to Ilpo Vattulainen, Department of Physics, Tampere University of Technology, P.O. Box 692, FI-33101 Tampere, Finland, +358 3 3115 3489, <Ilpo.Vattulainen@tut.fi>.

R. Faller et al. (eds.), *Biomembrane Frontiers: Nanostructures, Models, and the Design of Life*, Handbook of Modern Biophysics, DOI: 10.1007/978-1-60761-314-5_5,

In this contribution, we discuss the dynamics of lipids in particular. We first consider some of the most relevant dynamic processes where lipids are involved in biological membranes, and then move on to discuss some dynamic processes in detail, in particular those dealing with the formation of pores in membranes [10,11]. It turns out that the nature of dynamic phenomena taking place in membranes is largely *collective*. They cannot be described as a set of single-particle (single-lipid) processes that are largely independent from each other. Instead the actual dynamics of lipids involves large assemblies of molecules where the lipids move in unison in a collective manner. Several examples of dynamic phenomena discussed below show that this is a generic feature rather than an exception: dynamic membrane processes such as pore formation and flip-flops illustrate this view and especially highlight the importance of understanding the dynamics of lipids from a collective perspective.

5.2. DYNAMICS OF LIPIDS IN MEMBRANES

Membranes, like all biological and soft matter, are governed by weak interactions of the order of thermal energy, k_BT. In the same spirit, the dynamics of lipids in membranes is mainly driven by thermal fluctuations. This is truly exciting since it implies that Nature uses random walks in a variety of dynamical processes: the diffusion of individual lipids in a membrane is a random walk, the ordering and formation of membrane domains is partly related to random walks, and also membrane proteins move in a random walk-like manner.

It has to be emphasized that there is no specific timescale characterizing the dynamics of lipids. Instead, the timescales depend on the dynamic process in question as well as on the length scale involved. One of the fastest dynamic processes is the rotational diffusion of individual carbon–hydrogen bonds in CH_2 groups in lipid hydrocarbon chains. The timescale of these rotational motions is usually on the order of picoseconds in fluid-like membranes [12,13], though it can be considerably slower in the gel phase below the main transition temperature. The rotational motion of whole lipids around their principal axes of rotation is a slower process and takes place over a scale of nanoseconds [14,15]. Meanwhile, the timescale of lipid lateral diffusion depends on the length scale considered. For a given diffusion length $\ell_D = (2d\, D\, t)^{1/2}$, where d is the dimensionality, D the diffusion coefficient, and t the timescale of diffusion, it is obvious that the timescale and the range of diffusion given by ℓ_D are coupled. For a lipid in a fluid membrane, the diffusion coefficient is typically about 10^{-7} cm^2/s, thus it diffuses over a distance of its own size (about 0.8 nm) in about 15 ns. However, as biological membranes are heterogeneous and characterized by membrane domains, the more relevant timescale is the time needed to cross a domain. Assuming a domain to have a size of about 50 nm, a lipid would on average diffuse across it in about 60 μs. Further, given a cell with a radius of 10 μm, a lipid would diffuse to the opposing side in about a minute. In strongly ordered lipid rafts the timescales would be larger by a factor of 10, since in these systems $D \approx 10^{-8}$ cm^2/s [16,17]. These examples illustrate how the timescales of lateral diffusion increase with the length scale considered ($t \propto \ell_D^2$). Yet, there is reason to stress that diffusion provides an effective means for transport over cellular scales, since a timescale of the order of seconds is modest compared to times related to the folding of proteins, for instance. Additionally, as far as passive diffusion is concerned, no ATP is wasted: for a cell the motion of lipids comes for free, since the motion of lipids is driven by thermal fluctuations.

While the lateral diffusion of lipids is characterized by a wide range of timescales ranging from nanoseconds to minutes, there are considerably slower dynamic processes in a cell membrane, characterized by hours and even days. In this contribution, we consider one such phenomenon, the translocation (flip-flop) of lipids from one membrane leaflet to another. While it may sound surprising that atom-scale simulations can shed light on dynamic phenomena that take place on macroscopic timescales, it turns out that simulations can indeed complement experimental studies and provide added value in order to better understand the principles guiding complex dynamic processes in lipid membranes.

5.3. FLIP-FLOPS ASSOCIATED WITH ASYMMETRIC DISTRIBUTION OF LIPIDS IN MEMBRANES

One of the dynamical processes that we consider in more detail below is a flip-flop. In general, biological membranes of most animal cells are characterized by an asymmetric distribution of lipid molecules across a membrane [18,19]. This asymmetry is vital for numerous cellular functions and plays an important role, e.g., in membrane mechanical stability [20], membrane electrostatics [21–23], and modulation of the activity of membrane proteins [24]. Failure to maintain the asymmetric distribution of lipids can have dramatic consequences. For example, it has been shown that the externalization of anionic phosphatidylserines, which are normally localized in the inner leaflet of plasma membranes, is able to trigger programmed cell death [25].

To maintain the asymmetric transmembrane lipid distribution, the cell needs to be able to translocate lipid molecules from one membrane leaflet to another. One way of doing this is to employ special cellular mechanisms for active transport of lipids across biological membranes using specific membrane proteins — flippases [26,27]. In addition, cells also use passive transport mechanisms that facilitate the migration of lipids across a membrane; such passive translocations may take place with the help of proteins or without them [28].

There is reason to emphasize that the molecular mechanisms associated with passive transmembrane transport of lipids in protein-free membranes are exceptionally poorly understood. A thorough understanding of these mechanisms is one of the key challenges in cell biology [29]. Lipid flip-flops in the absence of flippases are rather difficult to characterize as they take place over molecular scales and often cannot be well resolved within existing experimental methods. On the computational side, the atomistic modeling of lipid flip-flops is usually considered to be prohibitive in terms of accessible timescales: lipid flip-flops in general are very slow processes characterized by an average waiting time ranging from several hours to several days [30,31]. In turn, the state-of-the-art atomic-scale simulations for systems of moderate size (20000–40000 atoms) are still limited to a timescale of about a microsecond.

However, there is strong experimental evidence that lipid translocation across a protein-free membrane is a pore-mediated process. It has been found that the activation energy for radioactive chloride flux across lipid membranes is close to the activation energy of lipid flip-flop [30], implying that the occurrence of pore defects is part of the flip-flop process. It has also been demonstrated that brief electric pulses are able to enhance the transbilayer mobility of phospholipids through electroporation of membranes [32]. Furthermore, recent computer simulations support the idea that pore formation and lipid flip-flop are coupled: when a lipid was dragged by an external force through the hydrophobic membrane interior, the formation of a small water pore was observed; the corresponding energies of the formation of the pore and of the lipid transmembrane translocation were found to be the same [33]. Taken together, these findings strongly suggest that a major fraction of lipid flip-flops takes place through water pores in membranes.

Remarkably, it turns out that formation of transient water pores is the rate-limiting step in the process of transmembrane lipid flip-flop [33,34]: once a pore has been formed, the subsequent lipid translocation occurs on timescales accessible through state-of-the-art computer simulations. Below, we demonstrate how this takes place. In particular, we use atomistic computer simulation data to discuss the complexity of lipid flip-flops, which turns out to be a highly collective process involving not only lipids but also their interplay with ions and water. Essentially, we will find that flip-flops take place through water pores generated by a locally strong transmembrane potential, which in turn is induced by an ion concentration gradient across a membrane. Without a doubt, this process is very appealing due to its collective character, which is likely a rule rather than an exception in biological systems in general.

5.4. FORMATION OF TRANSIENT WATER PORES IN LIPID MEMBRANES

We now discuss some practical applications of computer simulations for studying the dynamics of lipids in membranes. In particular, we highlight recent findings that have shed light on the formation of transient pores in membranes, and the consequent lipid flip-flops taking place through the pores.

On the computational side, various methods have been used to induce transient hydrophilic defects in lipid bilayers. In particular, pore formation has been observed under conditions where a bilayer has been subjected to an external electric field (electroporation) [35–37] or mechanical stress (surface tension) [35,38,39]. Here we consider a different scenario where pore formation is induced by a transmembrane ionic charge imbalance without any external forces [10,11]. The imbalance of ions across a membrane is an inherent feature in living cells. It is believed that the ion concentration differences determine the resting membrane potential in cells [40]. The average value of the potential is of the order of 100 mV. However, due to fluctuations its instantaneous value can deviate considerably from the average.

To characterize pore formation, let us consider recent results [10,11] based on atomic-scale molecular dynamics (MD) simulations on lipid bilayers comprised of zwitterionic dimyristoylphosphatidylcholine (DMPC) lipids. The temperature is 323 K, which is well above the main transition temperature of a DMPC bilayer (297 K) [41], meaning that the simulated DMPC bilayers are in the fluid phase.

There is reason to mention one practical issue that is relevant here. Because of periodic boundary conditions, it is not feasible to incorporate a transmembrane ionic charge imbalance into a conventional bilayer simulation (a single lipid bilayer in a box of water): salt ions on one side of the bilayer would diffuse to the other side within a nanosecond time span, thereby discharging the ionic imbalance. The way forward is to use a double-bilayer setup, that is, two lipid bilayers in the same simulation box [42,43] (see Fig. 5.1). This setup provides a means to control electrolyte compositions in the two water phases on the opposite sides of the membrane.

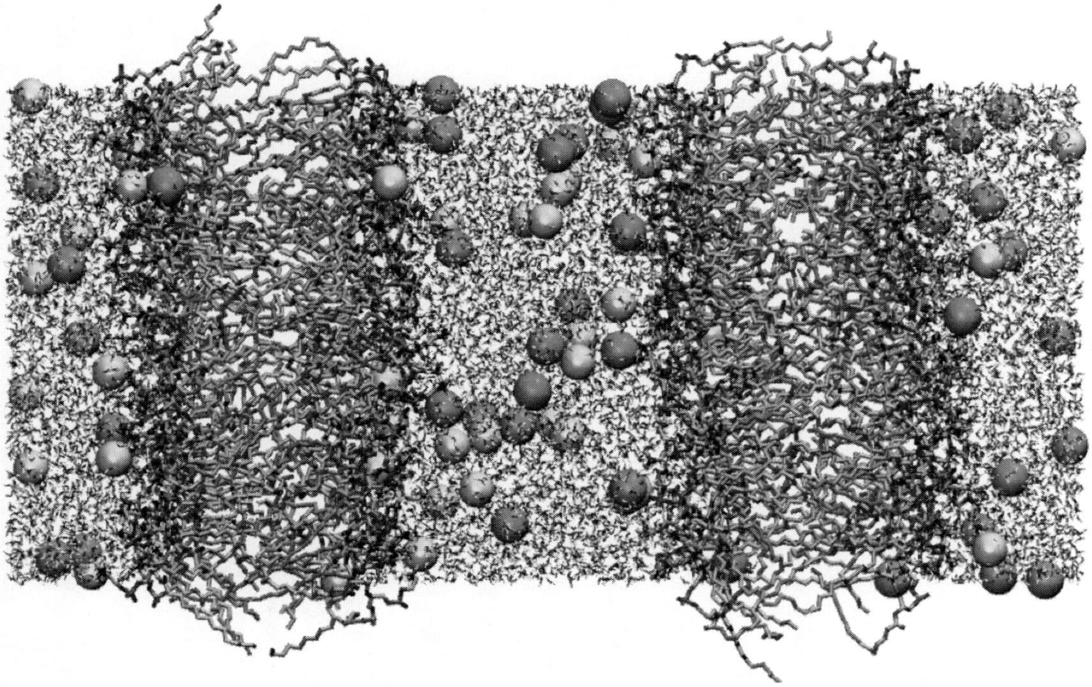

Figure 5.1. Initial structure of a double-bilayer system with a charge imbalance of 6 Na^+ ions per bilayer. The sodium ion concentration is larger in the "inner" water bath between two DMPC bilayers (see text for details). The two lipid and water regions are complemented by ions; sodium ions in light and chloride ions with darker color. Adapted with permission from [34]. Copyright © 2007, American Chemical Society. Please visit http://www.springer.com/series/7845 to view a high-resolution full-color version of this illustration.

In a double-bilayer system, the two bilayers separate the "inner" (between the bilayers) and "outer" water reservoirs. Here, the terms "inner" and "outer" are used for convenience only, as periodic boundary conditions are applied in all three directions. The next step is to create a transmembrane ionic charge imbalance. To do that, cations (Na^+ or K^+ ions) and anions (Cl^- ions) are added to the system, replacing randomly chosen water molecules. The number of anions is set to be the same in both water reservoirs (20 Cl^- ions). Meanwhile, an excess of cations is created in the "inner" water phase with respect to the "outer" one. The initial transmembrane imbalance of cations is chosen to be 6 cations per bilayer or 12 cations between the reservoirs, such that 26 and 14 cations are placed in the "inner" and "outer" water reservoirs, respectively (see Fig. 5.1). The condition of electroneutrality in the whole double-bilayer system is fulfilled.

To gain insight into characteristic times of pore formation and sensitivity of the whole process to the type of salt, 20 different runs simulated over 10 to 15 ns each have been performed by varying initial conditions (10 runs with NaCl and 10 runs with KCl) [10,11]. Figure 5.2 depicts the transmembrane electric field and electrostatic potential for one representative double-bilayer system with NaCl. We find that for this particular system the imbalance of 6 Na^+ ions per bilayer induces on average a transmembrane voltage of about 2.12 V across the membrane.

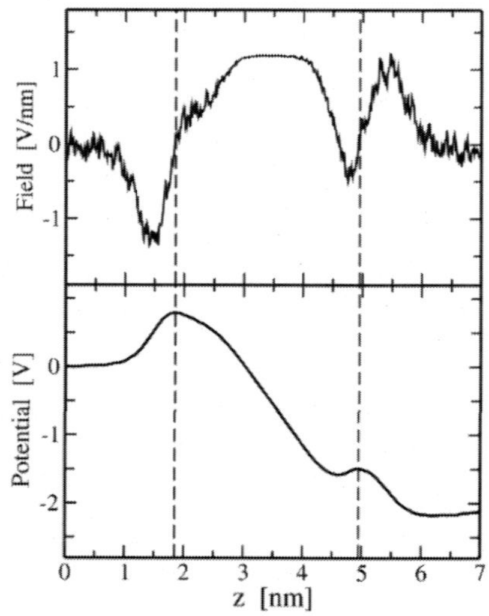

Figure 5.2. Electric field (top) and electrostatic potential (bottom) versus distance z from the middle of the "inner" water phase between the two bilayers. The field and the potential were calculated from the Poisson equation by integrating over charge densities that were determined directly from MD simulations; both quantities were chosen to be zero in the middle of the "inner" water phase ($z = 0$). Only the initial part of the trajectory prior to actual pore formation was used for the calculation. Dashed lines indicate average positions of phosphorous atoms of lipids for the two bilayer leaflets. Reprinted with permission from [11]. Copyright © 2007, The Biophysical Society.

Remarkably, the applied transmembrane imbalance of sodium ions gives rise to the formation of a water pore, which starts about 1.4 ns after the beginning of the simulation. The various stages of pore formation can be readily followed by considering water molecules in the hydrophobic core of a membrane (Figs. 5.3 and 5.4). The first step in pore formation is the creation of a single water defect spanning the entire membrane (Fig. 5.4B); this "chain" of water molecules expands rapidly on a nanosecond timescale (see Figs. 5.3 and 5.4C). The expansion of the pore is accompanied by a considerable redistribution of lipid headgroups from the water–lipid interface to the membrane interior; the hydrophilic headgroups line the pore and stabilize it. It is also interesting to note that the simulations did not express any signs of simultaneous formation of multiple pores. Finally, after most of the excess ions have migrated across the pore, the transmembrane field has become too weak to maintain the pore, which hence becomes metastable. Nonetheless, despite its metastable nature, the pore may remain open for a reasonably long period of time (see below).

The overall molecular picture of pore formation found here is very similar to the picture reported in MD simulations of electroporation in lipid membranes under the influence of an external (constant) electric field [35–37]. This similarity is due to the fact that in both situations a lipid membrane is exposed to a rather high transmembrane electric field. It should be emphasized that the electric field induced by a transmembrane ionic density gradient in the present case is stochastic in nature: the transmembrane electric field is determined by instantaneous positions of ions and is therefore subject to considerable fluctuations in time and space. Indeed, it has been found [11] that the average values of the electric field across a membrane (prior to pore formation) can scatter from 1.74 to 3.33 V.

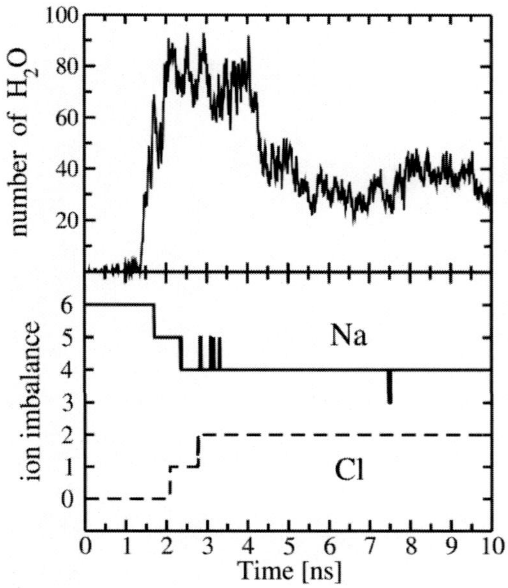

Figure 5.3. Time evolution for the number of water molecules in a pore. The number of water molecules is computed within 0.5 nm from the bilayer center (top) and for the transmembrane imbalance of Na^+ and Cl^- ions (bottom). Reprinted with permission from [11]. Copyright © 2007, The Biophysical Society.

The transmembrane potential of this magnitude exceeds the potentials typical for native plasma membranes by one order of magnitude and thus deserves some discussion. Due to obvious limitations of atomic-scale MD simulations regarding the length and timescales accessible, one has to promote pore formation by increasing the ionic charge imbalance across a membrane and, correspondingly, the transmembrane potential. By doing so, one essentially optimizes chances for observing the formation of water pores on timescales that are within reach by atomistic simulations. A similar approach has been used in simulations that mimic electroporation [35–37]. The choice of a reasonably large ion concentration gradient for simulation studies is also supported by the nature of biological matter, as it is driven by thermal fluctuations: in a macroscopic system, density fluctuations inevitably lead to fluctuations of ion densities that may locally be considerably larger than the average ones. Moving on, it is noteworthy that the overall picture of pore formation seems to be generic in the sense that the same stages of pore formation have also been found for a considerably smaller ion concentration gradient, though the characteristic timescales are then obviously longer [11]. On these grounds, it is reasonable to assume that the same pore formation mechanism would be found also with an ion imbalance that would match conditions in native cells.

The maximum size of a transient water pore varies considerably from one simulation to another. Assuming an ideal cylindrical shape for a pore, the radius of the inner part of a pore varies from 0.41 to 1.03 nm. Essentially, after the water pore has become large enough, the cations and anions permeate through the pore along the ion density gradient (cations and anions permeate in opposite directions). As an example, Figures 5.3 and 5.4 show permeation of Na and Cl ions across the membrane, discharging the transmembrane potential and driving the system to

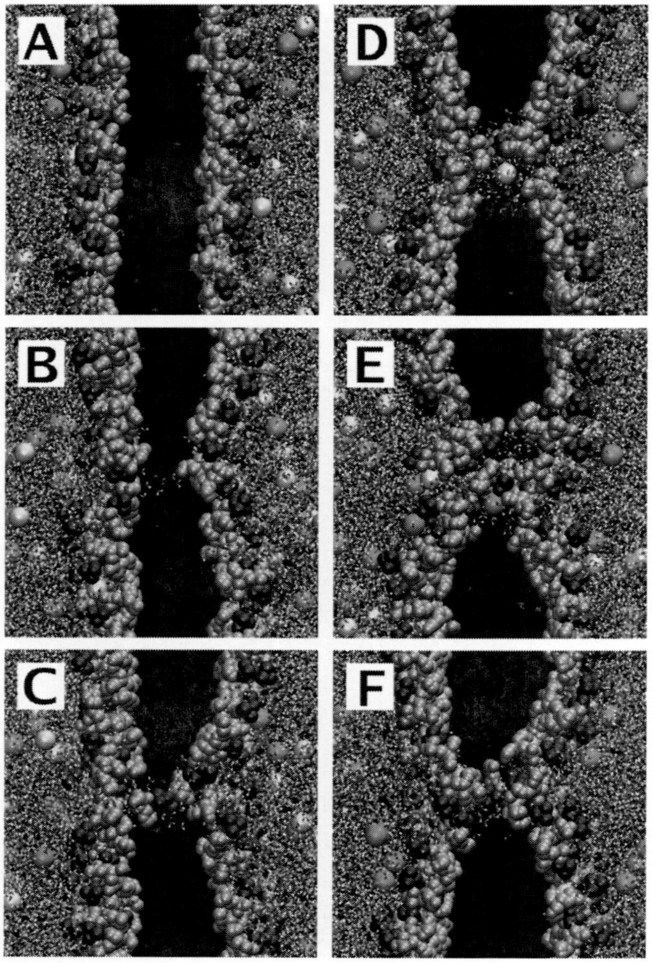

Figure 5.4. Pore formation and pore-mediated ion transport induced by a transmembrane imbalance of sodium ions. (A) 0 ps, the initial structure; (B) 1420 ps, appearance of a single water defect; (C) 1490 ps, redistribution of lipid headgroups toward the membrane interior; (D) 1710 ps, leakage of a sodium ion; (E) 2090 ps, leakage of a chloride ion; and (F) 10 ns, the metastable state of the water pore. Water is shown on the left- and right-hand sides of the figures, choline groups of lipid headgroups in dark, phosphate and glycerol groups using lighter color in the same region, and Na^+ (Cl^-) ions as light (dark) spheres. Nonpolar acyl chains of lipids are not shown. Excess of Na^+ ions is on the left-hand side. Reprinted with permission from [11]. Copyright © 2007, The Biophysical Society. The figure is adapted from (34). Please visit http://www.springer.com/series/7845 to view a high-resolution full-color version of this illustration.

equilibrium. It turns out that the size of the pore is controlled by the induced transmembrane potential: the transmembrane ionic leakage leads to a remarkable drop in the transmembrane potential difference and, as a result, to a considerable decrease in pore size (see Figs. 5.3 and 5.4). The drop in the ionic charge imbalance from 6 to 2 within the first 5 ns of simulations gives rise to a decrease in the number of water molecules by about 70% compared to its maximum value. In general, after a drop of the transmembrane voltage below some critical value due to ion leakage, the water pore becomes metastable, its lifetime expressing a broad distribution from 5 to more than 200 ns.

It is interesting to compare how the molecular picture of pore formation and subsequent ion leakage depend on the type of salt ions used to create a transmembrane ionic charge imbalance. It turns out that the sequence of events and most of the features of pore formation are very similar to that of NaCl and KCl salts; the maximal radius of the induced pores has been observed to be somewhat smaller in the case of potassium ions, as they require smaller pores for permeation to occur. However, ionic leakage through the pore is found to be sensitive to the type of cation: K^+ ions permeate through a lipid membrane much easier than Na^+ ions do. The origin of such a difference is most likely in the strong interactions of sodium ions with lipid headgroups lining the pore walls [43–47]. In contrast, potassium ions interact only weakly with the carbonyl regions of phospholipids [11,47].

5.5. FLIP-FLOPS OF LIPID MOLECULES ACROSS PROTEIN-FREE LIPID MEMBRANES

Given the above findings for pore formation, it is tempting to ask whether the pores could act as membrane defects, facilitating the occurrence of flip-flops. To consider this scenario, let us highlight some recent data [34] for 8 double-bilayer systems with preformed pores (4 systems with NaCl as well as KCl) that were simulated over 200 ns, or until the transient water pore closed.

In all the considered systems, one finds spontaneous pore-mediated translocation processes of lipid molecules from one leaflet to another. The overall process of lipid flip-flop is depicted in Figure 5.5 for one particular lipid of a DMPC bilayer system with NaCl. Starting from an intact lipid membrane (Fig. 5.5A), one first observes the formation of a water pore spanning the entire membrane. For the tagged lipid studied here, the pore is laterally located quite far away from the lipid (Fig. 5.5B), implying that a flip-flop cannot occur right after pore formation. However, after about 100 ns the lipid molecule has diffused laterally to the vicinity of the pore, allowing it to enter the pore and become part of it as the lipid's headgroup settles to line the pore (Fig. 5.5C). At this point, the spontaneous translocation of the lipid through the pore initiates: the translocation involves diffusion of a lipid headgroup along the pore walls coupled to the simultaneous desorption of lipid acyl chains out of the bilayer leaflet (Fig. 5.5D). As a result, one observes the appearance of the lipid in the opposite bilayer leaflet, accompanied by progressive reorientation of the lipid (Fig. 5.5E). For translocation to be successfully accomplished, the lipid should be irreversibly accommodated in the opposite leaflet. It turns out that such accommodation can be a rather slow process. For instance, more than 50 ns are required for the particular flip-flopped lipid considered here to settle down in the opposite leaflet (see Fig. 5.5E–G). Considering that the actual flip-flop takes about 30 to 100 ns to occur, the timescale of settling down is rather short and essentially limited by lateral diffusion: to avoid being displaced back to the original leaflet, the headgroup of the lipid being translocated should first be able to detach its headgroup out of pore "walls," followed by lateral diffusion of a lipid away from the pore site. Since diffusion in the plane of fluid membranes over a scale of molecular size takes roughly about 20 ns, it is obvious that about 50 ns are required to diffuse away from the immediate vicinity of the pore. We note, however, that the realization of a flip-flop can be facilitated by the closure of a pore; for the particular system in question, pore closure occurs at $t \cong 210$ ns (Fig. 5.5H).

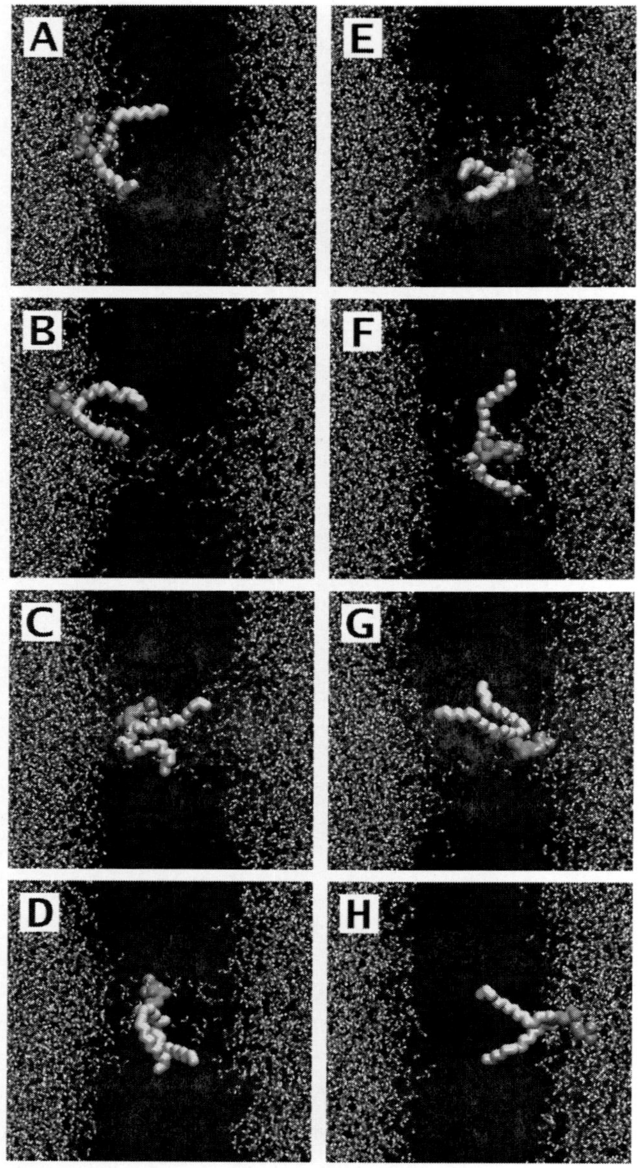

Figure 5.5. Pore-mediated lipid translocation across a DMPC membrane in aqueous solution with NaCl. (A) 0 ns, (B) 43.85 ns, (C) 118.9 ns, (D) 122.4 ns, (E) 152.7 ns, (F) 204.65 ns, (G) 208.9 ns, and (H) 215 ns. Lipids (except for the flip-flopped one) and ions are not shown; water is shown on the left- and right-hand sides of the figures, and the flip-flopped lipid is shown separately during various stages of the process. Reprinted with permission from [34]. Copyright © 2007, American Chemical Society. Please visit http://www.springer.com/series/7845 to view a high-resolution full-color version of this illustration.

To visualize lipid flip-flops across a membrane, it is instructive to follow the time evolution of the positions of centers of mass (CM) of lipid headgroups as well as the lipids' overall orientation measured through the angle between the bilayer normal and the vector directed from the

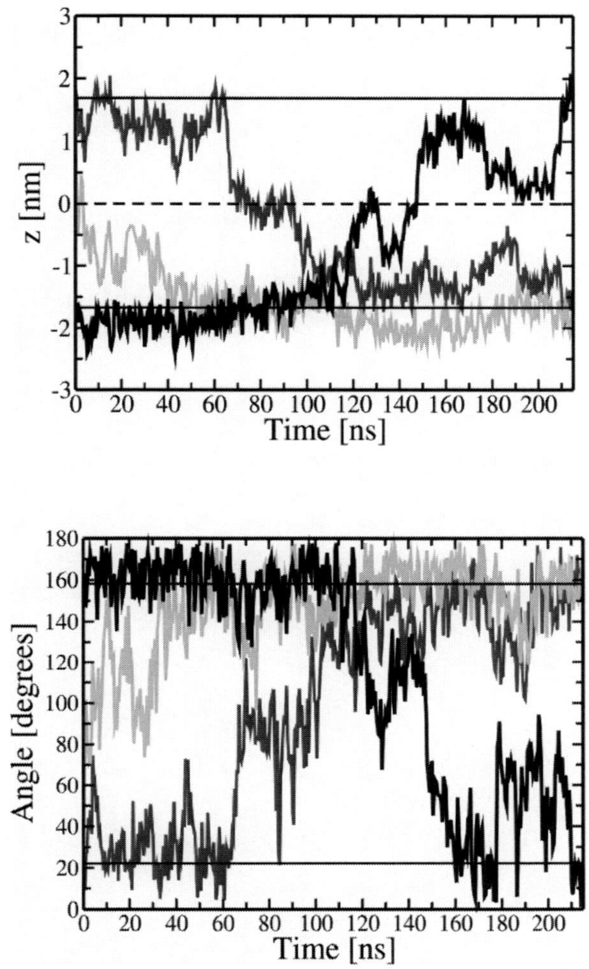

Figure 5.6. Time evolution of lipid positions and their orientations. Top: Time evolution of positions of the centers of mass (CM) of headgroups for four representative flip-flopped lipids of a DMPC bilayer system in an aqueous solution with NaCl. The $z = 0$ corresponds to the membrane center; solid black lines show the average positions of lipid headgroups in the two opposite leaflets, extracted from the intact membrane before pore formation. Bottom: Time evolution of the tail-to-head orientation for the same four flip-flopped lipids. The tail-to-head orientation is characterized as the angle between the bilayer normal and the vector directed from the CM of acyl chains to the CM of headgroups. Solid black lines again show typical values for the angle for lipids in the opposite leaflets. Adapted with permission from [34]. Copyright © 2007, American Chemical Society. Please visit http://www.springer.com/series/7845 to view a high-resolution full-color version of this illustration.

CM of lipid's hydrocarbon chains to the CM of its headgroup. Figure 5.6 presents trajectories for four representative flip-flopped lipids of a DMPC bilayer with NaCl [34]. Remarkably, the translocation of a lipid molecule across a membrane closely correlates with the overall reorientation of the lipid within a membrane: when a lipid molecule is accommodated in its initial membrane leaflet, the lipid's tail-to-head vector makes an average angle of 22 degrees (or 158 degrees depending on the leaflet) with the bilayer normal. During flip-flop (i.e., when the CM of a lipid headgroup crosses the membrane center, $z = 0$), a lipid changes its orientation and the corresponding angle lies between the two values typical to opposite leaflets. Interestingly, lipids tend to adopt an orientation perpendicular to the bilayer normal when they become close to the membrane center, which is clear for one of the lipids in Figure 5.6.

One can clearly distinguish two different types of lipid translocations [34]:

1. Flip-flops of lipid molecules that are directly involved in the initial formation of a water pore. These are very fast flip-flop processes whose onset coincides with the formation of a pore. They are also characterized by short times (10–20 ns) required for accommodation in the opposite leaflet (see the light curve in Fig. 5.6).
2. Considerably slower flip-flops of lipids that are either involved in initial pore formation but require much longer times to accomplish translocation (the dark curve in Fig. 5.6) or are initially remote from the pore site but diffuse to the pore with time (remaining curves in Fig. 5.6).

Overall, out of 50 successful flip-flop events observed, only 9 flip-flops are associated with category 1. The rest of the flip-flops (about 80%) belong to category 2. On the basis of these 50 flip-flop events resolved in atomic detail, one can estimate that the average time required for a lipid to translocate through a preformed water pore in a membrane is around 60 ns. Yet there is reason to mention that flip-flop times for individual lipids scatter considerably and have here been found to range from 10 to 130 ns. The reasonable statistics resulting from these events allows one to present the distribution of these flip-flop times (see Fig. 5.7). First, one finds that the distribution is indeed very wide. Second, one can easily distinguish two peaks that correspond to the two flip-flop categories outlined above: there is a clear peak around $t \cong 20$ ns (flip-flop category 1), a rather broad peak around the average flip-flop time $t \cong 60$ ns (category 2), and a long tail extending up to $t = 130$ ns corresponding to the slowest flip-flop observed.

Of the factors that affect the rate of lipid translocation, let us mention pore opening and membrane resealing. As discussed above, pore formation implies fast reorientation of some lipid headgroups toward the membrane interior [10,11,35–37] so that the lipids involved in pore formation are driven off from their equilibrium state. If desorption of lipid hydrocarbon chains occurs at the same time, one can observe a very fast lipid flip-flop. However, in most cases this scenario takes place rather rarely (see Fig. 5.7), because the time required for lipid accommodation in the opposite leaflet can be considerable. Pore closure, in turn, considerably speeds up the flip-flops of lipids that are partly translocated since irreversible membrane resealing simply makes it impossible for the lipids to move back to their initial leaflet. As an example, one can mention the translocation of a lipid shown by the dark curve in Figure 5.6 (note that pore closure for this bilayer system occurs at $t \cong 210$ ns).

Another factor whose role deserves some discussion is the type of cations used for pore formation. All together, 18 successful lipid flip-flops were found for DMPC bilayers with NaCl, while for their counterparts with KCl a number of 32 pore-mediated lipid translocations were observed. Therefore, despite the fact that the average times required for lipid flip-flops are similar in both cases (64 and 60 ns for bilayers with NaCl and KCl salt, respectively), the type of salt used does affect the probability of lipid flip-flop. In more practical terms, DMPC bilayer systems with NaCl and KCl were simulated for 485 and 515 ns, respectively; yet, the overall number of lipid flip-flops was almost twice as large in the case of KCl (see also Fig. 5.7). One can conclude that the probability of a spontaneous flip-flop event in a bilayer system with potassium ions is higher by a factor of about 1.7. This finding can readily be explained in terms of interactions of cations with lipid headgroups: sodium ions demonstrate much stronger interactions with zwitterionic phosphatidylcholine lipids than potassium ions [11,47]. More specifically, sodium ions are known to bind strongly to carbonyl oxygens of phosphatidylcholines, leading to the formation of tight complexes between neighboring lipids [11,43–47], thus weakening the desorption of lipids out of membrane leaflets toward the membrane interior.

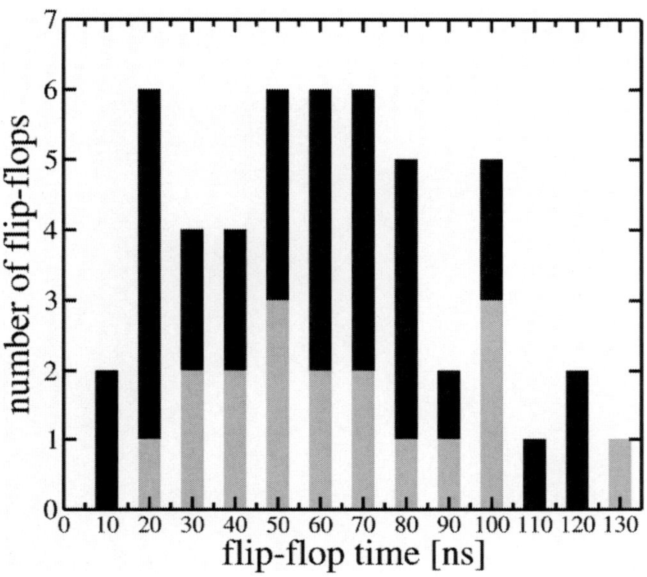

Figure 5.7. The distribution of flip-flop times. The distribution is shown for all eight DMPC bilayer systems considered (50 flip-flop events all together). The lipid flip-flops in the bilayer systems with NaCl are shown in light color, while the rest were observed under the influence of KCl. Adapted with permission from [34]. Copyright © 2007, American Chemical Society. Please visit http://www.springer.com/series/7845 to view a high-resolution full-color version of this illustration.

Overall, the molecular mechanism of lipid flip-flop discussed here consists of two steps [34]. The first step is the formation of a transient water defect in a membrane. In the present case this is achieved through the incorporation of an initial ionic concentration imbalance across the membrane; such an imbalance, being an inherent feature of plasma membranes of eukaryotic cells, initiates the formation of a water pore coupled to the subsequent fast ion leakage that quickly reduces the local field strength and makes the pore metastable. The second step is a fast translocation of lipid molecules through preformed water pores. Because lipid translocation through a pore is very fast (taking only 60 ns on average), one can observe considerable mixing of lipids between the opposite leaflets since the lifetime of water pores ranges from tens to several hundreds of nanoseconds (see Fig. 5.8).

The origin of the force that drives the pore-mediated lipid translocation across a membrane is central for understanding the molecular mechanism of lipid flip-flop. First, we recall that the PC lipids considered in this study are zwitterionic, i.e., neutral. Therefore, the transmembrane electrostatic potential can hardly play the main role. Indeed, the membrane-normal component of the total electrostatic force exerted on the lipids involved in flip-flops is found to fluctuate around zero [34], without a component that would drive translocation. Second, as already discussed, the electric field across a membrane is spatially inhomogeneous, as it is defined by instantaneous positions of ions. In general, the inhomogeneous electric field could contribute to translocation of zwitterionic lipids via dielectrophoretic motion. However, it has been found that the contribution of the dielectrophoretic force exerted on the dipolar lipid headgroup is the largest for a small number of lipids involved in flip-flops during the initial stages of pore formation, and even then the dielectrophoretic component is comparable to thermal energy [34]. At longer times, when the transmembrane electric field is almost fully discharged due to ion leakage, the dielectrophoretic force has been found to be an order of magnitude smaller than the force due to thermal fluctuations. Thus, we conclude that translocation of lipids through transient water pores is mainly *diffusive*, i.e., driven by thermal fluctuations.

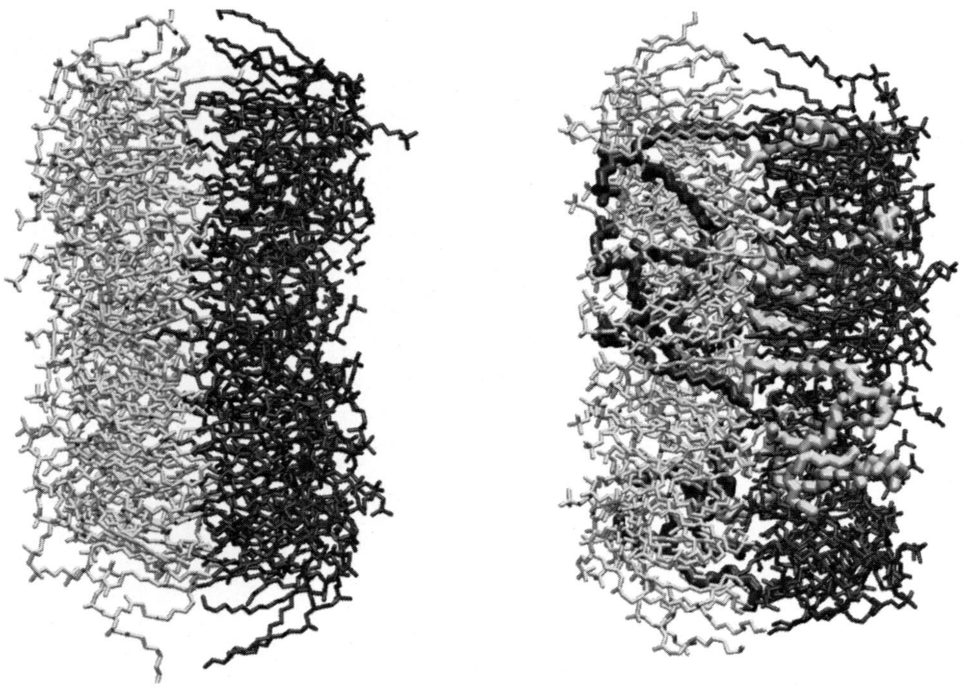

Figure 5.8. Flip-flop induced mixing up of the lipids which belong to the opposite leaflets of a DMPC bilayer system with KCl. Left: Intact membrane at $t = 0$ ns. Right: The same membrane with an opened pore at $t = 200$ ns. Lipids in opposite leaflets are shown in light and dark; for clarity, the size of flip-flopped lipids has been enlarged. Adapted with permission from [34]. Copyright © 2007, American Chemical Society. Please visit http://www.springer.com/series/7845 to view a high-resolution full-color version of this illustration.

Overall, the atomic-scale MD simulations demonstrate that the appearance of a water pore spanning a phospholipid membrane *inevitably* leads to diffusive transmembrane translocation of lipids through the pore. There is reason to especially emphasize that, while the formation of a water pore is induced here by a transmembrane ion density gradient, the precise way by which the pore is formed is irrelevant for the flip-flop mechanism discussed here. This has also been confirmed by very recent studies of chemically induced pore formation in phospholipid membranes: it was demonstrated that a small amphiphilic solute dimethylsulfoxide (DMSO) — which is widely employed in cell biology as an effective penetration enhancer [48], cryoprotectant [49], and cell fusogen [50] — is able to induce hydrophilic water pores in protein-free lipid membranes [51–53]. What is more, the appearance of DMSO-induced pores (formed without salt ions) was shown to lead to spontaneous pore-mediated lipid translocations [54], indicating that the formation of a water pore spanning the membrane is most likely the only required prerequisite for transmembrane lipid flip-flop.

Moving on, the above-discussed results provide concrete evidence for a view that the formation of water pores (defects) is the *rate-limiting step* in the lipid flip-flop processes that take place through pores. While the actual flip-flops driven by thermal fluctuations are very rapid processes, occurring spontaneously on a timescale of tens of nanoseconds, the total timescales of lipid flip-flops range from minutes to hours. The rate-limiting step is hence pore formation. Recent simulation studies and free energy calculations are in favor of this view [33,55,56].

However, we have to keep in mind that the above-discussed scheme is not necessarily the only possible mechanism for lipid translocation. Sterols have been observed to translocate without pores [57,58]. On the other hand, sterols are known to translocate at much larger rates compared to many other lipids, the timescales ranging over seconds rather than hours, which may explain why transient pores are not crucial for their translocation processes. Also, there is evidence for flip-flops that take place without transient membrane defects. The conditions where this has been reported to occur include high concentrations of glucose and a high transmembrane pH gradient [59].

Nonetheless, coming back to pore formation, let us summarize that the rate of lipid flip-flops can be significantly enhanced by spontaneous formation of water pores. In addition to their biological relevance, this view also offers a molecular-level framework for the development of ways to promote lipid transmembrane translocation and to make it possible to manipulate lipid distribution across cell membranes. This view also provides an explanation for the experimentally observed fact that the exposure of lipid membranes to electric field pulses drastically reduces the timescales where flip-flops take place [60]. Strong external electric fields induce electroporation of a lipid membrane, considerably increasing the number of pathways (defects) suitable for pore-mediated lipid flip-flops. Similarly, the presented pore-mediated mechanism for lipid flip-flops can be invoked to shed more light on how local anesthetics, amphiphilic compounds, and pore-forming peptides facilitate the occurrence of flip-flops [59].

5.6. SUMMARY

The results presented and discussed in this chapter illustrate the importance of understanding the complex interplay of lipids with water and ions. The dynamical processes associated with these systems are collective, involving a large number of molecules to function in unison. Pore formation and lipid flip-flops are excellent examples of this fact. Nonetheless, they are just a minor scratch on the surface of the issue since there are numerous membrane processes where the same view holds, too. For example, the fusion of membranes and endocytosis as well as exocytosis are large-scale processes that involve large amounts of lipids, whose joint dynamic interplay leads to realization of these events. Further, it has been shown very recently that lipid diffusion in the membrane plane occurs through the diffusion of locally correlated transient lipid clusters that move in unison, and the dynamical correlations associated with these motions range over tens of nanometers [61]. Also, as membranes are soft interfaces driven by weak interactions of the order of $k_B T$, it is not particularly surprising that membranes are characterized by undulations that are essentially collective motions of lipids in the bilayer normal direction. The elastic behavior of membranes exemplified by undulations also has a significant role to play in membrane protein functions, since it has been shown that the functionality of certain classes of proteins depends on elastic coefficients such as bending rigidity and area compressibility [62,63]. Concluding, the dynamics of lipids in membranes is an exceptionally appealing example of the dynamics of soft matter, and the *collective nature of lipid motions* has to be always accounted for if one desires to understand the physical laws that govern membrane dynamics.

ACKNOWLEDGMENTS

This work has been supported by the Russian Foundation of Basic Research through Grant No. 08-03-00150 (A.A.G.) and by the Academy of Finland (I.V.). The simulations were performed at the Finnish IT Center for Science and on the HorseShoe (DCSC) supercluster at the University of Southern Denmark.

PROBLEMS

5.1. In the chapter we discussed the overall process of pore-mediated lipid flip-flops and illustrated it for a dimyristoylphosphatidylcholine (DMPC) bilayer at $T = 323$ K. Make qualitative predictions regarding changes in the flip-flop rate when (a) temperature is decreased to e.g. the physiological value ($T = 310$ K); (b) DMPC lipids are replaced by dipalmitoylphosphatidylcholine (DPPC) lipids whose acyl chains are two hydrocarbons longer.

5.2. Assuming that the lipid lateral diffusion coefficient is $D = 1 \times 10^{-7}$ cm^2/s, compute the time that is needed to diffuse from one side of a cell to the opposing one, assuming the cell radius to be R = 10 micrometers.

FURTHER READING

For further reading on transmembrane translocations of lipids in cell membranes, one can recommend several recent review papers that summarize a contemporary view of related phenomena [26,64–67]. A thorough account of experimental observations of chemically induced lipid flip-flops (both pore-mediated ones and those that do not involve pores) can be found in [59]. A very recent computational study additionally highlights the overall process of pore formation in phospholipid membranes at atomistic resolution [68].

REFERENCES

1. Mouritsen, OG. 2005. *Life — as a matter of fat*. Berlin: Springer-Verlag.
2. Katsaras J, Gutberlet T, eds. 2001. *Lipid bilayers: structure and interactions*. Berlin: Springer-Verlag.
3. Yeagle PL, ed. 2005. *The structure of biological membranes*. Boca Raton, FL: CRC Press.
4. Simons K, Ikonen, E. 1997. Functional rafts in cell membranes. *Nature* **387**:569–572.
5. Edidin M. 2003. The state of lipid rafts: from model membranes to cells. *Annu Rev Biophys Biomol Struct* **32**:257–283.
6. Pike LJ. 2004. Lipid rafts: heterogeneity on the high seas. *Biochem J* **378**:281–292.
7. Hancock JF. 2006. Lipid rafts: continuous only from simplistic standpoints. *Nat Rev Mol Cell Biol* **7**:456–462.
8. Saxton MJ, Jacobson K. 1997. Single-particle tracking: applications to membrane dynamics. *Annu Rev Biophys Biomol Struct* **26**:373–399.
9. Vattulainen I, Mouritsen OG. 2005 Diffusion in membranes. In *Diffusion in condensed matter: methods, materials, models*, pp. 471–509. Ed P Heithans, J Kärger. Berlin: Springer-Verlag.
10. Gurtovenko AA, Vattulainen I. 2005. Pore formation coupled to ion transport through lipid membranes as induced by transmembrane ionic charge imbalance: atomistic molecular dynamics study. *J Am Chem Soc* **127**: 17570–17571.
11. Gurtovenko AA, Vattulainen I. 2007. Ion leakage through transient water pores in protein-free lipid membranes driven by transmembrane ionic charge imbalance. *Biophys J* **92**:1878–1890.
12. Feller SE, Gawrisch K, MacKerell Jr AD. 2001. Polyunsaturated fatty acids in lipid bilayers: intrinsic and environmental contributions to their unique physical properties. *J Am Chem Soc* **124**:318–326.
13. Doxastakis M, Sum AK, de Pablo JJ. 2005. Modulating membrane properties: the effect of trehalose and cholesterol on a phospholipid bilayer. *J Phys Chem B* **109**:24173–24181.
14. Klauda JB, Roberts MF, Redfield AG, Brooks BR, Pastor RW. 2008. Rotation of lipids in membranes: molecular dynamics simulation, 31P spin-lattice relaxation, and rigid-body dynamics. *Biophys J* **94**:3074–3083.
15. Moore PB, Lopez CF, Klein ML. 2001. Dynamical properties of a hydrated lipid bilayer from a multinanosecond molecular dynamics simulation. *Biophys J* **81**:2484–2494.
16. Niemelä P, Ollila S, Hyvönen MT, Karttunen M, Vattulainen I. 2007. Assessing the nature of lipid raft membranes. *PLoS Comput Biol* **3**:304–312.
17. Filippov A, Orädd G, Lindblom G. 2006. Sphingomyelin structure influences the lateral diffusion and raft formation in lipid bilayers. *Biophys J* **90**:2086–2092.
18. Gennis GB. 1989. *Biomembranes: molecular structure and function*. New York: Springer-Verlag.
19. Zachowski A. 1993. Phospholipids in animal eukaryotic membranes: transverse asymmetry and movement. *Biochem J* **294**:1–14.
20. Manno S, Takakuwa Y, Mohandas N. 2002. Identification of a functional role for lipid asymmetry in biological membranes: phosphatidylserine-skeletal protein interactions modulate membrane stability. *Proc Natl Acad Sci USA* **99**:1943–1948.
21. Latorre R, Hall JE. 1976. Dipole potential measurements in asymmetric membranes. *Nature* **264**:361–363.
22. Gurtovenko AA, Vattulainen I. 2007. Lipid transmembrane asymmetry and intrinsic membrane potential: two sides of the same coin. *J Am Chem Soc* **129**:5358–5359.
23. Gurtovenko AA, Vattulainen I. 2008. Membrane potential and electrostatics of phospholipid bilayers with asymmetric transmembrane distribution of anionic lipids. *J Phys Chem B* **112**:4629–4634.
24. Pomorski T, Hrafnsdottir S, Devaux PF, van Meer G. 2001. Lipid distribution and transport across cellular membranes. *Semin Cell Dev Biol* **12**:139–148.
25. Balasubramanian K, Schroit AJ. 2003. Aminophospholipid asymmetry: a matter of life and death. *Annu Rev Physiol* **65**:701–734.
26. Bevers EM, Comfurius P, Dekkers DWC, Zwaal RFA. 1999. Lipid translocation across the plasma membrane of mammalian cells. *Biochim Biophys Acta* **1439**:317–330.
27. Pomorski T, Menon AK. 2006. Lipid flippases and their biological functions. *Cell Mol Life Sci* **63**:2908–2921.
28. Raggers RJ, Pomorski T, Holthuis JCM, Kälin N, van Meer G. 2000. Lipid traffic: the ABC of transbilayer movement. *Traffic* **1**:226–234.
29. Holthuis JC, Levine TP. 2005. Lipid traffic: floppy drives and a superhighway. *Nat Rev Mol Cell Biol* **6**:209–220.

30. Kornberg RD, McConnell HM. 1971. Inside–outside transition of phospholipids in vesicle membranes. *Biochemistry* **10**:1111–1120.
31. Wimley WC, Thompson TE. 1990. Exchange and flip-flop of dimyristoylphosphatidylcholine in liquid-crystalline, gel, and two-component, two-phase large unilamellar vesicles. *Biochemistry* **29**:1296–1303.
32. Schwarz S, Haest CWM, Deuticke B. 1999. Extensive electroporation abolishes experimentally induced shape transformations of erythrocytes: a consequence of phospholipids symmetrization? *Biochim Biophys Acta* **1421**:361–379.
33. Tieleman DP, Marrink SJ. 2006. Lipids out of equilibrium: Energetics of desorption and pore mediated flip-flop. *J Am Chem Soc* **128**:12462–12467.
34. Gurtovenko AA, Vattulainen I. 2007. Molecular mechanism of lipid flip-flop. *J Phys Chem B* **111**:13554–13559.
35. Tieleman DP, Leotiadou H, Mark AE, Marrink SJ. 2003. Simulation of pore formation in lipid bilayers by mechanical stress and electric fields. *J Am Chem Soc* **125**:6382–6383.
36. Tieleman DP. 2004. The molecular basis for electroporation. *BMC Biochemistry* **5**:10.
37. Tarek M. 2005. Membrane electroporation: a molecular dynamics simulation. *Biophys J* **88**:4045–4053.
38. Leontiadou H, Mark AE, Marrink SJ. 2004. Molecular dynamics simulations of hydrophilic pores in lipid bilayers. *Biophys J* **86**:2156–2164.
39. Leontiadou H, Mark AE, Marrink SJ. 2007. Ion transport across transmembrane pores. *Biophys J* **92**:4209–4215.
40. Voet D, Voet JG. 2004. *Biochemistry*, 3rd ed. New York: John Wiley & Sons.
41. Cevc G, Marsh D. 1987. *Phospholipid bilayers: physical principles and models*. New York: John Wiley & Sons.
42. Sachs JN, Crozier PS, Woolf TB. 2004. Atomistic simulations of biologically realistic transmembrane potential gradients. *J Chem Phys* **121**:10847–10851.
43. Gurtovenko AA. 2005. Asymmetry of lipid bilayers induced by monovalent salt: atomistic molecular dynamics study. *J Chem Phys* **122**:244902.
44. Böckmann RA, Hac A, Heimburg T, Grubmüller H. 2003. Effect of sodium chloride on a lipid bilayer. *Biophys J* **85**:1647–1655.
45. Pandit SA, Bostick D, Berkowitz ML. 2003. Molecular dynamics simulation of a dipalmitoylphosphatidylcholine bilayer with NaCl. *Biophys J* **84**:3743–3750.
46. Gurtovenko AA, Miettinen M, Karttunen M, Vattulainen I. 2005. Effect of monovalent salt on cationic lipid membranes as revealed by molecular dynamics simulations. *J Phys Chem B* **109**:21126–21134.
47. Gurtovenko AA, Vattulainen I. 2008. Effect of NaCl and KCl on phosphatidylcholine and phosphatidylethanolamine lipid membranes: insight from atomic-scale simulations for understanding salt-induced effects in the plasma membrane. *J Phys Chem B* **112**:1953–1962.
48. Anchordoguy TJ, Carpenter JF, Crowe JH, Crowe LM. 1992. Temperature-dependent perturbation of phospholipid bilayers by dimethylsulfoxide. *Biochim Biophys Acta* **1104**:117–122.
49. Rall WF, Fahy GM. 1985. Ice-free cryopreservation of mouse embryos at –196°C by vitrification. *Nature* **313**:573–575.
50. Ahkong QF, Fisher D, Tampion W, Lucy JA. 1975. Mechanisms of cell fusion. *Nature* **253**:194–195.
51. Notman R, Noro M, O'Malley B, Anwar J. 2006. Molecular basis for dimethylsulfoxide (DMSO) action on lipid membrane. *J Am Chem Soc* **128**:13982–13983.
52. Gurtovenko AA, Anwar J. 2007. Modulating the structure and properties of cell membranes: the molecular mechanism of action of dimethyl sulfoxide. *J Phys Chem B* **111**:10453–10460.
53. Gurtovenko AA, Anwar J. 2007. Ion transport through chemically induced pores in protein-free phospholipid membranes. *J Phys Chem B* **111**:13379–13382.
54. Gurtovenko AA, Onike OI, Anwar J. 2008. Chemically-induced phospholipid translocation across biological membranes. *Langmuir* **24**:9656–9660.
55. Vernier PT, Ziegler MJ, Sun Y, Chang WV, Gundersen MA, Tieleman DP. 2006. Nanopore formation and phosphatidylserine externalization in a phospholipid bilayer at high transmembrane potential. *J Am Chem Soc* **128**:6288–6289.
56. Vernier PT, Ziegler MJ, Sun Y, Gundersen MA, Tieleman DP. 2006. Nano-pore facilitated, voltage-driven phosphatidylserine translocation in lipid bilayer — in cells and in silico. *Phys Biol* **3**:233–247.
57. Rog T, Stimson LM, Pasenkiewicz-Gierula M, Vattulainen I, Karttunen M. 2008. Replacing the cholesterol hydroxyl group by the ketone group facilitates sterol flip-flop and promotes membrane fluidity. *J Phys Chem B* **112**:1946–1952.

58. Marrink SJ, de Vries AH, Harroun TA, Katsaras J, Wassall SR. 2008. Cholesterol shows preference for the interior of polyunsaturated lipid membranes. *J Am Chem Soc* **130**:10–11.
59. Boon JM, Smith BD. 2002. Chemical control of phospholipid distribution across bilayer membranes. *Med Res Rev* **22**:251–281.
60. Dressler V, Schwister K, Haest CWM, Deuticke B. 1983. Dielectric breakdown of the erythrocyte membrane enhances transbilayer mobility of phospholipids. *Biochim Biophys Acta* **732**:304–307.
61. Falck E, Rog T, Karttunen M, Vattulainen I. 2008. Lateral diffusion in lipid membranes through collective flows. *J Am Chem Soc* **130**:44–45.
62. McIntosh TJ, Simon SA. 2006. Role of bilayer material properties in function and distribution of membrane properties. *Annu Rev Biophys Biomol Struct* **35**:177–198.
63. Lundbaek JA. 2006. Regulation of membrane protein function by lipid bilayer elasticity — a single molecule technology to measure the bilayer properties experienced by an embedded protein. *J Phys Cond Matter* **18**: S1305–S1344.
64. Holthuis JCM, van Meer G, Huitema K. 2003. Lipid microdomains, lipid translocation and the organization of intracellular membrane transport. *Mol Membr Biol* **20**:231–241.
65. Schekman R. 2004. Merging cultures in the study of membrane traffic. *Nat Cell Biol* **6**:483–486.
66. Holthuis JCM, Levine TP. 2005. Lipid traffic: floppy drives and a superhighway. *Nat Rev Mol Cell Biol* **6**:209–220.
67. Lenoir G, Williamson P, Holthuis JC. 2007. On the origin of lipid asymmetry: the flip side of ion transport. *Curr Opin Chem Biol* **11**:654–661.
68. Bockmann RA, de Groot BL, Kakorin S, Neumann E, Grubmuller H. 2008. Kinetics, statistics, and energetics of lipid membrane electroporation studied by molecular dynamics simulations. *Biophys J* **95**:1837–1850.

6

SPATIOTEMPORAL ORGANIZATION OF SPIN-COATED SUPPORTED MODEL MEMBRANES

Adam Cohen Simonsen
MEMPHYS—Center for Biomembrane Physics,
Department of Physics and Chemistry,
University of Southern Denmark, Odense

6.1. INTRODUCTION

All cells of living organisms are separated from their surroundings and organized internally by means of flexible lipid membranes. In fact, there is consensus that the minimal requirements for self-replicating life processes include the following three features: (1) information carriers (DNA, RNA), (2) a metabolic system, and (3) encapsulation in a container structure [1]. Therefore, encapsulation can be regarded as an essential part of life itself. In nature, membranes are highly diverse interfacial structures that compartmentalize cells [2]. While prokaryotic cells only have an outer plasma membrane and a less-well-developed internal membrane structure, eukaryotic cells have a number of internal membranes associated with the organelles and the nucleus. Many of these membrane structures, including the plasma membrane, are complex layered systems, but with the basic structure of a *lipid bilayer*. Biomembranes contain hundreds of different lipid species in addition to embedded or peripherally associated membrane proteins and connections to scaffolds such as the cytoskeleton. In vitro, lipid bilayers are spontaneously self-organized structures formed by a large group of amphiphilic lipid molecules in aqueous suspensions. Bilayer formation is driven by the entropic properties of the hydrogen bond network in water in combination with the amphiphilic nature of the lipids. The molecular shapes of the lipid constituents play a crucial role in bilayer formation, and only lipids with approximately cylindrical shapes are able to form extended bilayers.

Address correspondence to Adam Cohen Simonsen, MEMPHYS —Center for Biomembrane Physics, Department of Physics and Chemistry (IFK), University of Southern Denmark, Campusvej 55, DK-5230 Odense M, Denmark, +45 65503527 (office), +45 65504048 (fax), <http://memphys.sdu.dk/~adam>, <e-mail:adam@memphys.sdu.dk>

R. Faller et al. (eds.), *Biomembrane Frontiers: Nanostructures, Models, and the Design of Life,*
Handbook of Modern Biophysics, DOI: 10.1007/978-1-60761-314-5_6,

The bilayer structure of biomembranes was discovered by Gorter and Grendel in 1925 [3] using monolayer studies of lipid extracts from red blood cells. Later, a number of conceptual models were developed to rationalize the organization of lipids and proteins in biological membranes. One of the most celebrated is the fluid-mosaic model by Singer and Nicolson (1972) [4]. According to this model, the lipid bilayer component of biomembranes is randomly organized to facilitate membrane function. However, during the last 10–20 years it has become increasingly clear that the components of biomembranes are not randomly organized but that the lateral distribution is heterogeneous and time dependent. A picture is emerging where the interactions among membrane components and between membranes and external structures give rise to dynamically maintained domains with distinct sizes and compositions. Such domains are coupled to membrane function through their regulation of membrane-bound proteins. Experimentally, the investigation of domain structures in artificial and natural membranes have been enhanced by the proposition of the so-called *raft hypothesis* [5–7]. According to this hypothesis, rafts are nanoscale regions of biological membranes that are linked to important cellular processes and signaling pathways [8,9]. Rafts originated as insoluble membrane fragments upon treatment of cellular membranes with cold detergent [10].

To understand the role of individual membrane components, various types of simplified model membranes are useful. Currently, model membranes are providing a fundamental understanding of how domain formation can be regulated by the lipid components. It is essential that we fully understand the physics of domain formation in simple model membranes as a foundation for understanding lateral heterogeneity in complex biomembranes. Common model membrane types include vesicles with sizes from 50 nm to several micrometers as well as planar supported membranes. Supported membranes have a number of advantages in terms of the characterization techniques that can be applied; additionally, the planar geometry facilitates analysis of recorded images. Intended as a practical guide to supported model membranes, this chapter reviews common preparation techniques and provides a detailed introduction to our recently developed preparation method by spin coating. The last section contains examples of our recent applications of the spin-coating methodology to a number of systems probing the spatiotemporal structure of membranes.

6.2. METHODS FOR PREPARING SUPPORTED MEMBRANES

For fabrication of supported membranes, a range of preparation methods exists. Some of these are established and widely used, while others are less commonly used. Table 6.1 summarizes many of these methods and outlines the strengths and weaknesses of each technique.

One of the common methods for deposition of single or stacked multiple monolayers onto planar supports is by transfer of monomolecular films localized at the interface between air and an aqueous subphase [23]. This family of techniques was pioneered by Irving Langmuir, Katherine Blodgett, and Vincent Schaefer in the mid-1930s, and the experimental principles they devised are in use in present-day nano and surface science with only minor modifications [11,24,25,12]. The basic principle of the technique is that lipid is spread on the water surface from a hydrophobic, volatile organic solvent. Upon solvent evaporation, a monomolecular film is localized at the interface. This film is compressed by a barrier system while the lateral pressure is monitored. At a desired pressure the film is transferred to a solid support by translation of the support through the interface in a vertical (LB) or horizontal (LS) orientation. Bilayers are easily prepared by two consecutive strokes. But higher-order multilayers are difficult to prepare because previously deposited layers are easily desorbed. This class of techniques allows preparation of supported lipid monolayers. In addition, the lateral lipid density is strictly controlled. By replacement of the monolayer film between depositions it is also possible to prepare asymmetric supported bilayers.

Table 6.1. Overview of Selected Experimental Methods for Fabricating Supported Membranes

Method	Strengths	Weaknesses	Reference
Langmuir-Blodgett	Supported monolayers; asymmetric bilayers; lipid density controlled	Decoupling of membrane leaflets holes	[11] (1935)
Langmuir-Schäfer	Same as above; improved transfer in some cases	Same as above	[12] (1938)
Vesicle fusion	Membrane protein reconstitution; non-planar supports and microbeads; well tested and widely adopted	Hole formation is lipid and buffer specific	[13,14] (1984)
Solvent thinning	Membranes on metal supports; non-planar supports	Remaining solvent in membrane?	[15] (1989)
Painted lipid membranes	Metal support	Not true bilayers, but SAM + lipid monolayer	[16] (1993)
Spreading from a lipid reservoir	Requires attractive membrane-solid forces	Not tested for model studies	[17] (1995)
Self-assembly by chemical grafting	Membrane sensors; incorporation of gramicidin	Lateral diffusion possibly restricted by tethering	[18,19] (1997)
Self-assembly from micelles	Membrane protein reconstitution	Remaining detergent in membrane?	[20] (2002)
Spin coating	General lipid composition; single and multiple bilayers; membrane holes eliminated	Requires lipid solubility in solvent; requires a planar support	[21,22] (2004)

Perhaps the most widely used method for producing supported membranes is by adsorption and fusion of small unilamellar vesicles (SUV, 30–100 nm diameter) to the support. This technique was pioneered by McConnell [13,14] in 1984 and has since been widely adopted. Vesicle fusion is simple to perform and requires only the ability to produce SUVs by extrusion, sonication, or freeze thawing [26]. Exposure of the support to the vesicle solution ($\simeq$1 mM lipid) leads to spontaneous fusion and formation of a single-supported bilayer. The details of the adsorption and fusion events have been examined by number of groups using AFM, Quartz Crystal Microbalance (QCM), and Surface Plasmon Resonance (SPR) [27–32], but the process remains only partially understood. Some of the benefits of the vesicle fusion method are that it enables reconstitution of functional transmembrane proteins in supported membranes and that it allows formation of membranes on non-planar supports such as glass beads. However, it is also clear that a number of system-specific factors control membrane formation — e.g., lipid headgroup charge, electrolyte concentration, temperature, pH, and presence of divalent cations. The fundamental issue is that, although the final supported membrane may be stable, vesicle adsorption and fusion may require the vesicle to overcome a repulsive energy barrier when approaching the support (see §6.7). If thermal energy is not sufficient to pass this barrier, adsorption and fusion will not happen. This is also the most likely reason why supported membranes with high anionic lipid content (>20%) cannot be fabricated with vesicle fusion [29]. Membranes are formed by merging patches of individually ruptured vesicles, and for this reason complete coverage of the support without holes or minute defects is not always possible. With a few exceptions, the formation of a secondary membrane as depicted in Figure 6.4B is generally not possible using vesicle fusion.

Many alternative and less common preparation techniques are summarized in Table 6.1. A general shortcoming of most preparation techniques is that formation depends on the lipid composition and that double-supported bilayers cannot be prepared. For this reason we have developed a new and improved preparation method based on hydration of spin-coated lipid films. This technique is described below.

6.3. MEMBRANE SUPPORT MATERIALS

A variety of materials have been employed as supports for membranes. The proper choice of support material depends to a great extend on the intended application of the final membrane structure. Supported membranes as model systems for biomembranes should ideally preserve the features of freestanding membranes. At the same time, the ideal solid support should provide mechanical stability, optical transparency, flatness, and general low physical perturbation of membrane properties. For sensor applications, the support should also have good electrical conductivity as well as microporosity to allow for interrogation of ion transport events across the membranes and formation of a liquid buffer reservoir on both sides of the membrane.

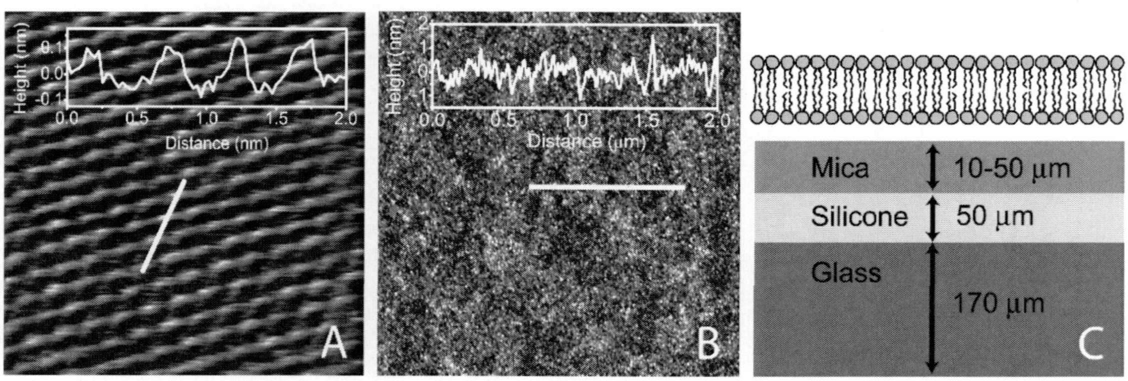

Figure 6.1. AFM morphology of two common membrane supports: mica (A) and glass (B). Mica is atomically planar with a hexagonal oxygen lattice that can be resolved atomically by contact-mode AFM (A). The lattice constant of mica (5.2 Å) is confirmed by the AFM images. Cleaned glass slides (B) have a pronounced microroughness with 1–3 nm height variations. Optical grade mica substrates (C) can be fabricated by sandwiching mica with glass cover slips using a silicone elastomer. Mica can subsequently be cleaved to ultrathin dimensions. Please visit http://www.springer.com/series/7845 to view a high-resolution full-color version of this illustration.

For model studies and characterization using scanning probe and fluorescence microscopy, only few known materials are suitable. One is mica (muscovite), a natural inorganic mineral with the chemical composition $KAl_2(AlSi_3O_{10})(FOH)_2$. It is semitransparent at visible wavelengths and is easily cleaved by scotch tape or a knife into hydrophilic and atomically planar crystal planes. Thin (<200 μm) mica sheets are mechanically flexible and can be cut with scissors. Mica crystal facets (0001) typically have macroscopic (mm) dimensions and are ideally suited as membrane supports for AFM studies. The surface of cleaved mica exposes a hexagonal oxygen lattice with K^+ ions coordinated centrally in the oxygen hexagons. In water, the K^+ ions are partially dissociated from the surface or replaced by other cations depending on the electrolyte concentration. Mica in water at neutral pH presents a net negative charge of approximately –0.0025 Coulombs per m^2 [33].

Fluorescence microscopy through the mica support is possible, but the quality of the fluorescence image may be negatively influenced by the semitransparent nature of mica. Signal intensity is reduced as a result of light absorption in mica, and image quality is compromised by light scattering from crystal defects in the natural mica mineral. These issues obviously depend greatly on the thickness of the mica plate. A partial solution to this problem is to sandwich mica with a glass cover slip using transparent glue. We have found that optically transparent and medical-grade silicone elastomers such as MED-6215 (Nusil Silicone Technology) can be used to sandwich for this purpose. Fixated mica can subsequently be cleaved to a very thin dimension (10–50 μm), which leads to a significant improvement in optical image quality [34].

The major advantage of glass as a membrane support is its superior optical transparency. For high-resolution and low-light fluorescence microscopy, glass is often the only viable option. Other advantages of glass include price and the wide variety of physical dimensions commercially available. However, before use as a membrane substrate, glass plates must be extensively cleaned to remove production impurities. Typical cleaning procedures involve boiling in detergent, boiling in an NH_3 (10%):H_2O_2 (10%) solution or piranha solution. Cleaned glass has a uniform microroughness with 1–3 nm height variations at short (<50 nm) length scales, as shown in Figure 6.1B. Depending on how closely the membrane follows the substrate topology, the glass structure will be superimposed on membrane structure and influence image quality. The negative surface charge density of glass depends on pH and electrolyte concentration and can be quantitatively described [35]. We have found that the roughness of glass and/or its reduced surface charge compared to mica results in significant perturbations of membrane properties. This perturbation is not easily screened by increasing electrolyte concentration, pointing to short-range van der Waals attractions.

A wide range of alternative membrane support materials has been employed, including single-crystalline Si wafers that are atomically planar, electrically conducting, but nontransparent. For sensor purposes, noble metal surfaces can be used as supports by chemical grafting of the membrane lipids via sulfur–metal bonds. This methodology has been demonstrated for gramicidin-based membrane sensors [18,19,36]. Soft materials such as polymer cushions are potentially ideal supports for membranes due to minimized membrane perturbations and the possibility to incorporate functional transmembrane proteins [37]. A typical design uses a thin hydrated polymer film between the membrane and support. However, the use of polymer supports requires careful adjustment of the wetting behavior between the individual layers. In particular, it is important to avoid attractive membrane–solid interactions that can destabilize the polymer film and cause the membrane to touch the solid surface. Moreover, polymer-supported membranes are not atomically planar and therefore not suited for investigation of membrane domains by AFM.

6.4. SPIN COATING

Spin coating [38–42] is a common technique for application of thin, uniform coatings to surfaces. The technique can be traced back about 100 years but was not quantified and treated theoretically until the 1950s. It has since been used extensively in fundamental thin-film studies as well as in various manufacturing processes. Polymer studies have benefited tremendously from the spin-coating technique that enables reproducible production of highly uniform polymer films with controlled thickness. In micro-fabrication and photolithography, spin coating plays a central role in deposition of a uniform light-sensitive polymer (photoresist) on the semiconductor wafer.

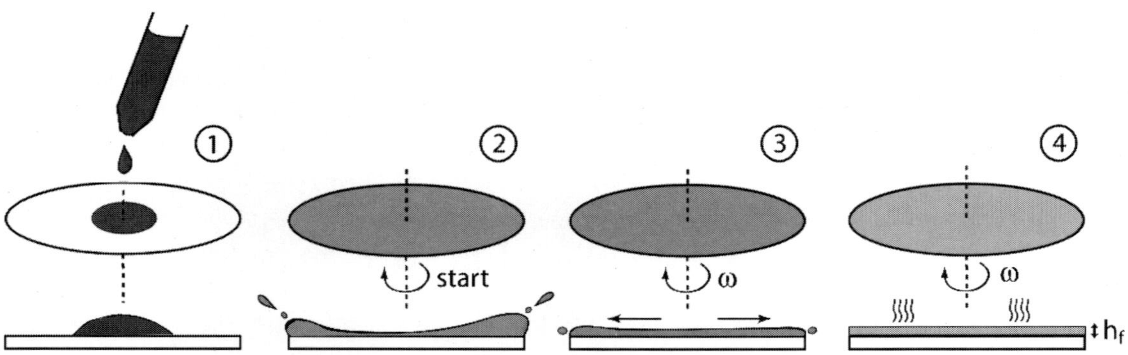

Figure 6.2. Principle of the spin-coating process and illustration of the characteristic time regimes during rotation and solvent evaporation. Step 1: placement of coating solution. Step 2: start of rotation (3000 rpm) and expulsion of solution. Step 3: laminar radial flow of solution and increase in viscosity. Step 4: evaporation of remaining solvent. Please visit http://www.springer.com/series/7845 to view a high-resolution full-color version of this illustration.

The very simple principle behind spin coating is shown schematically in Figure 6.2. The basic idea is that the solid coating material is dissolved in a solvent that wets the substrate. After placing a droplet on the substrate, rotation of the substrate while the solvent evaporates produces a uniform coating. Several characteristic time regimes can be identified during spin coating. In phase 1 a droplet of the coating solution is applied to the substrate surface. During subsequent rotation at the angular velocity ω, two time regimes (2 and 3) can be observed. During regime 2, the film becomes thinner due to radial outward flow of the coating solution. The radial flow stops when the viscosity has increased to a point where internal forces match the outward centrifugal force. In regime 3 the solvent evaporates from the film to the point where only the solid film is left on the substrate.

Several attempts have been made to make simple quantitative predictions of final film thickness based on known parameters for the coating process. This is a complex phenomenon to model accurately from basic principles, making it necessary to use a simplifying assumption regarding the properties of the coating solution. One of the more successful quantitative descriptions was that of Meyerhofer [41], who developed the following semi-empirical expression for the final film thickness, h_f:

$$h_f = \left(\frac{3}{2}\right)^{1/3} c_0(1-c_0)^{-1/3}\omega^{-2/3}v_0^{1/3}e^{1/3}, \tag{6.1}$$

where c_0 is the initial coating concentration (mass fraction units), ω is the angular rotation speed (s^{-1}), v_0 is the initial kinematic viscosity of the coating solution (m^2/s), and e is the evaporation rate (decrease of fluid thickness in m/s). This is the first and simplest description that takes into account solvent evaporation during the coating process. Based on Eq. (6.1) we can make a simple estimate of thickness of a lipid film deposited from a 1% hexane solution. We use $c_0 = 0.01$, $\omega = 314\ s^{-1}$, $v_0 = 6 \times 10^{-7}\ m^2/s$, and $e = 4 \times 10^{-7}$ m/s to obtain $h_f = 15$ nm. As found experimentally [43], this result is accurate to within a factor of 2 or better.

Spin coating is a versatile general method for producing uniform thin films on solid substrates. Successful coating can be obtained if some simple conditions are met: (1) the coating solvent must wet the solid substrate; if this is not fulfilled, 100% of the coating solution will be ejected in step 2 in Figure 6.2; (2) the coating material must be soluble in the solvent; and (3) the solvent must be volatile. When spin coating nonstandard coating materials such as biomolecules or nanoparticles, it is not trivial to fulfill all these conditions simultaneously. Despite its great potential, spin coating has not been widely used for the preparation of thin biomolecular films. This is likely to change due to the fact that this is a robust method for producing uniform thin films of materials such as polysaccharides, lipids, nanoparticles, and colloidal spheres.

6.5. LAMELLAR LIPID FILMS BY SPIN COATING

A wide range of membrane-forming lipids including phospholipids can be spin coated on solid surfaces suitable as substrates for supported membranes. This includes mica, silicon wafers, and glass. Spin coating of phospholipids has previously been applied to produce dry multimellar lipid films suitable for investigations with X-ray reflectivity [21,44]. Spin-coated films have also been used for protein adsorption studies [45]. The number of bilayers in spin-coated lipid films is found to scale in a linear fashion with the concentration of the coating solution, in agreement with Eq. (6.1) for low values of c_0. We have shown [22] that such dry lipid films can be used as a precursor structure to produce single- and double-supported model membranes by controlled hydration of the dry film.

In a typical preparation, spin-coated phospholipid films on mica are prepared using a 10-mM lipid stock solution in a hexane:methanol (97:3) solvent. This mixture dissolves most lipids and has good wetting properties on mica. In contrast, chloroform does not wet mica and is not suited as a coating solvent. A variety of other solvents are also suitable. Anionic lipids (phosphatidylglycerol) can be coated using the solvent mixture isopropanol:hexane:water (3:1:1). A droplet (30–100 μl) of the lipid stock solution is applied to the solid substrate with a size of roughly 10 × 10 mm and immediately thereafter spun on a spin coater at 3000 rpm for 40 s. Immediate acceleration is important due to the volatility of the solvent.

The morphology of several dry lipid films has been characterized with AFM [22]. Figure 6.3A,B shows the typical topography image and phase image of a dry POPC film spin coated from a solution of 5 mg/ml. The surface shows a lamellar structure with completely planar layers and an interlayer spacing that is determined by line scans to be 62 ± 5 Å. This thickness is considerably higher than for a hydrated fluid POPC membrane. However, POPC is known to undergo a lyotropic chain melting transition from the solid L_β to the fluid L_α state upon hydration, and the repeat distance in L_β POPC multilayers has previously been measured to around 58 Å by Binder et al. [46]. Since our spin-coated films have been stored under vacuum and the relative humidity (RH) during AFM measurements was between 20 and 30%, we conclude that the POPC films in Figure 6.3 are in the solid L_β state.

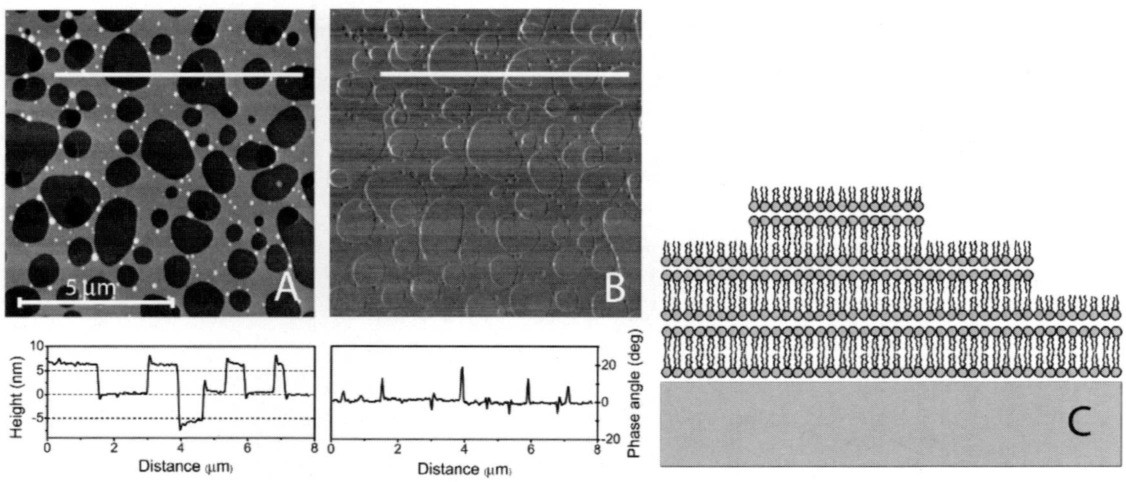

Figure 6.3. Typical AFM appearance of a dry POPC films spin coated on mica. The topography image (A) displays a lamellar organization of the film with a unique layer spacing of 6.2 nm. Conversely, the AFM phase image (B) shows an invariant phase contrast for all lipid layers, demonstrating that all regions have the same viscoelastic properties and that the solid support is not exposed. (C) Molecular configuration of the dry lipid film. Please visit http://www.springer.com/series/7845 to view a high-resolution full-color version of this illustration.

The phase image in Figure 6.3 exhibits no contrast except at the boundary lines of the individual layers. The absence of contrast in extended areas is a clear sign that the substrate is completely covered by lipids and that none of the holes in the lipid layers extend down to the mica substrate. The top individual lipid layers exhibit partial coverage and a characteristic dewetting pattern. The patterns formed in each lipid layer closely resemble theoretically and experimentally observed dewetting patterns observed in thin polymer films [47–49]. These patterns are most likely initiated by the thinning of the lipid film during the spin-coating process, which leads to rupture of the outer lipid layers, hole formation, and eventually film breakup. Once the solvent has evaporated, the patterns in the spin-coated dry films are rigid and stable over time. The outer layers of the film exhibit different degrees of film breakup depending on their proximity to the air interface.

A unique and striking property of the lipid films in Figure 6.3 is the fact that the lipids become organized in a perfectly lamellar structure during the coating process. The proposed molecular configuration of the dry film, as shown in Figure 6.3C, is based on AFM studies and force spectroscopy of the dry lipid film [22]. As shown in Figure 6.3C, the mica support is coordinated to the headgroups of a lipid monolayer. Depending on the coating concentration, a number of *inverse* bilayers are stacked on this primary monolayer. The total film structure is stable in air since all lipids in the outer surface will have their acyl chains pointing outward while the polar headgroups are interacting with either the polar substrate or with opposing lipid headgroups. An increase in the coating concentration effectively increases the number of stacked inverse bilayers. At sufficiently high coating concentrations, complete coverage of the support can be obtained. A thick multilamellar film will appear as a sequence of normal bilayer membranes resting on the support surface, while the outermost surface will have the lipid acyl chains pointing outward due to the hydrophobic air/vacuum region. Interestingly, structures similar to those shown in Figure 6.3C have been proposed for dried Langmuir-Blodgett lipid films [50].

The hydrophobic nature of the lipid-coated surface as compared to mica was confirmed by AFM force–distance measurements. The adhesion force between the AFM tip and the lipid surface was measured [22] and compared with results obtained on hydrophilic mica. The total attractive force existing between the silicon nitride tip and the surface is composed dominantly of capillary and van der Waals forces, but for a hydrophobic surface the capillary force will be absent, since a water meniscus cannot be formed [51]. This will generally give rise to a lowered adhesion force. We measure a 40% decrease in the adhesion for the lipid surface relative to mica. For comparison, measurements by Xiao et al. [51] on SiO_2 and on silanized SiO_2 (hydrophobic) show a similar decrease in adhesion force at the same relative humidity. This gives support to the structural model proposed in Figure 6.3C and confirms that the dry lipid surface is hydrophobic.

Bilayer formation by self-organization of lipids in fluid water is driven by the entropic properties of the hydrogen bond network in the water. The structure of the disordered hydrogen bond network in water is governed by a maximization of the configurational entropy. When small hydrophobic species are present in the water, the hydrogen bond network will tend to become more ordered around the solute, as described by the formation of clathrate cages around the solute [52]. This leads to an unfavorable reduction in entropy and is why purely hydrophobic molecules are normally insoluble in water. Membrane-forming lipids are amphiphatic and contain both a hydrophobic and an hydrophilic part in the same molecule. This amphiphatic property is what drives bilayer formation in water. The hydrophobic acyl chains are shielded from contact with the water by the bilayer structure, while the hydrophilic lipid headgroups are in contact with the water. The self-organization of a bilayer structure gives the system a lower free energy than uniform dispersion of the lipid monomers because the configurational entropy of the water is relatively unaffected by the presence of a bilayer.

Bilayer formation in lipid films dried on solid substrates cannot be driven by water entropy. Instead, weaker forces of the type lipid–lipid and lipid–solid become dominating and will govern the structure of the final film. The magnitude of these forces during spin coating is not characterized, and the exact mechanism for bilayer formation during spin coating is not known. We speculate that the planar solid surface acts as a template for the formation of an ordered planar lipid film. The lipid headgroups of phospholipids are attracted to the solid substrate via van der Waals and static dipole attractions. This explains the formation of a monolayer adjoining the solid support. Attractive headgroup–headgroup interactions between lipids and/or chain–chain attractions drive the formation of additional inverted bilayers, as shown in Figure 6.,3C. The overall result is the formation of a dry lamellar lipid film. Note that even much thicker dry lipid films retain the lamellar structure [22].

6.6. SUPPORTED MEMBRANES BY HYDRATION OF SPIN-COATED PRECURSOR FILMS

Vesicles are known to form during hydration and swelling of a dry lipid film [53,26] hydrated above the main phase transition temperature. Swelling of the lamellar film is accompanied by detachment from the surface of membrane structures in the form of tubes and vesicles in addition to highly complex membrane geometries. We have shown that such complex bilayer structures are also released from spin-coated lipid films [22]. The swelling and formation of membrane structures is a dynamic process that evolves over hours [54] and depends on such system parameters as lipid composition, electric charge of lipids, temperature, and buffer composition.

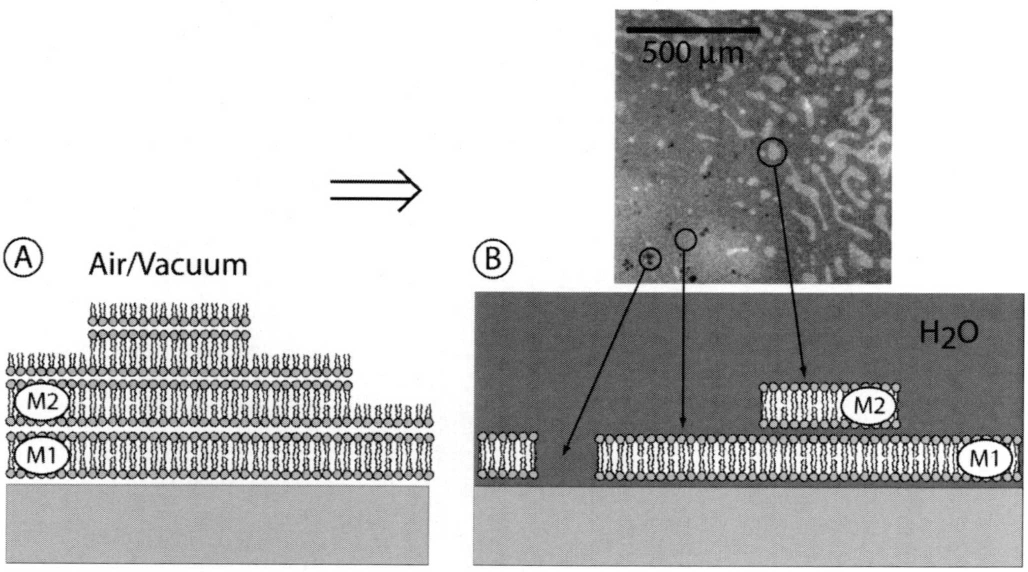

Figure 6.4. Hydration of the spin-coated precursor film (A) leads to release of membrane structures into solution and exposure of the primary (M1) or secondary (M2) supported membrane to the aqueous phase. Insert in (B) shows a fluorescence micrograph of a POPC film after hydration and identification of membrane holes as well as primary and secondary membrane regions. Please visit http://www.springer.com/series/7845 to view a high-resolution full-color version of this illustration.

As shown in Figure 6.4, swelling of the lipid film can be monitored live by fluorescence microscopy (see §6.10). The release of membranes into solution is induced by generating a liquid flow toward the interface with a suction pipette. By monitoring the response of the lipid film to the flow, the removal of lipid layers can be accurately controlled. Typically, a central spot on the sample is flushed, and this completely removes all but the primary bilayer in this region (see insert in Fig. 6.4B). Some distance away from the flushed region, lipid multilayers are present, while in the transition zone one can routinely localize secondary membrane islands situated on the primary membrane. This method is a fast and easy way to produce large batches of samples with a consistently high quality.

Supported membranes prepared as explained in Figure 6.4 present a number of advantages. Large (mm-sized) regions with 100% coverage by the primary membrane (M1) can easily be achieved since full coverage by M1 is only a question of increasing the concentration of the coating solution. In comparison, achieving full membrane coverage by vesicle fusion is difficult to control, and such membranes will often contain minute defects and holes that are below the optical resolution limit. Patches or extended regions of a secondary membrane (M2) are also easily prepared. In comparison, preparation of secondary membranes using alternative preparation methods is typically successful only with certain lipid species at low salt conditions. As shown later, secondary membranes also preserve many features of freestanding membranes that are influenced by the support in the primary membrane. Spin coating offers great flexibility regarding the lipid composition of the supported membrane. Many classes of membrane-forming lipids have been successfully tested — including PC, PG, PE, ceramides, sphingolipids, and membranes with a high (>50%) sterol content. An experimental protocol describing the practical details of fabricating spin-coated supported membranes is included at the end of this chapter.

6.7. MEMBRANE SUPPORT AND MEMBRANE–MEMBRANE INTERACTIONS

An important issue of supported membranes used as model systems is the possible influence of the solid support on the properties of the membrane. Such important characteristics as thermotropic phase behavior [55], lateral lipid mobility, and the structural coupling between bilayer leaflets [56] have been reported to be adversely influenced by an adjoining solid support. Many aspects of such membrane–solid interactions are still poorly understood. A proper description of these effects in model systems may hold the key to a better understanding of such important biological questions as the physical interactions between cell membranes and the surroundings and the possible influence of the cytoskeleton on lateral organization of the plasma membrane. In addition, the many possible technological applications of supported membranes depend on an accurate quantitative description of membrane–solid interactions.

The interactions between a supported membrane and the solid can be assigned to a number of fundamental forces that are caused by electrostatic effects, van der Waals forces, and several entropy-driven forces, including undulation and protrusion forces. The interactions between charged and uncharged surfaces in electrolyte solutions is simplified by the so-called DLVO (Derjaguin, Landau, Verwey, Overbeek) theory [52,57]. According to this theory the interaction potential is approximately given by the sum of two interactions: the van der Waals (W_{vdW}) and the electrostatic double layer energy (W_{DL}). All other interactions are neglected. Interaction potentials are defined as the energy difference per unit area between completely separated surfaces and surfaces with spacing D. The attractive van der Waals potential is caused by random spatial fluctuations in the charge density of the surfaces. For two planar surfaces it is given as

$$W_{vdW} = -\frac{H_A}{12\pi D^2}, \tag{6.2}$$

with H_A being the Hamaker constant, which has a value around 1×10^{-20} J for hydrocarbon materials in water. The repulsive double layer potential is caused by the diffuse layer of counterions outside the surface (the double layer), as determined by solution to the Poisson-Boltzmann equation. When two surfaces approach, the ions outside the surfaces are confined to a smaller and smaller space, whereby the entropy of the double layer is reduced. This leads to a repulsive force. The double layer repulsion between two planar, negatively charged surfaces is

$$W_{DL} = \frac{2\sigma_m \sigma_s \lambda_d}{\varepsilon \varepsilon_0} e^{-D/\lambda_d}, \tag{6.3}$$

with σ_m and σ_s being the charge per unit area (C/m^2) of the membrane and the support, respectively. The Debye screening length λ_d is determined by the electrolyte concentration and the ion valency. For monovalent ions it is given as $\lambda_d = 0.304/\sqrt{C_{NaCl}}$ (nm), where $[C_{NaCl}]$ is expressed in moles/liter.

The total interaction potential $W_{tot} = W_{vdW} + W_{DL}$ between a membrane and the support can be estimated if the respective surface charges are known. Mica has a negative surface density of approximately –0.0025 C/m^2 [33], while the membrane charge depends on the lipid composition and the content of anionic lipids. It should be noted that even membranes of zwitterionic lipids will typically exhibit an effective negative charge due to adsorption of salt ions [58]. The membrane–solid interaction potential is determined by a delicate balance between the contributing forces. At low separation distances (<1 nm), van der Waals attraction will dominate. For low salt and large separations, double-layer repulsion will dominate. At intermediate distances and intermediate salt concentrations (50–500 mM), a secondary potential minimum around D = 3–4 nm is possible. A supported membrane in such a secondary potential minimum is desirable as a model system for a freestanding membrane since the membrane is stabilized several nanometers from the solid. Other contributions to the interaction forces include hydration forces due to repulsion from the water layer at the membrane surface [59] and undulation forces due to thermal bending modes of the membrane [60]. The total interaction potential depends entirely on the relative magnitude of these forces, which is generally not known. At present, no accurate quantitative theory of membrane–solid interactions exists since DLVO theory at best only provides a rough guideline.

Recently, the first atomistic molecular dynamics simulation of a lipid bilayer interacting with a silica support was reported [61]. Despite the apparent simplicity of this problem, it is not trivial to accurately parametrize the force fields in a system combining a crystalline solid with soft matter and water. For DPPC, an equilibrium bilayer–solid distance of 3.5 nm and an adhesion energy of 30 mJ/m^2 were reported. However, this interaction energy is measured experimentally to be around 0.5 mJ/m^2 [59], whereas the equilibrium distance is typically 0.5–2 nm (see below). More accurate simulations would have to include ions, and to expand the size of the simulation box to allow for membrane bending modes.

Experimentally, the distance between the membrane and the solid support is measured to be around 0.5–2 nm [62–64] for membranes on glass. Some of these measurements have been performed with membranes supported by colloidal glass beads, which are easier to measure spectroscopically (NMR). A short separation distance (~1 nm) indicates that van der Waals forces dominate. Recently, direct TEM imaging of membranes supported by glass beads have been performed [65]. These measurements confirmed the short interaction distance ($\simeq$1 nm) and revealed that adsorbed membranes faithfully follow the detailed topography of the glass. Therefore, this study is also pointing to significant short-range attractive interactions between the membrane and solid.

Membrane–membrane interactions are also very important since this type of interaction is relevant for the secondary membrane (M2) situated on the primary supported membrane (M1) (see Fig. 6.4B). The interaction between M1 and M2 determines such physical characteristics of M2 as lateral mobility, cooperativity, and coupling between bilayer leaflets. It has been found [66] that M2 in many of these aspects resembles a freestanding membrane much more than M1. This implies that the M1–M2 interaction is weaker than the M1–solid interaction. Membrane–membrane interactions are easier to probe experimentally, and membrane multilayers have been examined with such techniques as X-ray scattering. An excellent recent study provides a unified quantitative description of the interaction between zwitterionic membranes in salt solutions [58]. It was demonstrated that salt influences membrane interactions in two ways: (1) by binding of polarizable anions to the zwitterionic lipid headgroups, and (2) by reduction of van der Waals forces through a reduction of the Hamaker constant. According to these results, a significant increase in intermembrane spacing should result when using large polarizable anions — e.g., bromide (Br^-) — instead of chlorine (Cl^-).

6.8. IMAGING THE MEMBRANE MICROSTRUCTURE

Several microscopic imaging techniques are available for visualizing the microstructure of supported model membranes. Two of these techniques are fluorescence microscopy (FM) and atomic force microscopy (AFM). These two techniques are highly complementary both in terms of resolved length scales and in terms of type of information obtained. Due to this complementarity, it is highly desirable to integrate both techniques in a single experimental setup, thus allowing concurrent imaging of the same sample region. Such integrated instruments are becoming commercially available from a number of vendors. In order to be practically useful, any type of microscopy must provide two essential features: (1) *spatial resolution* and (2) *contrast*. The spatial resolution h of a microscope is defined as the ability to discriminate point-like sample features with a spacing larger than h. Conversely, the contrast is the general intensity difference that exists between sample features and the background. This can be expressed in terms of the following definition:

$$C = \frac{I - I_{BG}}{I_{BG}}, \tag{6.4}$$

where C is contrast, I is sample intensity, and I_{BG} is background intensity. In order for a microscope image to provide usable information, it is necessary that *both* spatial resolution and contrast be present at a sufficient level.

Another critical issue in the characterization of model membranes is the temporal resolution of the microscope setup. While contrast and spatial resolution are parameters characterizing the static properties of single image frames, temporal resolution is given by the ability of the instrument to acquire sequential images at the highest possible rate (frame rate) while minimizing time spent on acquiring single image frames (acquisition time). In general, the highest frame rate and shortest acquisition time will provide the best temporal resolution. Below a discussion is presented of Atomic Force Microscopy and Fluorescence Microscopy with an emphasis on characterization of supported model membranes.

6.9. AFM FOR IMAGING SUPPORTED MEMBRANES

In 1986, Binnig, Quate, and Gerber [67] proposed the basic principle of the Atomic Force Microscope (AFM) and demonstrated in preliminary measurements that a spatial resolution at the Ångstrom-level was possible. Shortly thereafter, the first AFM prototypes were developed by Paul Hansma and put into commercial production. Since its invention, AFM has undergone rapid technological development from a prototype instrument with applications in solid-state physics to an instrument used routinely in the life sciences for investigations of biomolecular interfaces and live cells in their native aqueous environment. The current focus is development of more user-friendly instruments that can be handled by non-expert users. Other major goals are the development of fast scanners ("video-rate AFM") and integration of AFM with complementary optical techniques.

In the atomic force microscope, repulsive interactions between the probe and sample induce a bending of a microscopic cantilever that in its end has an atomically sharp tip (the probe). Cantilever bending is detected by deflection of a laser beam reflected from the back side of the cantilever and into a quadrant photodiode. By recording the four signals from this photodiode, cantilever bending and shear motion (twisting) can be measured. For optimal sensitivity, the laser should be properly aligned in the center where the four segments of the detector meet. Cantilever bending will be proportional to the difference signal S from the photodiode:

$$S \propto (I_1 + I_4) - (I_2 + I_3)\,, \tag{6.5}$$

where $I_1 \ldots I_4$ is the signal from the individual sensor elements. The cantilever/detector system is capable of detecting forces on the pN level, a force regime below the unbinding force of many single biomolecular bonds. To understand how this force sensitivity comes about, we must consider the Hookian mechanics of the cantilever and the geometrical amplification of the laser signal.

The spring constant of a cantilever is typically in the range κ = 0.01–0,5 N/m. From Hooke's law we have that, if a force of say F = 100 pN is to be measured with a spring with force constant κ = 0.01 N/m, this will require a detecting a deflection of

$$|\Delta y| = \frac{F}{\kappa} = 10 \text{ nm}\,. \tag{6.6}$$

This deflection is going to be detected by the detection system shown in Figure 6.5. By theoretically calculating the shape of a cantilever subjected to a force F at the tip, the following relationship can be derived between the deflection of the cantilever end (Δy) and the deflection in the photodiode (Δz):

$$\Delta z = \frac{3L}{l}\Delta y\,. \tag{6.7}$$

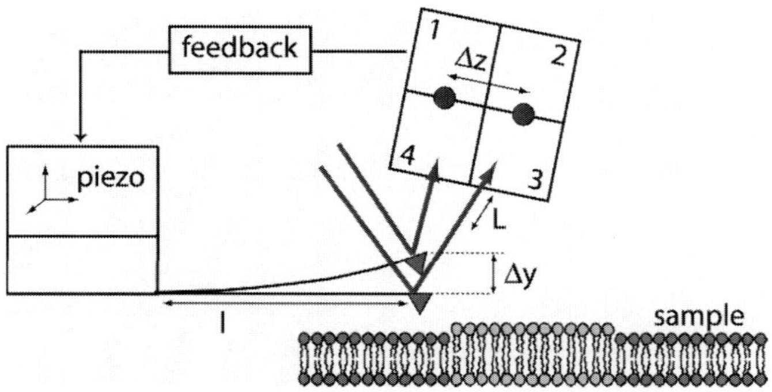

Figure 6.5. Principle of scanning AFM using the tip-scanner design. Symbols are explained in the text. Please visit http://www.springer.com/series/7845 to view a high-resolution full-color version of this illustration.

If we enter a typical cantilever length $l = 100$ μm and a cantilever–detector distance $L = 2$ cm, we find $\Delta z = 6$ μm, which is easily detectable by the photodiode. This demonstrates that the geometrical amplification of the AFM is sufficient to amplify cantilever deflections at the pN level into signals detectable by the photodiode. This amplification is the primary reason for the extremely good height resolution (typically below 1 Å) in scanning AFM. It also accounts for the excellent force sensitivity in single-molecule force spectroscopy experiments. In fact, the detection limit of small cantilever deflections is typically given by thermal noise (Brownian motion) of the cantilever rather than by limitations in the physical detection system.

Imaging with Atomic Force Microscope is based on lateral scanning of the cantilever across the sample surface. The tip is brought into close contact with the sample surface and subsequently lateral xy raster scanning takes place while the height (z) is controlled by a feedback loop connected to the photodiode sensor. The feedback loop adjusts the z-height according to a specific criterion. This criterion is typically that the tip–sample force is constant (contact mode) or that the cantilever vibration amplitude is constant (tapping mode). Whenever the tip encounters a change in surface topography, the feedback circuit adjusts the height accordingly. The height recorded at each point on the surface results in a dataset $z(x,y)$ that reflects the sample topography.

AFM performed on such soft biomolecular specimens as supported membranes puts strong demands on the performance of the AFM instrument. A critical issue is the ability to avoid sample damage by minimizing the tip-sample contact force and the energy deposited into the sample during scanning. Model membranes in the fluid state are delicate structures that are easily perturbed by interactions with the AFM tip. Even if bilayer integrity is not compromised by scanning, more subtle properties of the bilayer such as domain patterns or lateral diffusion processes can be strongly influenced by repeated scanning over the same sample region.

There are at least two imaging modes suitable for imaging supported model membranes: (1) contact-mode scanning with ultrasoft cantilevers and (2) magnetically actuated tapping mode, also known as MAC-mode [68,69] (Agilent™). Currently, the softest cantilevers available commercially have a spring constant of $\kappa = 0.01$ N/m (Veeco, MSCT, C-cantilever). Such cantilevers enable contact-mode scanning of supported membranes with minimal perturbation of the membrane structure. One drawback of such cantilevers is that they are highly sensitive to temperature variations in the fluid cell that cause bending of the cantilever that can result in a change of the proper force set-point for scanning.

When imaging in tapping mode a simple estimate can be given of the energy deposited into the sample during scanning. Since the energy of a harmonic oscillator is $E = \frac{1}{2}\kappa A^2$, where A is the amplitude, the energy deposited into the sample during scanning is of the order of $\Delta E = \frac{1}{2}\kappa(A_0^2 - A_S^2)$, with A_0 and A_S being the free amplitude and scan amplitude, respectively. In conventional tapping mode (TM), the cantilever is brought into oscillations by mechanically vibrating the sample holder at the cantilever resonance frequency. With some typical numbers for conventional tapping mode (TM) (A_0^{TM} = 30 nm, A_S^{TM} = 27 nm, κ = 2.8 N/m), the energy difference becomes $\Delta E^{TM} \approx$ 1500 eV. In MAC mode, special cantilevers are used that carry permanent magnetization. The vibration of these cantilevers is excited through the effect of an oscillating magnetic field from a coil placed near the cantilever. This excitation principle dramatically reduces the free amplitude. Using typical numbers for MAC mode (A_0^{MAC} = 4.8 nm, A_S^{MAC} = 3.8 nm, κ = 2.8 N/m), the energy difference now becomes ΔE = 75 eV, or roughly 20 times smaller than conventional tapping mode. MAC mode is therefore an excellent low-perturbation method for AFM imaging of supported membranes. The drawbacks include the price of magnetic cantilevers and the complexity of the magnetic excitation system.

The tip of an AFM instrument has a finite sharpness and an (often unknown) specific apex shape. The result is that the recorded surface topography in AFM is not identical to the true surface morphology. Rather, the resulting image contains combined information on the tip and the sample shape. The transformation from true to recorded topography is described formally using concepts from the discipline of mathematical morphology. It can be shown [70] that the recorded image I is the *dilation* of the surface shape S by the tip shape P:

$$I = S \bullet P. \tag{6.8}$$

Conversely, it can be shown that, given that the detailed shape of the tip is known, the best possible reconstruction S_r of the surface features from the image is *erosion* of the recorded image by the tip:

$$S_r = I \circ P. \tag{6.9}$$

The effect of tip shape on sample features schematized in Figure 6.6. AFM imaging of membrane domains in supported membranes resembles this schematic situation with $B \gg H$. Typical numbers for membrane domains are B = 1–10 μm, H = 1 nm, and R = 10 nm. In terms of resolution, this is an ideal situation and means that broadening will constitute only a small fraction of the true domain width ($\Delta B \ll B$). At the same time, the domain height H will typically be measured accurately since the background is planar. A different question is whether or not the measured domain heights are equal to the difference in bilayer thickness across domains. For membranes that are closely attached to the support this is likely to be the case, whereas for less strong membrane–support interactions thickness variations will be distributed to both sides of the membrane.

A final remark concerns the temporal resolution of AFM measurements. This is limited by the acquisition time of a single image frame, which is typically in the range from 30 s to 10 min. The main reason for this relatively slow speed is that the piezo feedback loop has to be adjusted in each image pixel. Strong efforts are currently being devoted to development of much faster AFMs with frame rates on the order of 10–30 Hz.

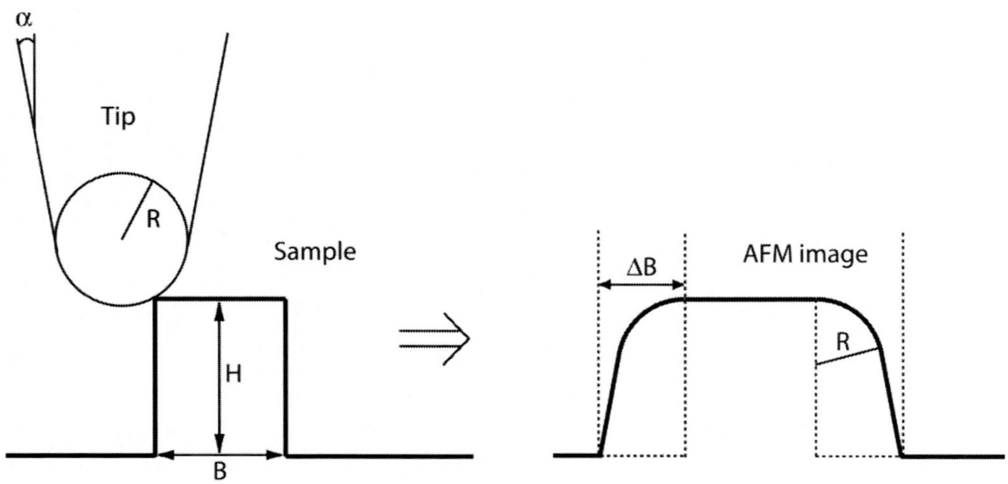

Figure 6.6. Effect of tip broadening illustrated by a rectangular sample structure and an idealized tip shape. The broadening ΔB of the structure depends on H, R, and α. The measured height (H) is accurate if the sample features are isolated from each other on a planar background topography.

6.10. FLUORESCENCE MICROSCOPY FOR IMAGING SUPPORTED MEMBRANES

Optical characterization by means of epifluorescence microscopy provides imaging of domain structures in supported membranes in a way that is complementary to AFM. This complementarity is reflected both in the length and timescales covered by these two techniques as well as in the physical mechanisms that give rise to image contrast. Fluorescence microscopy uses labeling of the sample with fluorescent probes as the mean to achieve contrast. For membranes this is conveniently achieved using lipid analogues (lipid probe <1 mole%) carrying the fluorescent group. These lipid analogues are typically distributed uniformly in a homogenous bilayer structure. If the bilayer has microstructure in the form of domains or other lateral variations, the lipid probe will typically partition differently among the lipid phases, thereby providing contrast to the fluorescent image. However, it is important to bear in mind that there is no general correlation between the lipid phase state and the partitioning of specific probes [71]. Rather, the local chemical environment in the membrane is the determining factor.

The principle of image formation in epifluorescence microscopy is illustrated in Figure 6.7. A light source (mercury, xenon, or laser) provides the excitation light. This light is directed to the filter cube in which the excitation filter (ex) selects the wavelength band for excitation of the probe and the dichroic mirror (dm) directs it to the sample. Light emitted by the probe is collected by the imaging objective and transmitted through the dichroic mirror and filtered through the emission filter (em) before reaching the CCD camera. The exposure time of the CCD camera must be long enough so that the pixel intensities are above the noise level of the camera. Depending on the performance of the CCD chip, this may require exposure times in the range from milliseconds to seconds.

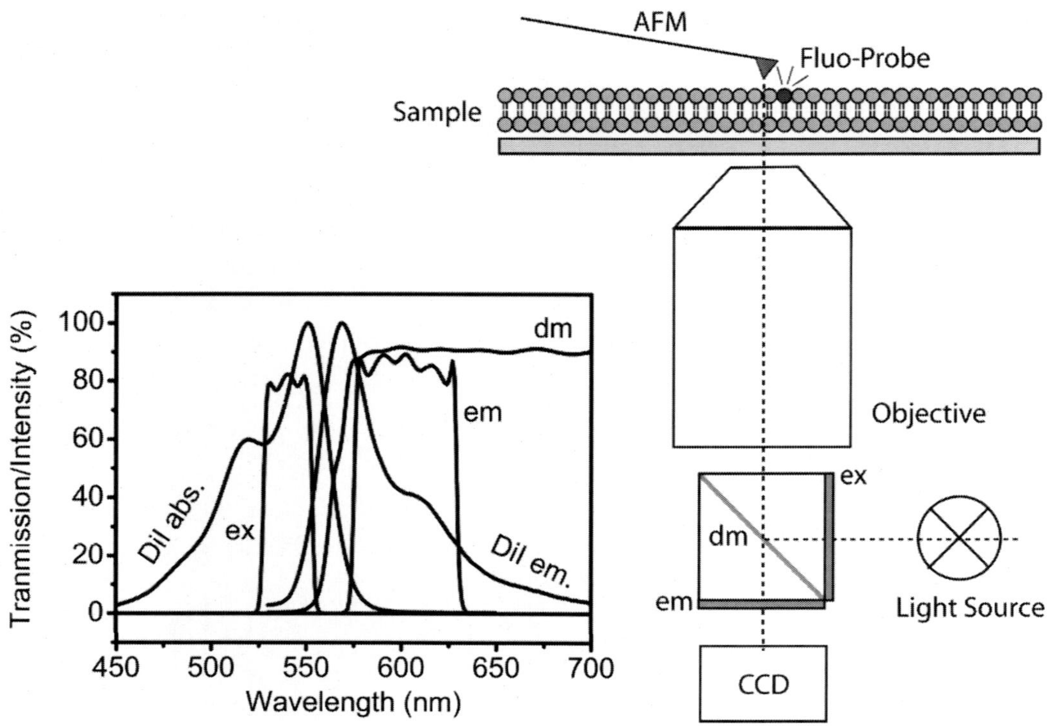

Figure 6.7. Principle of fluorescence microscopy combined with AFM for characterization of supported membranes. Insert shows transmission spectra of the filter cube elements as well as absorption and emission spectra of a typical lipid probe (DiI–C_{18}). Please visit http://www.springer.com/series/7845 to view a high-resolution full-color version of this illustration.

The spatial resolution h in conventional epifluorescence microscopy is defined by the Rayleigh criterion and is given by the following expression:

$$h = \frac{0.61\lambda}{NA}, \tag{6.10}$$

where λ is the emission wavelength and $NA = n \sin \alpha$ is the numerical aperture of an objective with a collecting angle of α. The objectives with the highest numerical apertures ($NA \approx 1.4$) have a typical working distance (WD) of 0.2 mm and require immersion liquid to be used. Under such conditions the best obtainable resolution is around 250 nm. Note that, if thick membrane supports (thicker glass slides) are used, this may require different objectives with a longer working distance and a resulting lower NA and poorer resolution. For studies involving temperature control of supported membranes, immersion objectives will often act as undesired heat sinks and are better replaced by air objectives for better thermal isolation of the sample and better temperature control.

While epifluorescence constitutes the most simple and basic type of fluorescence microscopy, it is eminently suited to investigation of supported membranes. There are several more advanced versions of fluorescence microscopy that restrict the depth of field (z-range), such as TIRFM and confocal microscopy. In TIRFM (total internal reflection fluorescence microscopy), excitation light is coupled to the sample via an evanescent field from the support. This restricts the fluorescence excitation depth to <100 nm from the substrate surface. Confocal microscopy uses a pinhole system to limit the collected light solely to the focal plane. These techniques improve vertical resolution and reduce background signal, but since a supported membrane is an inherently two-dimensional structure where background light can be eliminated during the sample preparation, these features are only a minor advantage for such samples. A different matter concerns imaging of a number of alternative optical parameters, such as environment-sensitive two-photon lipid probes [72–74], second harmonic generation (SHG) [75], lifetime imaging (FLIM), and resonance energy transfer (FRET) [76]. These techniques are being used increasingly to characterize supported model membranes and provide new information, but a more detailed treatment is beyond the scope of this text.

The time resolution in fluorescence image recording is determined by the exposure time necessary to obtain pixel signals above the noise level. During the past 8 years, a new generation of camera sensors, denoted EMCCDs (electron multiplying charge coupled devices), have revolutionized the field of fluorescence microscopy. These sensors have an on-chip amplification stage that amplifies the signal before readout. The result is sensors with a quantum efficiency above 90% and strongly reduced exposure times. This has enabled video-rate fluorescence recording of biomolecular specimens over extended time periods. It has also enabled sharp imaging of domain structures in supported membranes and made it possible to follow changes in domain features over extended periods.

6.11. PHASE TRANSITIONS AND DOMAINS IN MODEL MEMBRANES

One of the main challenges in the biomembrane field is to understand lateral protein/lipid organization in a quantitative way. To advance our understanding of the fundamental mechanisms that contribute to lateral membrane organization, simplified model membrane systems with a well-known composition are useful. It is obviously much easier to make quantitative physical and thermodynamic models of the behavior of simple membranes than that of complex biomembranes with poorly characterized compositions. Importantly, concepts from thermodynamics and statistical mechanics are powerful tools for rationalizing the behavior of simple model membranes, and these insights can often subsequently be applied to more complex membranes.

In general, self-organized lipid structures belong to the large family of soft condensed matter systems. Specifically, bilayer membranes are classified as *smectic* liquid crystals, which are liquid crystals composed of separated molecular layers with the molecules oriented perpendicular to the layers. Lipid assemblies such as membranes can undergo several different types of phase transitions when physical parameters are varied. These transformations can be discontinuous (first order) or continuous (second order) and can be induced by a number of changes — such as varying the water content at constant temperature (lyotropic) or by varying temperature at full hydration (thermotropic). With the purpose of modeling lateral domain formation in biological membranes, we will focus our attention on thermotropic phase transitions in fully hydrated bilayers. In this class of phase transitions, the bilayer structure is intact, but the internal packing and conformation of the lipid constituents is changed when the temperature is varied. The simplest possible example is provided by single component membranes. The

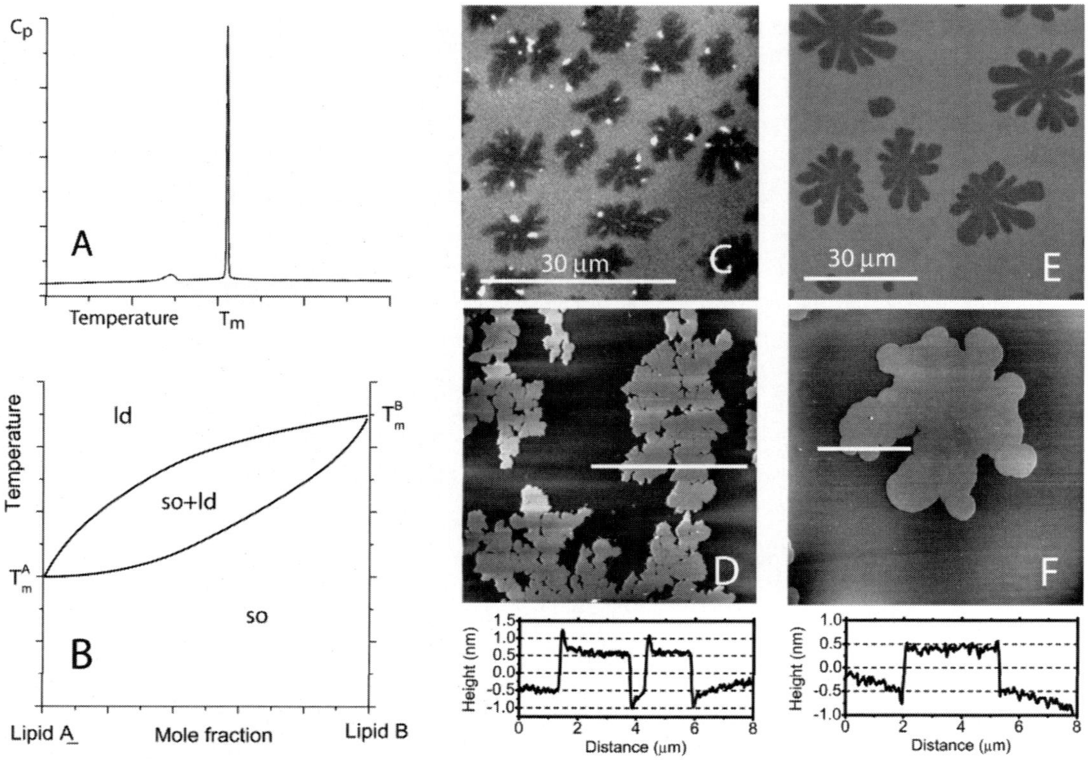

Figure 6.8. Thermotropic phases in single-component and binary membranes. (A) Schematic calorimetric data of a one-component phospholipid membrane showing the variation of heat capacity with temperature. (B) Schematic phase diagram for a binary membrane composed of two lipids with different melting points. Phases are solid ordered (so) and liquid disordered (ld). (C) Fluorescence and corresponding (D) AFM images of a supported POPC:DPPC (1:1) membrane at 20°C exhibiting coexistence of so and ld domains. Please visit http://www.springer.com/series/7845 to view a high-resolution full-color version of this illustration.

heat capacity of multilamellar phospholipids membranes such as DPPC can be measured in a differential scanning calorimeter. A typical result, depicted in Figure 6.8A, shows a sharp peak at the main phase transition temperature T_m. The peak area corresponds to the latent heat of fusion for the phase transition. Above T_m, the membrane is in the *liquid disordered* state (ld), where lipid acyl chains have large conformational freedom and entropy. The ld state is not only associated with acyl chain disorder but also with large lateral mobility of the lipids, hence the designation of this state as liquid disordered. Well below T_m, the membrane is in the *solid-ordered* state (so), where the lipid acyl chains are conformationally ordered and the lateral mobility is low. In the vicinity of T_m, the membrane displays coexistence of ld and so phases, but for single-component membranes this domain coexistence only happens within a narrow temperature interval.

6.12. BINARY MODEL MEMBRANES

For membranes composed of two lipids, the phase behavior becomes slightly more complex. Imagine a binary membrane composed of two phospholipids with different melting temperatures. Due to differences in the attractive forces between the two components, a demixing takes place and the membrane will phase separate into so and ld domains over a temperature interval between the two melting points and over a wide range of compositions. This is formally described by a coexistence region in the binary phase diagram shown in Figure 6.8B. The well-known formalism for interpreting phase diagrams such as the lever rule also applies to such membrane phase diagrams. Phase diagrams are thus very powerful tools for rationalizing phase separation in multicomponent membranes. Next, we will give some examples of the characterization of domains in different types of binary model membrane systems.

We have investigated domain structures formed in supported membranes of binary mixtures exhibiting phase coexistence [22]. Membranes exhibiting solid/fluid phase coexistence with micrometer-sized domains have been observed with several binary lipid mixtures in both free-standing GUV [77] and supported membranes [78–80]. We show results for two binary mixtures — POPC+DPPC and POPC+Brain ceramide (BC). These mixtures were chosen for their physiological relevance and because some previous information on phase separation is available. Thus, the POPC:DPPC system is known to phase separate over a large range of compositions, as is evident from the phase diagram [81] and as observed in GUVs [82,83]. We have investigated a sample with a 1:1 mixture, for which a membrane should be located centrally in the coexistence region of the phase diagram at 20°C. Single-supported membranes were prepared by spin coating (as described above). The dry spin-coated film was hydrated in water followed by annealing for 1 hour (60°C) above the gel–fluid phase transition temperature and finally re-cooled to 20°C. As observed in the fluorescence and AFM images of Figure 6.8C,D, the supported membrane has phase separated into a characteristic gel–fluid pattern, as evident from the irregular boundaries of the bright (tallest) regions. These gel domains have a size of several micrometers and a height relative to the fluid phase of 1.0–1.5 nm.

Important physiological functions have been associated with the presence and generation of ceramide in cell membranes. These effects include the role of ceramide as a messenger in apoptosis [84,85] and as a main structural component in the lamellar lipid structures found in the outer skin (stratum corneum) [86]. Several of these specific biological functions are speculated to be linked to formation of ceramide-enriched domains or patches in the bilayers. A simplified model system for this phase separation would be membranes of the binary lipid mixture POPC+Brain ceramide [87]. Hsueh et al. [88] have constructed a partial phase diagram for pure palmitoyl ceramide and POPC based on NMR data and showed that a gel/fluid coexistence region covers a wide range of temperatures and compositions. We provide microscopic evidence of domains in mixed ceramide/phospholipid membranes. Figure 6.8E,F shows fluorescence and AFM images obtained with the mixture POPC+16% BC. The domains are more rounded in shape than the DPPC/POPC gel domains, and have a height difference relative to the fluid phase of 10 Å. The domain height difference of 1.2 nm found in Figure 6.8E,F supports the notion that the coexisting phases are gel and fluid in type. In general, we find that formation of solid domains upon cooling is initiated by random nucleation of domains when crossing the fluidus line. These domains grow in what appears to be a diffusion-controlled manner. The final domain shape is given by a competition between diffusion limited growth that promotes branching and line tension at the domain boundary, which promotes round domain shapes. Since the lipid composition influences line tension and diffusion properties, domain shapes are coupled to lipid composition. Hence the differences observed for the two mixtures in Figure 6.8.

6.13. TERNARY MODEL MEMBRANES

There is growing evidence that cholesterol is a key lipid for maintaining a particular lateral structure of biomembranes, as has been emphasized in a recent study of the pulmonary surfactant system [89]. In particular, the so-called liquid-ordered (lo) phase is speculated to be important in relation to the formation of functional microdomains (rafts) in membranes. The lo phase is enriched in saturated lipids and cholesterol and is characterized by ordered acyl chains and lateral lipid mobility corresponding to the fluid disordered state [90]. Ternary model membranes constitute a simple class of relevant model systems used for modeling biological membrane domain formation. Such ternary model membranes are typically composed of an unsaturated or branched phospholipid with a low melting point and cholesterol, as well as a saturated phospholipid or sphingolipid that colocalizes with the sterol. Specifically, the ternary phase diagram of DOPC, DPPC, and Chol is known to exhibit a two-phase coexistence region where the membrane displays two macroscopic fluid phases corresponding to the lo and the ld phases (Fig. 6.9A). It is known from studies of freestanding membranes (Giant Unilamellar Vesicles [GUVs]) that the membrane domains are 5–20 μm in diameter and circular due to line tension acting at the domain interfaces [71]. On the other hand, many studies of supported ternary membrane reveal much smaller (100–1000 nm) domains with very irregular shapes [91–93]. This suggests that the membrane support may have a critical influence on observed domain shapes.

We have systematically investigated the characteristic domain shapes that arise in supported ternary membranes following hydration of the spin-coated lipid film and annealing/cooling to ambient temperature. Figure 6.9 shows the membrane domains in a membrane with composition DOPC:DPPC:Chol 2:2:1 that was chosen to be centrally located in the lo/ld coexistence region of the ternary phase diagram. For a single-supported membrane hydrated in pure water, the domain pattern shown in Figure 6.9B clearly displays two distinct phases with a height difference of 0.6–0.8 nm (line scan). This value is in agreement with previously measured height differences for the lo/ld phases and is smaller than the typical height difference between solid and fluid domains (so/ld) of 1.2 nm. The membrane domains are in this case small (100–1000 nm), with highly branched and irregular shapes. The domains are static and do not coarsen with time due to strong membrane–support interactions that restrict lateral lipid mobility. This situation is changed if the membrane is hydrated in buffer under 150-mM NaCl and using double-supported membranes, as illustrated in Figure 6.9C. The presence of salt will screen membrane–solid and membrane–membrane interactions and release lateral lipid mobility. This is evident from the fluorescence image in Figure 6.9C, which demonstrates much larger domains. In particular, the second bilayer forms large and round domains that are closely similar to the domains found in freestanding GUV membranes [71]. This indicates that double-supported membranes prepared in physiological buffer are sufficiently decoupled from the support to mimic the domain shapes of freestanding membranes. In Figure 6.9C, domains in the first and second membrane are overlapping. The domains in bilayer 1 are round and well defined but have domains much smaller

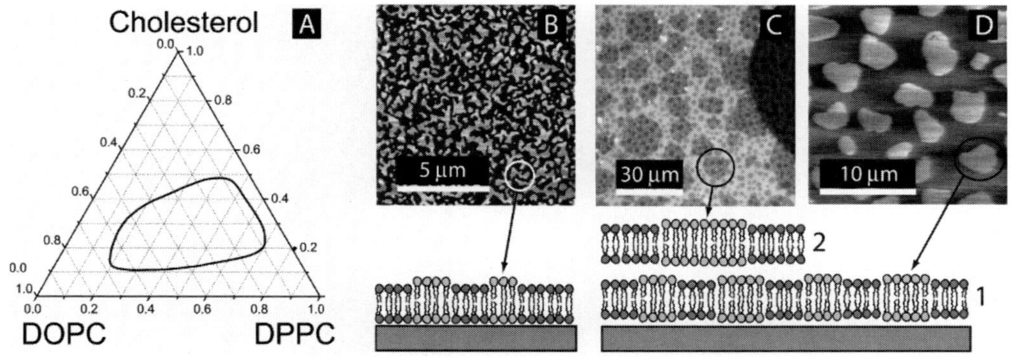

Figure 6.9. (A) Single- and double-supported ternary membranes with the composition DOPC:DPPC:Chol (2:2:1) corresponding to the lo/ld coexistence region (marked) of the membrane phase diagram. (B) AFM scan of a single-supported membrane hydrated in water showing branched nanoscale domains. (C) Fluorescence image of a double-supported membrane system (150-mM NaCl) with large circular liquid-ordered domains in bilayer 2 overlapping with smaller domains in bilayer 1. The domains in bilayer 1 have been imaged with AFM in (D), which confirms the absence of nanoscale domains in this case. Please visit http://www.springer.com/series/7845 to view a high-resolution full-color version of this illustration.

than those in bilayer 2. This was confirmed by AFM imaging of bilayer 1 (see Fig. 6.9D), which also confirms that the branched nanoscale domains in Figure 6.9B are absent when the aqueous phase contains 150-mM NaCl. AFM imaging of bilayer 2 is possible when employing magnetically excited tapping mode (images not shown). However, due to the weak coupling between membrane 2 and the support, the topographical variations across domains are much smaller than when scanning membrane 1. Moreover, this topography is subject to fluctuations from membrane undulations, which complicates image acquisition. These results demonstrate that the second bilayer of a double-supported membrane system is the most suitable membrane model system which to a high degree reproduces the domain features observed in freestanding membranes.

Domain coarsening, or Ostwald ripening [94], is one of the fundamental physical mechanisms that can alter domain size in many types of phase separation in nature. It describes the common observation that, after nucleation of domains, large domains grow at the expense of smaller ones because the system seeks to minimize overall interface free energy. In fluid two-dimensional systems, the driving force for domain coarsening is line tension, which seeks to minimize total perimeter length around domains. Experimental measurements of domain coarsening in supported membranes have until now been complicated by membrane–support interactions, which can affect coarsening dynamics. Using the second membrane of the double-supported membrane system as described above, we have obtained a model system in which the membrane is sufficiently decoupled from the support to allow coarsening phenomena to be observed and quantified [66]. We have performed an analysis of domain coarsening after a rapid quench of the membrane from above the phase transition and into the two-phase region of the ternary phase diagram. Since the first membrane is affected by interactions with the solid support, domain coarsening takes place more rapidly in the second membrane. This means that for short times after the quench domains in the first membrane stay below the optical resolution

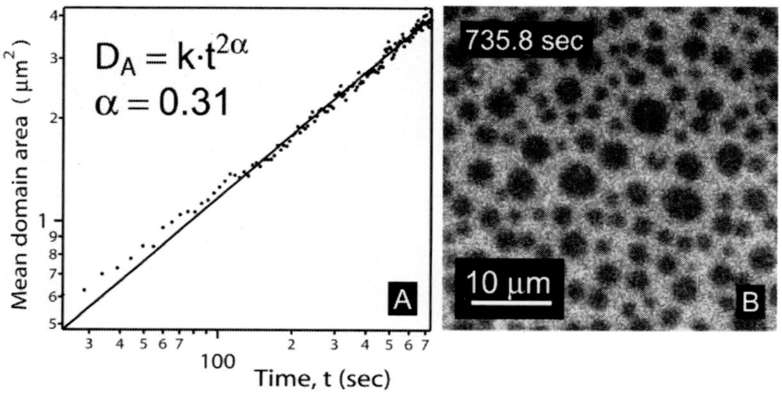

Figure 6.10. (A) Domain coarsening in the second membrane of a double-supported ternary membrane (DOPC:DPPC:Chol, 2:2:1) following a quench of the system into the two-phase coexistence region at time $t = 0$ s. The mean domain area (circles) approaches a power law behavior (solid line) with growth exponent $\alpha = 1/3$. (B) Fluorescence micrograph demonstrating liquid-ordered domains (dark) formed after the thermal quench.

limit while resolvable domains develop in the second one. In this initial time window, we can therefore avoid overlapping domains, and we have quantified the average domain size in the second membrane using image analysis. Figure 6.10A shows a log–log plot of the mean domain area versus time. For times longer than 100 s, the datapoints fall on a straight line corresponding to a power law with a growth exponent 2α, with $\alpha = 0.31$. This value is close to the classical Lifshitz exponent of 1/3 [95] and also close to previous experiments on fluid domains in binary lipid monolayers ($\alpha = 0.28$ [96]). These results demonstrate that a freestanding ternary membrane displaying coexistence of two fluid phases will undergo coarsening and develop micrometer-sized domains within a few minutes after the quench. The small nanoscale domains previously observed with AFM by us and others are found only when strong interactions with the support surface restrict lateral lipid mobility. However, there is growing evidence that functional microdomains in biological cells do in fact have nanoscale dimensions [8]. We speculate that in biological systems the interaction between the membrane and scaffolds such as the cytoskeleton may be a crucial factor that limits domain size by a similar mechanism as that found for solid supports.

ACKNOWLEDGMENTS

I would like to thank my colleagues at MEMPHYS for creating and nurturing a very stimulating research environment. Specifically, I would like to thank the following for their direct or indirect contributions to the work mentioned in this chapter: Luis Bagatolli, Eliza Morris, Mikkel Jensen, Uffe Bernchou Jensen, Esben Thormann, Per Lyngs Hansen, and Ole G. Mouritsen. I am grateful to The Danish National Research Foundation for support via a grant to MEMPHYS—Center for Biomembrane Physics.

PROBLEMS

6.1. Derive the cantilever amplification factor in the AFM detection system,

$$\Delta z = \frac{3L}{l}\Delta y\,, \tag{6.11}$$

where Δz is deflection of the laser spot when reaching the photodiode and Δy is the physical deflection of the cantilever end (see also §6.9).

6.2. Consider a situation where a fluorescence image is recorded of a double-membrane system with resolvable and superimposed domain features from membranes 1 and 2. The difference in partitioning of the fluorescence probe between the membrane domains generates contrast and gives rise to two intensities — high (H) and low (L) — from the two domains types, respectively. If we are only interested in analyzing the domain pattern in membrane 2, there is a potential problem in detecting these domain features without interference from membrane 1 domains. We will consider if and how this problem can be resolved.

a. Assume that the fluorescent intensity of the two domains is the same (H.L) in membrane 1 and 2. Will it be possible to detect domains in membrane 2 by simple threshold filtering (intensity cut-off)?

Figure 6.11.

b. Assume that the fluorescent intensity of the two domains is different (H_1,L_1,H_2,L_2) in membranes 1 and 2. Will it be possible to detect domains in membrane 2 by simple threshold filtering (intensity cutoff)? Will it be possible to detect the contour of islands in membrane 2? If yes, write down the restrictions on H_1,L_1,H_2,L_2 that enable these detections.

c. Discuss alternative experimental or computational tricks to improve domain detection in the above case.

FURTHER STUDY

Mouritsen OG. 2005. *Life — as a matter of fat: the emerging science of lipidomics.*Heidelberg: Springer-Verlag.

Heimburg T. 2007. *Thermal biophysics of membranes.* New York: Wiley VCH.

Israelachvili J. 1991. *Intermolecular and surface forces*, 2nd ed. New York: Academic Press.

Evans DF, Wennerström H. 1999. *The colloidal domain. where physics chemistry, biology and technology meet.* New York: Wiley-VCH.

REFERENCES

1. Rasmussen S, Chen LH, Deamer D, Krakauer DC, Packard NH, Stadler PF, Bedau MA. 2004. Transitions from nonliving to living matter. *Science* **303**(5660):963–965.
2. Mouritsen OG. 2005. *Life — as a matter of fat: the emerging science of lipidomics*. Heidelberg: Springer-Verlag.
3. Gorter E, Grendel F. 1925. On bimolecular layers of lipoids on chromatocytes of blood. *J Exp Medicine* **41**:439–443.
4. Singer S, Nicolson GL. 1972. The fluid-mosaic model of cell membranes. *Science* **172**:720–730.
5. Simons K, Ikonen E. 1997. Functional rafts in cell membranes. *Nature* **387**:569–572.
6. Edidin M. 2003. The state of lipid rafts: from model membranes to cells. *Ann Rev Biophys Biomol Struct* **32**:257–283.
7. London E. 2005. How principles of domain formation in model membranes may explain ambiguities concerning lipid raft formation in cells. *Biochim Biophys A* **1746**(3):203–220.
8. Jacobson K, Mouritsen O, Anderson RGW. 2007. Lipid rafts: at a crossroad between cell biology and physics. *Nature Cell Biol* **9**(1):7–14.
9. Mayor S, Rao M. 2004. Rafts: Scale-dependent, active lipid organization at the cell surface. *Traffic* **5**(4):231–240.
10. Brown DA, Rose JK. 1992. Sorting of GPI-anchored proteins to glycolipid-enriched membrane subdomains during transport to the apical cell-surface. *Cell* **68**(3):533–544.
11. Blodgett KB. 1935. Films built by depositing successive monomolecular layers on a solid surface. *J Am Chem Soc* **57**:1007–1022.
12. Langmuir I, Schaefer VJ. 1938. Activities of urease and pepsin monolayers. *J Am Chem Soc* **60**:1351–1360.
13. Brian AA, McConnell HM. 1984. Allogenic stimulation of cytoxic t cells by supported planar membranes. *Proc Natl Acad Sci USA* **81**:6159–6163.
14. Watts TH, Brian AA, Kappler JW, Marrack P, Mcconnell HM. 1984. Antigen presentation by supported planar membranes containing affinity-purified I-A^{d}. *Proc Natl Acad Sci USA* **81**:7564–7568.
15. Tien HT, Salamon Z. 1989. Formation of self-assembled lipid bilayers on solid substrates. *Bioelectrochem Bioelectr* **22**(3):211–218.
16. Florin EL, Gaub HE. 1993. Painted supported lipid membranes. *Biophys J* **64**:375–383.
17. Rädler J, Strey H, Sackmann E. 1995. Phenomenology and kinetics of lipid bilayer spreading on hydrophilic surfaces. *Langmuir* **11**:4539–4548.
18. Cornell BA, Braachmaksvytis VLB, King LG, Osman PDJ, Raguse B, Wieczorek L, Pace RJ. 1997. A biosensor that uses ion-channel switches. *Nature* **387**(6633):580–583.
19. Raguse B, Braach-Maksvytis V, Cornell BA, King LG, Osman PDJ, Pace RJ, Wieczorek L. 1998. Tethered lipid bilayer membranes: formation and ionic reservoir characterization. *Langmuir* **14**:648–659.
20. Karlsson OP, Lofas S. 2002. Flow-mediated on-surface reconstitution of g-protein coupled receptors for applications in surface plasmon resonance biosensors. *Anal Biochem* **300**(2):132–138.
21. Mennicke U, Salditt T. 2002. Preparation of solid-supported lipid bilayers by spin-coating. *Langmuir* **18**:8172–8177.
22. Simonsen AC, Bagatolli LA. 2004. Structure of spin-coated lipid films and domain formation in supported membranes formed by hydration. *Langmuir* **20**:9720–9728.
23. Petty MC. 1996. *Langmuir-Blodgett films*. Cambridge: Cambridge UP.
24. Blodgett KB. 1934. Monomolecular films of fatty acids on glass. *J Am Chem Soc* **56**:495.
25. Blodgett KB, Langmuir I. 1937. Built-up films of barium stearate and their optical properties. *Phys Rev* **51**:964–982.
26. Torchilin VP, Weissig V. 2003. *Liposomes*, 2nd ed. Oxford: Oxford UP.
27. Reviakine I, Brisson A. 2000. Formation of supported phospholipid bilayers from unilamellar vesicles investigated by atomic force microscopy. *Langmuir* **16**:1806–1815.
28. Richter R, Mukhopadhyay A, Brisson A. 2003. Pathways of lipid vesicle deposition on solid surfaces: a combined qcm-d and afm study. *Biophys J* **85**(5):3035–3047.
29. Richter RP, Berat R, Brisson AR. 2006. Formation of solid-supported lipid bilayers: an integrated view. *Langmuir* **22**(8):3497–3505.

30. Keller CA, Glasmästar K, Zhdanov VP, Kasemo B. 2000. Formation of supported membranes from vesicles. *Phys Rev Lett* **84**:5443–5446.
31. Reimhult E, Höök F, Kasemo B. 2002. Temperature dependence of formation of a supported phospholipid bilayer from vesicles on SiO_2. *Phys Rev E* **66**:051905.
32. Reimhult E, Hook F, Kasemo B. 2003. Intact vesicle adsorption and supported biomembrane formation from vesicles in solution: Influence of surface chemistry, vesicle size, temperature, and osmotic pressure. *Langmuir* **19**(5):1681–1691.
33. Muller DJ, Amrein M, Engel A. 1997. Adsorption of biological molecules to a solid support for scanning probe microscopy. *J Struc Biol* **119**(2):172–188.
34. Simonsen AC. 2008. Activation of phospholipase A(2) by ternary model membranes. *Biophys J* **94**:3966–3975.
35. Behrens SH, Grier DG. 2001. The charge of glass and silica surfaces. *J Chem Phys* **115**(14):6716–6721.
36. Krishna G, Schulte J, Cornell BA, Pace RJ, Osman PD. 2003. Tethered bilayer membranes containing ionic reservoirs: selectivity and conductance. *Langmuir* **19**:2294–2305.
37. Tanaka M, Sackmann E. 2005. Polymer-supported membranes as models of the cell surface. *Nature* **437**(7059):656–663.
38. Emslie AG, Bonner FT, Peck LG. 1958. Flow of a viscous liquid on a rotating disc. *J Appl Phys* **29**:858–862.
39. Acrivos A, Shah MJ, Petersen EE. 1960. Momentum and heat transfer in laminar boundary-layer flows of non-Newtonian fluids past external surfaces. *AICHE J* **6**(2):312–317.
40. Acrivos A, Shah MJ, Petersen EE. 1960. On the flow of a non-Newtonian liquid on a rotating disk. *J Appl Phys* **31**(6):963–968.
41. Meyerhofer D. 1978. Characteristics of resist films produced by spinning. *J Appl Phys* **49**(7):3993–3997.
42. Flack WW, Soong DS, Bell AT, Hess DW. 1984. A mathematical model for spin coating of polymer resists. *J Appl Phys* **56**:1199–1206.
43. Unpublished ellipsometry data of spincoated phospholipid films in collaboration with P. Thiesen, Nanofilm Technologie GmbH.
44. Perino-Gallice L, Fragneto G, Mennicke U, Salditt T, Rieutord F. 2002. Dewetting of solid-supported multilamellar lipid layers. *Eur Phys J E* **8**:275–282.
45. Malmsten M. 1995. Protein adsorption at phospholipid surfaces. *J Colloid Interface Sci* **172**:106–115.
46. Binder H, Gawrisch K. 2001. Dehydration induces lateral expansion of polyunsaturated 18:0–22:6 phosphatidylcholine in a new lamellar phase. *Biophys J* **81**:969–982.
47. Müller-Buschbaum P. 2003. Dewetting and pattern formation in thin polymer films as investigated in real and reciprocal space. *J Phys Cond Matter* **15**:R1549–R1582.
48. Sharma A, Reiter G. 1996. Instability of thin polymer films on coated substrates, rupture, dewetting, and drop formation. *J Colloid Interface Sci* **178**:383–399.
49. Sharma A, Khanna R. 1999. Pattern formation in unstable thin liquid films under the influence of antagonistic short and long-range forces. *J Chem Phys* **110**:4929–4936.
50. Solletti JM, Botreau M, Sommer F, Brunat WL, Kasas S, Duc TM, Celio MR. 1996. Elaboration and characterization of phospholipid Langmuir-Blodgett films. *Langmuir* **12**:5379–5386.
51. Xiao X, Qian L. 2000. Investigation of humidity-dependent capillary force. *Langmuir* **16**:8153–8158.
52. Israelachvili JN. 1991. *Intermolecular and surface forces: with applications to colloidal and biological systems*. San Diego: Academic Press.
53. Lasic DD. 1988. The mechanism of vesicle formation. *Biochem J* **256**:1–11.
54. Bagatolli LA, Parasassi T, Gratton E. 2000. Giant phospholipid vesicles: comparison among the whole lipid sample characteristics using different preparation methods. *Chem Phys Lipids* **105**:135–147.
55. Faiss S, Schuy S, Weiskopf D, Steinem C, Janshoff A. 2007. Phase transition of individually addressable microstructured membranes visualized by imaging ellipsometry. *J Phys Chem B* **111**(50):13979–13986.
56. Keller D, Larsen NB, Møller IM, Mouritsen OG. 2005. Decoupled phase transitions and grain-boundary melting in supported phospholipid bilayers. *Phys Rev Lett* **94**:025701.
57. Evans DF, Wennerström H. 1999. The colloidal domain: where physics, chemistry, biology and technology meet, 2nd ed. New York: Wiley-VCH.
58. Petrache HI, Zemb T, Belloni L, Parsegian VA. 2006. Salt screening and specific ion adsorption determine neutral-lipid membrane interactions. *Proc Natl Acad Sci USA* **103**(21):7982–7987.
59. Marra J, Israelachvili J. 1985. Direct measurements of forces between phosphatidylcholine and phosphatidylethanolamine bilayers in aqueous-electrolyte solutions. *Biochemistry* **24**(17):4608–4618.

60. Israelachvili JN, Wennerstrom H. 1992. Entropic forces between amphiphilic surfaces in liquids. *J Phys Chem* **96**(2):520–531.
61. Heine DR, Rammohan AR, Balakrishnan J. 2007. Atomistic simulations of the interaction between lipid bilayers and substrates. *Mol Simul* **33**(4–5):391–397.
62. Bayerl TM, Bloom M. 1990. Physical-properties of single phospholipid bilayers adsorbed to micro glass-beads — a new vesicular model system studied by H2-nuclear magnetic-resonance. *Biophjys J* **58**(2):357–362.
63. Johnson SJ, Bayerl TM, Mcdermott DC, Adam GW, Rennie AR, Thomas RK, Sackmann E. 1991. Structure of an adsorbed dimyristoylphosphatidylcholine bilayer measured with specular reflection of neutrons. *Biophys J* **59**(2):289–294.
64. Koenig BW, Kruger S, Orts WJ, Majkrzak CF, Berk NF, Silverton JV, Gawrisch K. 1996. Neutron reflectivity and atomic force microscopy studies of a lipid bilayer in water adsorbed to the surface of a silicon single crystal. *Langmuir* **12**(5):1343–1350.
65. Mornet S, Lambert O, Duguet E, Brisson A. 2005. The formation of supported lipid bilayers on silica nanoparticles revealed by cryoelectron microscopy. *Nano Lett* **5**(2):281–285.
66. Jensen MH, Morris EJ, Simonsen AC. 2007. Domain shapes, coarsening and random patterns in ternary membranes. *Langmuir* **23**:8135–8141.
67. Binnig G, Quate CF, Gerber C. 1986. Atomic force microscope. *Phys Rev Lett* **56**(9):930–934.
68. Han WH, Lindsay SM, Jing TW. 1996. A magnetically driven oscillating probe microscope for operation in liquids. *Appl Phys Lett* **69**(26):4111–4113.
69. Revenko I, Proksch R. 2000. Magnetic and acoustic tapping mode microscopy of liquid phase phospholipid bilayers and dna molecules. *J Appl Phys* **87**(1):526–533.
70. Villarrubia JS. 1997. Algorithms for scanned probe microscope, image simulation, surface reconstruction and tip estimation. *J Res Natl Inst Stand Technol* **102**:425–454.
71. Bagatolli LA. 2006. To see or not to see: Lateral organization of biological membranes and fluorescence microscopy. *Biochim Biophys A* **1758**(10):1541–1556.
72. Bagatolli LA, Gratton E. 2000. A correlation between lipid domain shape and binary phospholipid mixture composition in free standing bilayers: a two-photon fluorescence microscopy study. *BiophysJ* **79**(1):434–447.
73. Bagatolli LA, Parasassi T, Gratton E. 2000. Giant phospholipid vesicles: comparison among the whole lipid sample characteristics using different preparation methods — a two-photon fluorescence microscopy study. *Chem Phys Lipids* **105**(2):135–147.
74. Bagatolli LA, Gratton E. 2000. Two photon fluorescence microscopy of coexisting lipid domains in giant unilamellar vesicles of binary phospholipid mixtures. *Biophys J* **78**(1):290–305.
75. Moreaux L, Sandre O, Mertz J. 2000. Membrane imaging by second-harmonic generation microscopy. *J Opt Soc Am B* **17**(10):1685–1694.
76. Wallrabe H, Periasamy A. 2005. Imaging protein molecules using fret and flim microscopy. *Curr Opin Biotechnol* **16**(1):19–27.
77. Bagatolli LA, Gratton E. 2000. A correlation between lipid domain shape and binary phospholipid mixture composition in free standing bilayers: a two-photon fluorescence microscopy study. *Biophys J* **79**:434–447.
78. Dietrich C, Bagatolli LA, Volovyk ZN, Thompson NL, Levi M, Jacobson K, Gratton E. 2001. Lipid rafts reconstituted in model membranes. *Biophys J* **80**:1417–1428.
79. Burns AR. 2003. Domain structure in model membrane bilayers investigated by simultaneous atomic force microscopy and fluorescence imaging. *Langmuir* **19**:8358–8363.
80. Ianoul A, Burgos P, Lu Z, Taylor RS, Johnston LJ. 2003. Phase separation in supported phospholipid bilayers visualized by near-field scanning optical microscopy in aqueous solution. *Langmuir* **19**:9246–9254.
81. Curatolo W, Sears B, Neuringer LJ. 1985. A calorimetry and deuterium NMR study of mixed model membranes of 1-palmitoyl-2-oleoylphosphatidylcholine and saturated phosphatidylcholines. *Biochim Biophys Acta* **817**:261–270.
82. Bagatolli LA, Gratton E. 2000. Two photon fluorescence microscopy of coexisting lipid domains in giant unilamellar vesicles of binary phospholipid mixtures. *Biophys J* **78**:290–305.
83. Shoemaker SD, Vanderlick TK. 2003. Material studies of lipid vesicles in the l_α and l_α-gel coexistence regimes. *Biophys J* **84**:998–1009.
84. Hannun YA. 1996. Functions of ceramide in coordinating cellular responses to stress. *Science* **274**:1855–1859.
85. Venkataraman K, Futerman AH. 2000. Ceramide as a second messenger: sticky solutions to sticky problems. *Trends Cell Biol* **10**:408–412.

86. Bouwstra JA, Honeywell-Nguyen PL, Gooris GS, Ponec M. 2003. Structure of the skin barrier and its modulation by vesicular formulations. *Prog Lipid Res* **42**:1–36.
87. Fidorra M, Duelund L, Leidy C, Simonsen AC, Bagatolli LA. 2006. Absence of fluid-ordered/fluid-disordered phase coexistence in ceramide/POPC mixtures containing cholesterol. *Biophys J* **90**(12):4437–4451.
88. Hsueh YW, Giles R, Kitson N, Thewalt J. 2002. The effect of ceramide on phosphatidylcholine membranes: a deuterium NMR study. *Biophys J* **82**:3089–3095.
89. de la Serna J, Perez-Gil J, Simonsen AC, Bagatolli LA. 2004. Cholesterol rules: direct observation of the coexistence of two fluid phases in native pulmonary surfactant membranes at physiological temperatures. *J Biol Chem* **279**(39):40715–40722.
90. Ipsen J, Karlström G, Mouritsen O, Wennerström H, Zuckermann M. 1987. Phase-equilibria in the phosphatidylcholine–cholesterol system. *Biochim Biophys A* **905**(1):162–172.
91. van Duyl BY, Ganchev D, Chupin V, de Kruijff B, Killian JA. 2003. Sphingomyelin is much more effective than saturated phosphatidylcholine in excluding unsaturated phosphatidylcholine from domains formed with cholesterol. *FEBS Lett* **547**:101–106.
92. Lawrence JC, Saslowsky DE, Edwardson JM, Henderson RM. 2003. Real-time analysis of the effects of cholesterol on lipid raft behavior using atomic force microscopy. *Biophys J* **84**(3):1827–1832.
93. Henderson RA, Edwardson JM, Geisse NA, Saslowsky DE. 2004. Lipid rafts: feeling is believing. *News Physiol Sci* **19**:39–43.
94. Zinke-Allmang M. 1999. Phase separation on solid surfaces: nucleation, coarsening and coalescence kinetics. *Thin Solid Films* **346**:1–68.
95. Lifshitz I, Slyozov V. 1961. The kinetics of precipitation from supersaturated solid solutions. *J Phys Chem Sol* **19**(1–2):35–50.
96. Seul M, Morgan NY, Sire C. 1994. Domain coarsening in a 2-dimensional binary mixture: growth dynamics and spatial correlations. *Phys Rev Lett* **73**:2284–2287.

APPENDIX

Protocol: How to Prepare Spin-Coated Supported Membranes for Microscopy

This is a simple practical recipe for preparing single- or double-supported model membranes from lipid stock solutions. You need the following experimental equipment:

1. Membrane support (freshly cleaved muscovite mica or cleaned glass). Roughly 1 × 1 cm^2.
2. Lipid stock solution with a concentration of 10 mM. Usable lipid solvents include n-hexane, n-hexane + MeOH (volume ratio 97:3), or isopropanol + n-hexane + water (volume ratio 3:1:1). For fluorescence observations the lipid stock must contain 0.1–1% (mole fraction of lipids) of fluorescent lipid dye.
3. Spin coater and vacuum pump (such as Chemat Technology KW-4A).
4. Epifluorescence microscope with sensitive camera for live, video-rate readout.
5. Transparent liquid cell sample holder for microscopy. Temperature control in the fluid cell or, alternatively, an external oven (adjustable to below 100°C).
6. Accessories (Hamilton syringes, pipettes, hydration buffer, etc.).

The method includes two steps: (1) fabrication of a dry spin-coated lipid film followed by (2) controlled hydration/annealing of the dry precursor film. To prepare the dry spin-coated lipid film, deposit a droplet (30–100 μl) of the lipid stock solution onto a clean substrate surface that has been placed on the spin coater. The lipid stock solution must wet the substrate completely to ensure uniform quality of the final film. Immediately after lipid deposition, the sample is rotated on the spin coater at 3000 rpm for 40 s, or long enough for the solvent to evaporate (visual inspection). This creates a dry multilayered lipid film that is placed under vacuum in a desiccator for >10 hours to ensure complete evaporation of the solvent. To prepare single- or double-supported membranes, the sample is hydrated by immersion in an aqueous buffer with the desired ionic composition and strength. After aqueous immersion, the sample must be kept hydrated at all times to avoid disruption of the bilayer structure. The hydrated sample is heated to above the main phase transition temperature for 1–2 hours, typically 50–60°C. While at elevated temperature, the sample is transferred to the fluorescence microscope for imaging of the lipid film. The sample is now flushed with a pipette to remove excess lipid bilayers. For this step use an adjustable pipette set to 500 μl and induce a vertical downward flow directed at the center of the sample. This completely removes all but the lowest (primary) membrane in this region. Some distance away from the center, multilayers are present, and in the transition zone one can routinely localize double (secondary) bilayer islands situated on the primary bilayer. Depending on whether the primary or patches of the secondary membrane is the objective, the washing step can be fine tuned. Finally, the sample can be cooled to the target temperature for the final phase of membrane investigation.

NANOPORE ANALYSIS OF NUCLEIC ACIDS: SINGLE-MOLECULE STUDIES OF MOLECULAR DYNAMICS, STRUCTURE, AND BASE SEQUENCE

Felix Olasagasti and David W. Deamer
Department of Chemistry and Biochemistry,
University of California Santa Cruz

7.1. INTRODUCTION

Nucleic acids are linear polynucleotides in which each base is covalently linked to a pentose sugar and a phosphate group carrying a negative charge. If a pore having roughly the cross-sectional diameter of a single-stranded nucleic acid is embedded in a thin membrane and a voltage of 100 mV or more is applied, individual nucleic acids in solution can be captured by the electrical field in the pore and translocated through by single-molecule electrophoresis. The dimensions of the pore cannot accommodate anything larger than a single strand, so each base in the molecule passes through the pore in strict linear sequence. The nucleic acid strand occupies a large fraction of the pore's volume during translocation and therefore produces a transient blockade of the ionic current created by the applied voltage. If it could be demonstrated that each nucleotide in the polymer produced a characteristic modulation of the ionic current during its passage through the nanopore, the sequence of current modulations would reflect the sequence of bases in the polymer.

According to this basic concept, nanopores are analogous to a Coulter counter that detects nanoscopic molecules rather than microscopic [1,2]. However, the advantage of nanopores is that individual macromolecules can be characterized because different chemical and physical properties affect their passage through the pore. Because macromolecules can be captured in the pore as well as translocated, the nanopore can be used to detect individual functional complexes that form between a nucleic acid and an enzyme. No other technique has this capability.

Address all correspondence to Felix Olasagasti, PhD, Department of Chemistry and Biochemistry, University of California Santa Cruz, 1156 High Street, Santa Cruz, CA 95064-1077 USA, 831 459-5157, <olasagas@chemistry.ucsc.edu>.

R. Faller et al. (eds.), *Biomembrane Frontiers: Nanostructures, Models, and the Design of Life*,
Handbook of Modern Biophysics, DOI: 10.1007/978-1-60761-314-5_7,

Over the past decade, nanopore science has become a burgeoning field in which both biological nanopores such as α-hemolysin are being used, as well as a variety of synthetic nanopores. A recent review has described this progress in detail [3]. Here we will focus on recent observations from our laboratory that employ the hemolysin nanopore, as well as related studies by our colleagues.

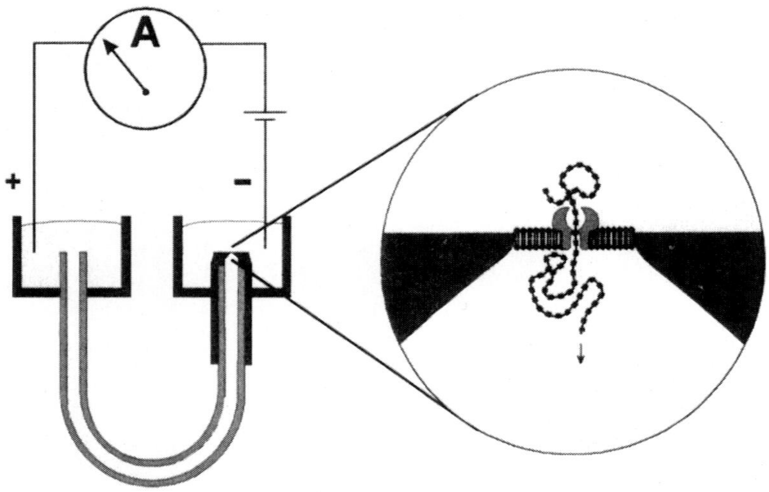

Figure 7.1. Bilayer apparatus. A U-shaped Teflon patch tube connects two 70-μl baths milled into a Teflon support (left). The baths and the Teflon tube are filled with KCl buffer. The baths are connected to an Axopatch 200B amplifier by Ag–AgCl electrodes. One end of the Teflon patch tube has a conical tip that narrows abruptly to a 25-mm conical aperture (right). Diphytanoyl phosphatidylcholine/hexadecane bilayers are formed across this aperture, and one or more α-hemolysin channels are inserted into the bilayer. Nucleic acids are driven through the α-hemolysin channel by applying a positive voltage. Reprinted with permission from [5]. Copyright © 1999, Biophysical Society.

7.2. THE NANOPORE APPARATUS

Biological nanopores are prepared for study by inserting a single protein pore in a planar lipid bilayer membrane. Most analytical nanopore research has been carried out with α-hemolysin pores, but several other interesting biological pores have also been investigated [4]. The approach used by Akeson et al. [5] is representative of the devices used to support a bilayer membrane with a hemolysin pore (Fig. 7.1). The apparatus consists of a U-shaped Teflon tube, which connects two baths of ~70 μl each milled into a single Teflon support piece. The Teflon tube has a conical shape that narrows to an aperture of 25 μm. This shape is made by shrinking a double-walled PTFE tube around a thin conical steel mold. After the mold is removed, the tube is cut to produce an aperture of the desired diameter that supports the lipid bilayer. In order to form this bilayer, the aperture is first coated with phospholipid, and then a small amount of lipid dissolved in an organic solvent is manually "painted" over it. Typically, a mixture of diphytanoyl phosphatidylcholine/hexadecane is used, but other lipid mixtures can also be employed [6–9].

The tube and the baths are filled with a buffer containing an appropriate KCl concentration for the experiments to be performed. KCl provides ions for the conduction of current through the solution. The current amplitude and the detection of analytes depend on the concentration of the salt [10] as well as the temperature, which also affects the current [11]. When the lipid bilayer is formed, it isolates the two baths and there is no electrical current. Then α-hemolysin is added to the 70 μl volume, which is designated the *cis* side of the bilayer, and the monomers spontaneously assemble into a heptamer with a stem that penetrates the bilayer. The insertion is detected as a stepped increase in current that corresponds to 1 pA per 1 mV applied in between 60 and 300 mV using a buffer containing 1 M KCl [11]. α-Hemolysin pores have a minimum diameter of about 1.5 nm [12]. After insertion of a single α-hemolysin channel is detected, the excess protein must be promptly removed by perfusion of 2–3 ml of fresh buffer in order to avoid further insertions. In order to record the electrical signal, the conductance is monitored by an Axopatch 200B amplifier using Ag–AgCl electrodes. Electrical noise from the environment can be a significant problem when pA magnitude currents are measured, and proper grounding of the system allows reducing this noise [11]. After a single channel is detected and appears to be stable, macromolecules or macromolecular complexes can be analyzed. These molecules and complexes are usually negatively charged, so they are added to the *cis* chamber, so that the applied potential produces electrophoretic movement from the *cis* to the *trans* bath through the nanopore.

7.3. DETECTION OF SINGLE MACROMOLECULES

The nanopore system has been envisioned as a device that could facilitate quick and inexpensive nucleic acid sequencing, and most of the research has been directed toward analysis of nucleic acids. Kasianowicz et al. [13] observed that when homopolymers of nucleic acids were added on the *cis* side of a lipid bilayer membrane containing an α-hemolysin pore, numerous short current blockades appeared (Fig. 7.2). They found that the blockades reduced the current by 85–95%, depending on the nature of the homopolymer, and that they lasted from hundreds to thousands of microseconds. These millisecond-scale blockades consistently appeared in the presence of a variety of nucleic acid molecules — such as poly[A], poly[C], poly[dT], and poly[dC] — as well as with single-stranded synthetic DNA composed of 150 nt of poly[dA,dT,dC]. The initial work demonstrated that the blockades observed in the electrical current were actually due to translocation of nucleic acids through the nanopore rather than experimental artifacts. For instance, the number of blockades was proportional to the molar concentration of the polymer, and blockade duration was directly proportional to the length of the polymer and inversely proportional to applied voltage (Fig. 7.3). The critical experiment was to analyze the DNA present in the *trans* side after running an experiment in which both single- and double-stranded DNA were present on the *cis* side. The DNA in the *cis* and *trans* chambers was then amplified by PCR and analyzed. The results showed that only the single-stranded DNA appeared on the *trans* side. Furthermore, the number of blockades matched the number of translocated DNA molecules, as determined by quantitative PCR. This result provided unequivocal evidence that ionic current blockades were in fact produced by translocation, rather than simply by collisions of nucleic acid molecules with the pore.

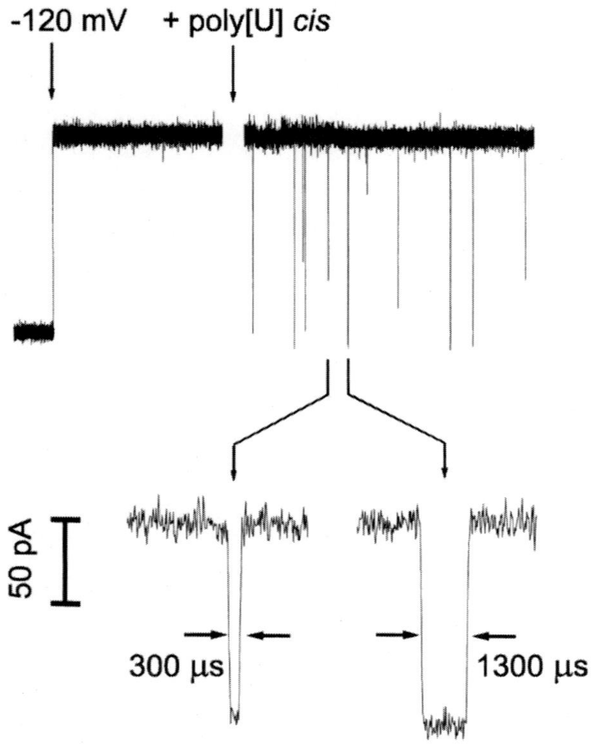

Figure 7.2. Current blockades. Oligomers of poly[U] caused transient blockades in the α-hemolysin single-channel current. At the first arrow, a potential of –120 mV was applied across the membrane (*cis* side negative). This voltage caused a continuous current of –120 pA to flow. At the second arrow, poly[U] of mean length 210 bases was stirred into the *cis* compartment to a final concentration of 0.1 mg/ml. The polynucleotides caused short-lived current blockades. The inset (expanded timescale) shows two typical blockades with lifetimes of 300 and 1300 ms. For the purposes of illustration, the low-time-resolution current recordings (a total of 4 sec is shown here) were digitally filtered. Reprinted with permission from [13]. Copyright © 1996, National Academy of Sciences.

Akeson et al. [5] went on to determine whether the hemolysin nanopore could resolve different homopolymers. This test was essential if nanopores were expected to resolve individual bases, as required for sequencing applications. Single-stranded ribonucleic acid homopolymers of known length were studied and a statistical analysis carried out on the resulting current blockades. Polyadenylic acid (polyA) produced three kind of blockades: (1) short (<200 ms) current blockades of 35–50% of the open-channel current; (2) indefinitely long blockades of 55% of the open-channel current; and (3) 1–4 ms blockades of 85% of the open-channel current. The lower-amplitude blockades (35–50 and 55% of the open-channel current) were interpreted as collisions or partial entries of the polymers into the pore vestibule without translocation. This finding was consistent with the previous work by Kasianowicz et al., who had shown that only the blockades that were strand length dependent represented translocations through the pore [13]. These vestibule-associated events have been recently studied in detail by Butler et al. [14].

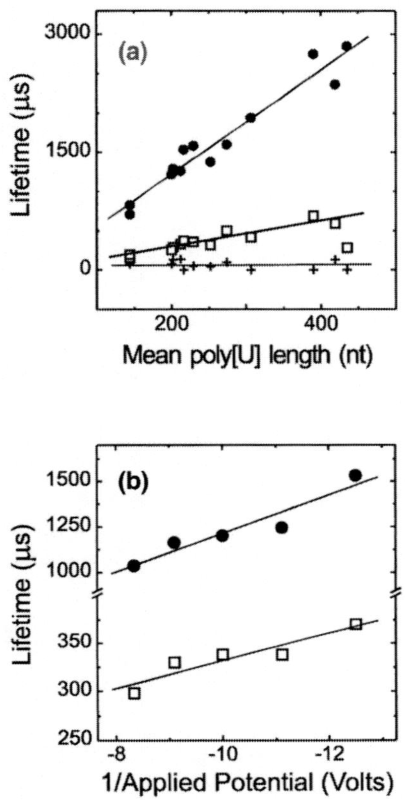

Figure 7.3. Current blockades. Poly[U]-induced channel blockade lifetimes were (a) proportional to mean polymer length and (b) inversely proportional to applied voltage. The plots show lifetimes for different blockade populations in experiments using $V = -120$ mV with (a) 13 different-size-selected poly[U]s and (b) poly[U] of mean length 215 nt at the indicated voltages. Reprinted with permission from [13]. Copyright © 1996, National Academy of Sciences.

When polycytidylic acid (polyC) was added, the pattern of blockades was clearly distinguishable from that of polyA. PolyC reduced the current of the channel by 91 and 95% and rarely produced the lower-amplitude blockades that were common in polyA. It was initially surprising to see that the blockades produced by polyA had a lower amplitude than those produced by polyC, because the larger purine bases in polyA were expected to obstruct the current more than the smaller pyrimidines in polyC. It was proposed that the helical structures of polyC and polyA accounted for this unexpected finding. PolyC forms single-stranded helices 1.3 nm in diameter, which readily enter the pore. The polyA helix has a larger diameter that must be stretched by the electric field to enter the pore, and the extended chain has less effect on ionic current. PolydC forms less stable helices than polyC and also produces blockades of lesser amplitude.

The effects of secondary structure in translocation phenomena were further analyzed by Meller et al. in their studies of the temperature dependence of event characteristics [15]. Akeson et al. could observe these differences even in mixed samples composed of polyA and polyC, where individual blockades produced by either molecule could be identified. PolyU presented current blockade amplitudes similar to polyA, but approximately tenfold faster in translocation velocity. These characteristic properties allowed polyA, polyC, and polyU to be distinguished.

It was encouraging to find that homopolymers composed of different bases could be differentiated in the nanopore, so the next step was to see whether different bases could be distinguished in a single molecule. To this end, a synthetic RNA was prepared that contained a block of 70 C nucleotides followed by a block of 30 A nucleotides. As expected, bilevel blockades were observed in which one component of the blockade reduced the current by 95% of the open-channel current (consistent with polyC), and the other one by 85% (consistent with polyA). The translocation patterns indicated that the polyC end entered the nanopore first, and this observation was recently extended by Butler et al. [16]. In addition to confirming the previous results, Butler et al. also saw evidence of orientation-dependent differences in translocation duration by analyzing copolymers with alternating location of the polyA and polyC segments in the 5′ and 3′ ends of the nucleic acids. Mathé et al. [17] also studied DNA orientation phenomena in the pore and found significant differences depending on whether the 3′ or 5′ end threaded through the pore first. The fact that segments of differing composition could be detected within single molecules supported the possibility of resolving each base within a given nucleic acid strand. However, Akeson et al. detected segments that were 30 nucleotides in length and found that single-stranded helical structures had a major role in determining the amplitude level. So it was still doubtful whether single-nucleotide resolution could be achieved with the nanopore system. Vercoutere et al. [18] addressed this question by combining the nanopore detector with machine-learning algorithms. These algorithms can classify individual current blockade traces, and ambiguous current traces can be automatically discarded after a period of training with an additional set of independent data. By employing this approach, Vercoutere et al. studied DNA hairpins in the nanopore and observed that DNA hairpin molecules captured by the nanopore caused partial blockades that lasted hundreds of milliseconds, followed by a rapid downward spike. These characteristic traces were called "shoulder-spike" signatures. The signature could be explained if the long partial blockade represented the capture of a hairpin stem in the vestibule of the α-hemolysin nanopore. The hairpin would stay there due to the fact that the duplex stem cannot fit through the limiting 1.5-nm diameter aperture of the pore. The second part of the blockade, where a downward spike appeared, was interpreted as spontaneous dissociation of the base pairing in the stem of the hairpin (unzipping), which would allow extended single-strand molecules to traverse the channel.

This hypothesis was tested with blunt-ended DNA hairpins having stems of different length. With each additional base pair, the blockade shoulder lifetime increased substantially. Furthermore, the increase in median shoulder lifetime was directly correlated with the calculated ΔG^0 for the formation of each hairpin (Fig. 7.4). The shoulder current amplitude was reduced from 68% of the open-channel current for hairpins with a 3 base-pair stem, to 32% for a 9-bp stem, which was consistent with an increased obstruction of the ionic current as the hairpin stem extended farther into the vestibule of the pore.

After observing that the addition of a single base pair could be detected in the nanopore system, Vercoutere et al. analyzed the effect of single-nucleotide differences between otherwise identical DNA hairpins. In order to do this, they took a hairpin with 3 bases in the hairpin loop instead of the 4-base loop used in the previous studies. This resulted in an increase in the amplitude of the blocked current and a threefold reduction in the median time with respect to the

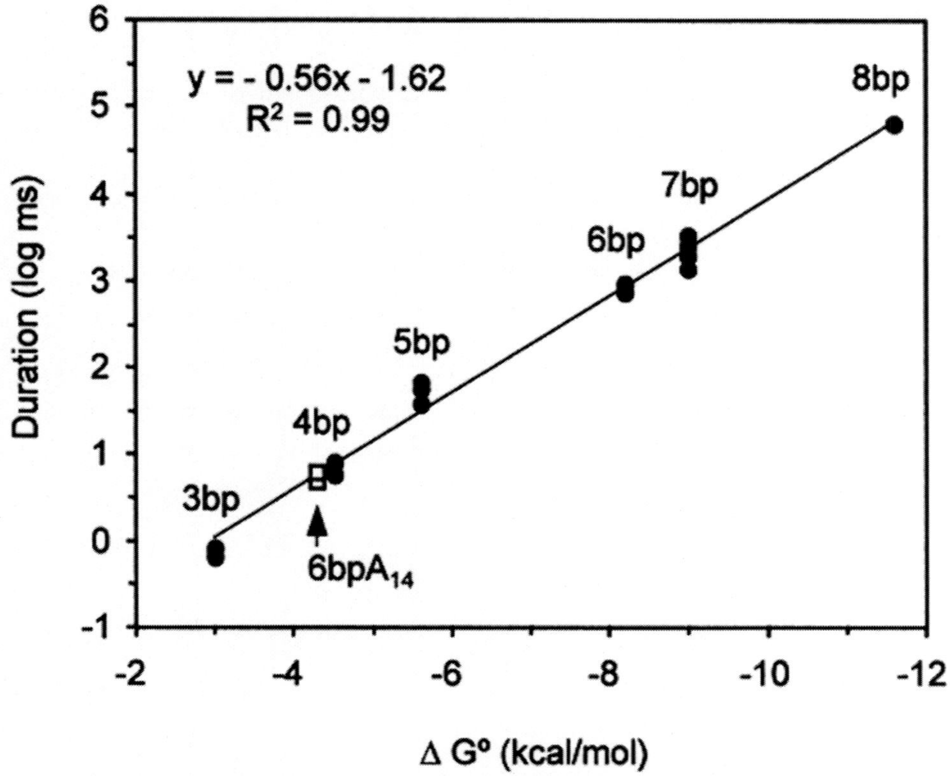

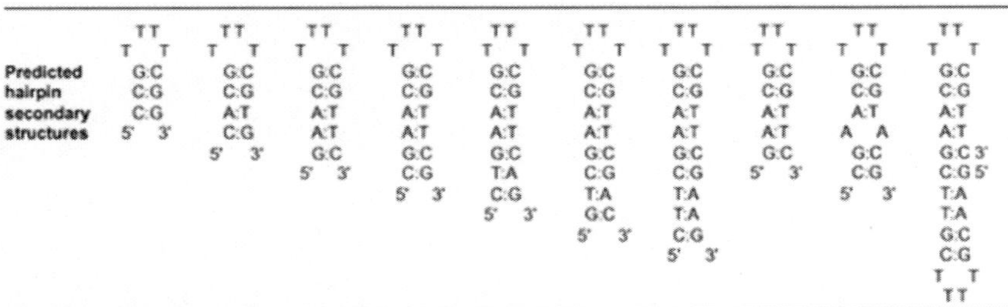

Identity	3bp	4bp	5bp	6bp	7bp	8bp	9bp	5bp3dT	6bpA14	Dumbbell
ΔG° (kcal/mol)[a]	−3.0	−4.5	−5.6	−8.2	−9.0	−11.4	−12.8	−4.2	−4.3	−11.3
I/I_0 (%)[b]	68	64	60	52	47	35	32	62	53	NA

[a] ΔG° values for hairpin formation were calculated using the DNA mfold server (http://bioinfo.math.rpi.edu/~folder/dna/form1.cgi) based on data from SantaLucia[17]. Assume 22°C and 1 M KCl.
[b] I is the current average for an event shoulder (in pA). I_0 is the current average for the open channel (in pA).

Figure 7.4. Standard free energy of hairpin formation versus shoulder-blockade duration. Standard free energy of hairpin formation correlated with median duration of hairpin shoulder blockades (solid circles). Each point represents the median blockade duration for a given hairpin length acquired using a separate α-hemolysin pore on a separate day. Median blockade durations and ΔG^0 for the equivalent of the 6-bp hairpin with a single mismatch (6bpA14) are represented by open squares. Reprinted with permission from [18]. Copyright © 2001, Nature America Publishing.

hairpin with a 4-base loop. After this, they also tested a 6-bp hairpin with a single-bp mismatch; interestingly, this single-nucleotide difference caused an approximately 100-fold decrease in the median blockade shoulder duration with respect to the hairpin with no mismatch. This change in duration is consistent with the difference in the calculated ΔG^0 between the mismatched and the well-matched hairpins. In this example, while both populations could be manually separated in 90% of cases, automated separation using the machine-learning algorithms allowed a separation with 97.6% sensitivity and 99.9% specificity.

Vercoutere et al. observed that longer molecules produce longer blockades, as expected, but this also resulted in a smaller number of blockades, so the statistical significance of the acquired data was reduced. In order to circumvent this problem, a data acquisition protocol was devised that alternated the voltage applied to the system from +120 mV for 249.5 ms to –40 mV for 0.1 ms. This allowed individual hairpin signals to be captured during the initial time and then ejected. By using this method coupled with automated data analysis, signals for individual hairpins with 7, 8, and 9 base pairs could be distinguished with an average sensitivity of 99% and specificity of 96%.

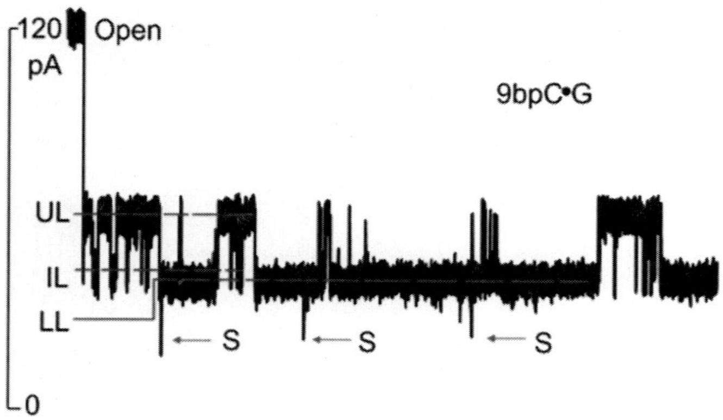

Figure 7.5. Representative blockade of ionic current caused by a 9-bp DNA hairpin (9bpC dG). Open-channel current was typically 120 pA at 120 mV and 23.0°C, but this could vary by 63 pA. Capture of a DNA hairpin caused a rapid current decrease. In the case of 9-bp hairpins, the residual current transitions between four levels: an upper conductance level (UL), an intermediate level (IL), a lower level (LL) and a transient downward spike (S). Reprinted with permission from [19]. Copyright © 2003, Oxford University Press. Please visit http://www.springer.com/series/7845 to view a high-resolution full-color version of this illustration.

In a second paper Vercoutere et al. [19] demonstrated the resolution of single-bp differences by using a kinetic analysis. They tested a series of 9-bp hairpin molecules with all possible single-bp differences at the terminus of the hairpin duplex stem. Analysis of the ionic current traces showed four conductance states (Fig. 7.5). These included an intermediate level (IL) that initiates all 9-bp hairpin events, an upper conductance level (UL), and a lower conductance level (LL) that must be preceded by the upper level, and spikes down from the lower level, which indicates close proximity of the terminal base pair to the pore-limiting aperture. To account for these states, it was proposed that the 9-bp hairpin can transiently bind to the amino acids in the vestibule wall near the limiting aperture of α-hemolysin. The IL state is caused by orientation and immobilization of the hairpin terminus due to electrostatic interactions between the terminal base pair and residues in the vestibule wall. The fact that the average dwell time for this intermediate state was largely independent of base pair identity or orientation also supports the idea that the interaction is electrostatic. The IL state invariably gave way to the upper conductance state (UL), which, according to the model, corresponds to desorption of the terminal base pair from the protein wall and thermal motion of the hairpin stem, allowing higher ion current to flow through the limiting aperture. After the UL state, the hairpin could return to IL or move on to LL. It is interesting that the residence time in this state is dependent on the terminal base pair identity and orientation (e.g., C–G and G–C would produce different residence times). In this state, both terminal nucleotides of the duplex could be adsorbed near the pore-limiting aperture and would strongly interact with the ionic current. When the duplex end is freed from the bound LL, one strand may extend and penetrate the aperture, producing the observed transient current spikes.

The characteristic current traces observed for different base pairs allowed discrimination between Watson and Crick base pairs at the termini of individual DNA hairpin molecules. The average dwell time in the LL level and the frequency of the spikes were highly dependent on the presence of a stable base pair at the duplex terminus. This principle was tested by using hairpins with dangling nucleotides in the 5′ or 3′ ends, as well as blunt-ended hairpins with non–Watson-Crick bases, and none of these hairpins showed any stable blockade signatures. By looking at the kinetic parameters (especially at the frequency of the spikes) and using the pattern recognition method previously described by Winters-Hill et al. [20], they were able to distinguish between A–T, T–A, G–C, and C–G.

Other investigators have also examined hairpin molecules in the hemolysin pore. Mathé et al. [21] studied the unzipping kinetics of individual DNA hairpins under constant force or constant loading rate (also called nanopore force spectroscopy or NFS). They used a relatively high voltage to increase the probability of capturing hairpins in the pore. The presence of the hairpin was detected by an abrupt decrease in current, and the hairpin was allowed to remain in the pore long enough to have the hairpin fully lodged in the vestibule. After this step, the voltage was changed to a lower potential that would hold the hairpin in the vestibule, but would not induce unzipping. Unzipping voltage or loading rate was then applied depending on whether constant force or constant loading rate was to be measured. It was found that the unzipping time of the hairpins decays exponentially with voltage, and this decay is independent of hairpin sequence.

Mathé et al. [22] further analyzed the unzipping process using NFS and found that the critical unzipping voltage is proportional to the logarithm of the voltage ramp at high ramp rates, while at low ramp values the dependence on ramp rate is weaker. This phenomenon could be understood in terms of a two-state model in which there is quasi-equilibrium unzipping at low-voltage ramps and irreversible unzipping at high ramp rates.

These examples clearly demonstrate the potential of a nanopore sensor to be used as an analytical tool. Its potential can also be exploited to characterize polymers other than nucleic acids. For instance, Robertson et al. [23] investigated poly(ethylene glycol) (PEG) molecules as analytes and observed that they produce mass-dependent conductance states with characteristic mean times. These authors used monodisperse PEG to calibrate the current intensities they obtained for different masses of PEG. As expected, they observed that higher PEG molecular masses blocked the current more than low-mass molecules.

Rajamani et al. [24] recently pioneered the use of a nanopore system in a basic research application. The aim was to determine whether RNA-like polymers could be synthesized from mononucleotides under conditions in which phospholipids produced an organizing matrix that promoted formation of phosphodiester bonds. The yields were expected to be very low, so it was essential to use a highly sensitive method capable of detecting individual molecules. Reaction products were presented to the hemolysin nanopore, and blockade signals were observed that could only be due to linear strands of polymers up to 100 nucleotides in length. The presence of long RNA-like strands was afterward confirmed with standard enzymatic labeling methods using ATP-32. As nanopore instruments become more widely available to investigators, it seems likely that such basic research applications will be increasingly employed.

7.4. DETECTION OF MACROMOLECULAR COMPLEXES

The first indication that a nanopore could detect enzyme activity was reported by Kasianowicz et al. [13], who used ribonuclease A to hydrolyze polyU molecules and observed that the number of blockades was dramatically increased as the enzyme produced multiple strands of shorter RNA. Other researchers have attempted to detect and characterize interactions of nucleic acids with macromolecules by taking advantage of the single-molecule sensitivity of nanopores. These approaches focus mainly on characterizing nucleic acid properties in a given molecular complex. For instance, Nakane et al. [25] developed a clever hybridization test that allowed detection of specific base sequences in the nanopore. In order to achieve this, a DNA strand was biotinylated at the 5′ end and a nucleotide sequence was designed to non-covalently bind a specific single-stranded oligonucleotide sequence at the 3′ end. The DNA strand was then bound to avidin by the biotin residue in the 5′ end, and the 3′ end of the DNA strand was captured in a hemolysin nanopore as described earlier. However, because the avidin was bound to the DNA strand, the latter could not translocate completely through the pore and instead was held in place until it paired with the complementary oligonucleotides in solution on the *trans* side of the pore. After binding, the applied potential was reversed, which after a time interval could force dissociation of the bound oligonucleotide. A longer time required for dissociation of the oligonucleotide duplex indicated that a duplex had in fact formed, and capture of the DNA itself was detected as a decrease to ~25% of the open-channel current. When the decrease was detected, the potential was then reduced to 10 mV. If the DNA strand was bound to an oligonucleotide on the *trans* side of the pore, its exit was prevented and the whole probe remained in the pore. Otherwise, it diffused away from the pore. After binding of the oligonucleotide to the probe was detected, the potential was reversed to –60 mV, which caused the probe to withdraw from the pore. The time required for the whole molecular complex to leave the pore was measured, and statistical analysis of many dissociation event lifetimes at different reverse potentials yielded identifying characteristics of the oligonucleotides. This technique proved to be sensitive enough to discriminate between molecules differing by a single nucleotide.

Hornblower et al. [26] explored protein–DNA interactions by applying electrical force to individual ssDNA–exonuclease I complexes. This force was sufficient to dissociate the complex when the electrical potential was applied. When a complex formed by streptavidin or avidin and biotinylated ssDNA is captured in the pore, the polynucleotide cannot translocate through the pore because the protein bound to the ssDNA is larger than the diameter of the pore. This effect produces current blockades that can be seconds in duration. Hornblower et al. [26] proposed that weaker protein–DNA complexes would dissociate in the pore under the applied voltage, resulting in shorter translocation times than in the case of avidin-biotynilated DNA. The use of Exonuclease I allowed them to test this hypothesis. When an ssDNA 64-mer by itself was captured in the pore, the blockades had a characteristic (most probable) translocation time of 66 ± 10 μs. However, when Exonuclease I was added, a new population appeared with a characteristic time of 380 ± 50 μs that was due to the capture and retention of the ssDNA–Exonuclease I complex until it dissociated.

The activity of the Exonuclease I is Mg^{2+} dependent. Upon addition of 5 mM Mg^{2+} to the medium, the frequency of translocation events was markedly reduced. However, when a 3′-phosphorylated DNA oligomer was used, the frequency of translocations did not change, consistent with the previous data showing that phosphorylation of the 3′ end prevents Exonuclease I from hydrolyzing the DNA. In addition, when Mg^{2+} was added to the ssDNA modified this way, a new population characterized by a dwell time of around 10 ms appeared, indicating a significantly stronger binding of DNA to the active site of Exonuclease I.

Hornblower et al. [26] also used nanopore force spectroscopy to probe the dissociation kinetics of the Exonuclease I–ssDNA. When an Exonuclease I–ssDNA complex was captured, a voltage ramp was applied to the pore as previously described by Mathé et al. [21,22]. At a given voltage value, the Exonuclease I–ssDNA complex dissociated (Fig. 7.6) and the distribution of voltage values at which dissociation occurred was obtained. These measurements could also be used to calculate the dissociation equilibrium constant as well as the rate constants for association and dissociation of the ssDNA to the Exonuclease I.

In a similar approach, Astier et al. [27] used α-hemolysin pores to study the RNA-binding ATPase P4, a viral packaging motor from bacteriophage ϕ8. They employed an oligoribonucleotide that produced blockades with an average duration of 0.5 ms. When the ATPase P4 was added, events appeared that were one or two orders of magnitude longer in duration. Furthermore, the long events disappeared when ATP was added, indicating that the P4–RNA complex dissociated as the DNA moved through the P4 and the nanopore. The frequency of events depended on the concentration of P4 as well as on the length of the oligoribonucleotide.

Benner et al. [28] studied the effect of the Klenow fragment (KF) of *Escherichia coli* DNA polymerase I on nucleic acids. These investigators were able to discriminate among unbound DNA, binary DNA/polymerase complexes, and ternary DNA/polymerase/deoxynucleotide triphosphate complexes, each producing blockades having characteristic duration times, amplitudes, and modulation within the blockade signal. The DNA was a hairpin with a 2′–3′ dideoxycytidine terminus. The most striking result relates to the ability of the nanopore system to distinguish between the nonspecific nucleotides and nucleotides that were complementary to the template in the catalytic site of the KF. The correct dNTPs would have produced the extension of a regular DNA template, but since the hairpin DNA contained a dideoxy terminus it could not be extended. However, addition of the correct dNTP resulted in the formation of a ternary complex with higher stability, which produced blockades with a median duration much longer than the duration corresponding to the molecular complex with incorrect dNTPs.

Most recently, Cockroft and coworkers [29] have demonstrated that the hemolysin nanopore has sufficient resolution to attain base-by-base resolution of the action of a DNA polymerase. They used a DNA–PEG construct containing a terminal biotinyl group on the PEG

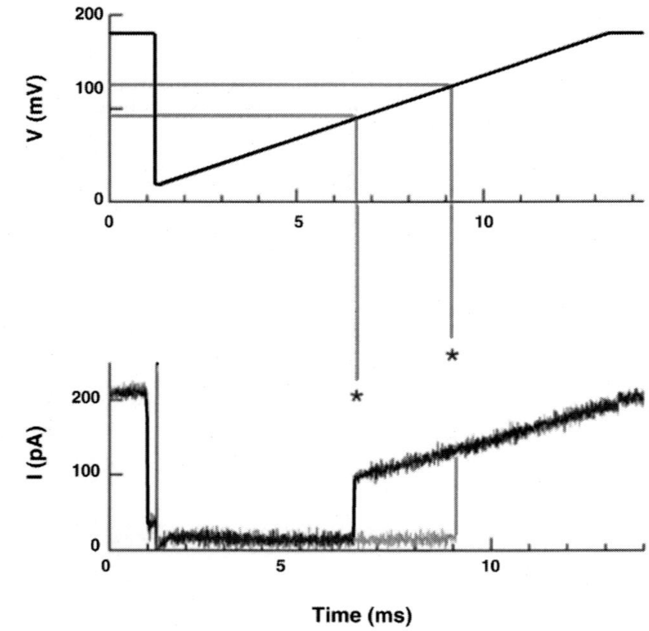

Figure 7.6. Current and voltage profiles during two typical events in a nanopore voltage-ramp force spectroscopy experiment. Plotted are the time-dependent voltage applied to each individual complex (top) and the measured pore current (bottom). Dissociation of the complexes results in an abrupt rise of the current (asterisks) to the open-pore level. Reprinted with permission from [26]. Copyright © 2007, Nature Publishing Group.

side. After threading the construct through the α-hemolysin nanopore, they formed an interlocked complex with the addition of streptavidin on one side and a primer complementary to the DNA segment on the other. The I–V traces were calibrated to establish the response of the system to changes in the primer length and used the voltage that maximized the difference in current intensity. It was then possible to monitor the DNA polymerase as it catalyzed base-by-base elongation of the primer. The voltage was alternated from positive, where the length of the primer was estimated according to the calibration, to negative, where extension of the primer by the DNA polymerase was allowed. This alternation was required because the terminal 3′-OH of the primer, which serves as the site for DNA polymerase-catalyzed incorporation of deoxynucleotide triphosphates (dNTPs), was concealed within the pore vestibule when the positive voltage was applied and the double-stranded DNA was sitting on the pore.

The results described above show that integration of DNA processing enzymes with nanopore analysis of DNA significantly enhances the potential applications of nanopore analytical systems. Enzymes can detect and discriminate in real time between different mononucleotides at the single-molecule level of resolution, and this can be exploited for next-generation sequencing technology.

7.5. CONCLUSIONS AND FUTURE PROSPECTS

Nanopore technology has advanced considerably since Kasianowicz et al. [13] published the first paper suggesting that nanopores could be used for sequencing DNA. Progress since that time has demonstrated that single-nucleotide and single-bp differences can be detected in individual nucleic acid molecules. The motion of single nucleic acid molecules can be monitored in real time, including transient unzipping of blunt-ended dsDNA captured in the vestibule of the pore. Hairpins that differ by a single base pair can be distinguished, as well as molecules having single nucleotide mismatches.

In addition, the use of enzymes that interact with nucleic acids in the nanopore has provided the means to extract more information about single nucleic acids as well as enzyme–nucleic acid complexes. The activity of the enzymes can be followed in real time as the nucleic acids that they process translocate through the pore. This approach may eventually provide the approach that will lead to nanopore sequencing of DNA and RNA, because many nucleic acid processing enzymes do actually discriminate between different single nucleotides as demonstrated by Benner et al. [28]. Once this is achieved, individual genomes may routinely be sequenced and a new horizon will open for biomedical research and personal diagnostic procedures in health care.

7.5.1. Limitations of nanopore biosensors

Despite the progress toward achieving direct sequencing with a nanopore instrument, there are still several significant problems that need to be resolved. Although single-nucleotide differences can be detected by the nanopore, it is still not possible to obtain sequential information about the nucleotides of a given nucleic acid molecule. This is in part due to the fact that nucleic acid molecules translocate too fast through the nanopore to be resolved by modulations in ionic current blockades. Another limitation concerns the stability and reproducibility of the hemolysin pore. The hemolysin heptamer is robust, but the supporting lipid bilayer is not. Considerable effort is now being made by several laboratories to increase stability. For instance, Henry White's group at the University of Utah has produced bilayers on silanized glass supports that are stable for days [39]. The supporting apertures are a micron or less in diameter, and the small size appears to be an important factor in increasing overall stability.

7.5.2. Future Prospects

Future research should be directed toward remedying the problems described in the previous section, and particularly those that limit single-base resolution. A promising approach to increasing resolution of a nanopore is to modify its structure. For instance, Astier et al. [30] fabricated a mutant α-hemolysin pore containing an arginine ring that can accommodate a cyclodextrin adapter in the lumen. The cyclodextrin ring constrains the pore diameter, thus reducing the current going through but also allowing detection and identification of single mononucleotides. This finding can be extended to detection of enantiomers of chiral drug molecules such as ibuprofen and thalidomide [31], and recently, the cyclodextrin adapter has been covalently bound to the α-hemolysin [32]. The covalent bond has the advantage of preventing dissociation of the adapter from the pore, which will permit continuous detection of mononucleotides required for sequencing.

Although the hemolysin pore has led the way for developing nanopore analysis, the fact remains that biological pores are stable only under certain conditions of pH, ionic strength, and temperature. For this reason, several laboratories are exploring solid state nanopores as plausible alternatives. (See Dekker [33] for a comprehensive review.) Synthetic nanopores can be produced by drilling nanometer-sized holes through very thin membranes of SiN and SiO_2 with a focused ion beam [33]. The diameter of the pores is then reduced by exposing them to Ar^+ ion beams [34]. Although synthetic nanopores have their own limitations, research is progressing toward solving critical issues like noise reduction [35]. Solid-state nanopores can already detect dsDNA and ssDNA [36] as well as aid in determining the conformation and base number of nucleic acid polymers [37].

Synthetic nanopores do not depend on stochastic processes to form the pore, because they are physical holes in a solid matrix that should in principle be stable for an indefinite period of time. It may even be possible that a hybrid nanopore could be fabricated in which a solid-state pore is chemically functionalized [38] to allow insertion of a hemolysin heptamer.

An increasingly important aspect would be to incorporate electronic control circuits that would allow a single molecule to be manipulated in the nanopore. One such computational approach was described recently by Benner et al. [28], and in earlier papers by Bates et al. [41] and by Gershaw and Golovchenko [39] for solid-state nanopores as well. It seems increasingly obvious that active control systems will be required to achieve sequencing, as well as providing a novel method for investigating a variety of substrates by nanopore analysis of single-molecule structure and function.

PROBLEMS

7.1. Why is ionic current reduced when a single-stranded nucleic acid molecule traverses a nanopore?

7.2. A homopolymer molecule such as polydeoxycytidylic acid, 100 bases in length, requires an average of 0.2 ms to traverse the pore.
   a. How much time does an individual base spend in the pore?
   b. Given that this time is too short for identification of individual bases, what can be done to slow down the translocation velocity?

FURTHER STUDY

Bockris JO'M, Reddy AKN. 1998. *Modern electrochemistry*. Dordrecht: Kluwer Academic.
Saenger, W. 1984. *Principles of nucleic acid structure*. Berlin: Springer-Verlag.
Sakmann B, Neher E. 1995. *Single-channel recording*. New York: Plenum Press.
Deamer D, Branton D. 2002. Characterization of nucleic acids by nanopore analysis. *Acc Chem Res* **35**:817–825.

REFERENCES

1. Nakane J, Akeson M, Marziali A. 2002. Evaluation of nanopores as candidates for electronic analyte detection. *Electrophoresis* **23**:2592–2601.
2. Bezrukov SM. 2000. Ion channels as molecular Coulter counters to probe metabolite transport. *J Membr Biol* **174**:1–13.
3. Healy K. 2007. Nanopore-based single-molecule DNA analysis. *Nanomedicine* **2**(4):459–481.
4. Halverson KM, Panchal RG, Nguyen TL, Gussio R, Little SF, Misakian M, Bavari S, Kasianowicz JJ. 2005. Anthrax biosensor, protective antigen ion channel asymmetric blockade. *J Biol Chem* **280**(40):34056–34062.
5. Akeson M, Branton D, Kasianowicz JJ, Brandin E, Deamer DW. 1999. Microsecond time-scale discrimination among polycytidylic acid, polyadenylic acid, and polyuridylic acid as homopolymers or as segments within single RNA molecules. *Biophys J* **77**:3227–3233.
6. Chandler EL, Smith AL, Burden LM, Kasianowicz JJ, Burden DL. 2004. Membrane surface dynamics of DNA-threaded nanopores revealed by simultaneous single-molecule optical and ensemble electrical recording. *Langmuir* **20**:898–905.
7. Shenoy DK, Barger WR, Singh A, Pnachal RG, Misakian M, Standford VM, Kasianowicz JJ. 2005. Functional reconstitution of protein ion channels into planar polymerizable phospholipid membranes. *Nano Lett* **5**(6):1181–1185.
8. Malmstadt N, Nash MA, Purnell RF, Schmidt J. 2006. Automated formation of lipid-bilayer membranes in a microfluidic device. *Nano Lett* **6**(9):1961–1965.
9. Pastoriza-Gallego M, Oukhaled G, Mathé J, Bénédicte T, Jean-Michel B, Auvraya LC, Pelta J. 2007. Urea denaturation of alpha-hemolysin pore inserted in planar lipid bilayer detected by single nanopore recording: loss of structural asymmetry. *FEBS Lett* **581**:3371–3376.
10. Bonthuis DJ, Zhang J, Hornblower B, Mathé J, Shklovskii BI, Meller A. 2006. Self-energy-limited ion transport in subnanometer channels. *Phys Rev Lett* **97**:128104.
11. Meller A, Branton D. 2002. Single molecule mesurements of DNA transport through a nanopore. *Electrophoresis* **23**:2583–2591.
12. Song L, Hobaugh MR, Shustak C, Cheley S, Bayley H, Gouaux EJ. 1996. Structure of staphylococcal α-hemolysin, a heptameric transmembrane pore. *Science* **274**(5294):1859–1866.
13. Kasianowicz JJ, Brandin E, Branton D, Deamer DW. 1996. Characterization of individual polynucleotide molecules using a membrane channel. *Proc Natl Acad Sci USA* **93**:13770–13773.
14. Butler TZ, Gundlach JH, Troll MA. 2007. Ionic current blockades from DNA and RNA molecules in the α-hemolysin nanopore. *Biophys J* **93**:3229–3240.
15. Meller A, Nivon L, Brandin E, Golovchenko J, Branton D. 2000. Rapid nanopore discrimination between single polynucleotide molecules. *Proc Natl Acad Sci USA* **97**(3):1079–1084.
16. Butler TZ, Gundlach JH, Troll MA. 2006. Determination of RNA orientation during translocation through a biological nanopore. *Biophys J.* **90**:190–199.
17. Mathé J, Aksimentiev A, Nelson DR, Shulten K, Meller A. 2005. Orientation discrimination of single-stranded DNA inside the α-hemolysis membrane channel. *Proc Natl Acad Sci USA* **102**(35):12377–12382.
18. Vercoutere W, Winters-Hilt S, Olsen H, Deamer D, Haussler D, Akeson M. 2001. Rapid discrimination among individual DNA hairpin molecules at single-nucleotide resolution using an ion channel. *Nat Biotechnol* **19**:248–252.
19. Vercoutere WA, Winters-Hilt S, DeGuzman VS, Deamer D, Ridino SE, Rodgers JT, Olsen HE, Marziali A, Akeson M. 2003. Discrimination among individual Watson-Crick base pairs at the termini of single DNA hairpin molecules. *Nucleic Acids Res* **31**(4):1311–1318.
20. Winters-Hilt S, Vercoutere W, DeGuzman VS, Deamer D, Akeson M, Haussler D. 2003. Highly accurate classification of Watson-Crick basepairs on termini of single DNA molecules. *Biophys J* **84**:967–976.
21. Mathé J, Visram H, Viasnoff V, Rabin Y, Meller A. 2004. Nanopore unzipping of individual DNA hairpin molecules. *Biophys J* **87**:3205–3212.
22. Mathé J, Arinstein A, Rabin Y, Meller A. 2006. Equilibrium and irreversible unzipping of DNA in a nanopore. *Europhys Lett* **73**(1):128–134.
23. Robertson JWF, Rodrigues CG, Standford VM, Rubinson KA, Krasilnikov OV. 2007. Single-molecule mass spectrometry in solution using a solitary nanopore. *Proc Natl Acad Sci USA* **104**(20):8207–8211.

24. Rajamani S, Vlassov A, Benner S, Coombs A, Olasagasti F, Deamer D. 2008. Lipid-assisted synthesis of RNA-like polymers from mononucleotides. *Orig Life Evol Biosph* **38**:57–74.
25. Nakane J, Wiggin M, Marziali A. 2004. A nanosensor for transmembrane capture and identification of single nucleic acid molecules. *Biophys J* **87**:615–621.
26. Hornblower B, Coombs A, Whitaker RD, Kolomeisky A, Picone SJ, Meller A, Akeson M. 2007. Single-molecule analysis of DNA-protein complexes using nanopores. *Nat Methods* **4**(4):315–317.
27. Astier Y, Kainov DE, Bayley H, Tuma R, Howorka S. 2007. Stochastic detection of motor protein-RNA complexes by single-channel current recording. *ChemPhysChem* **8**:2189–2194.
28. Benner S, Chen RJA, Wilson NA, Abu-Shumays R, Hurt N, Lieberman KR, Deamer DW, Dunbar WB, Akeson M. 2007. Sequence-specific detection of individual DNA polymerase complexes in real time using a nanopore. *Nat Nanotechnol* **2**:718–724.
29. Cockroft SL, Chu J, Amorin M, Ghadiri MR. 2008. A single-molecule nanopore device detects DNA polymerase activity with single-nucleotide resolution. *J Am Chem Soc* **130**(3):818–820.
30. Astier Y, Braha O, Bayley H. 2006. Toward single molecule DNA sequencing: direct identification of ribonucleoside and deoxyribonucleoside 5′-monophosphates by using an engineered protein nanopore equipped with a molecular adapter. *J Am Chem Soc* **128**:1705–1710.
31. Kang X-F, Cheley S, Guan X, Bayley H. 2006. Stochastic detection of enantiomers. *J Am Chem Soc* **128**:10684–10685.
32. Wu H-C, Astier Y, Maglia G, Mikhailova E, Bayley H. 2007. Protein nanopores with covalently attached molecular adapters. *J Am Chem Soc* **129**:16142–16148.
33. Li J, Stein D, McMullan C, Branton D, Aziz MJ, Golovchenko J. 2001. Ion-beam sculpting at nanometre length scales. *Nature* **412**:166–169.
34. Mitsui T, Stein D, Kim Y-R, Hoogerheide D, Golovchenko JA. 2006. Nanoscale volcanoes: accretion of matter at ion-sculpted nanopores. *Phys Rev Lett* **96**:036102.
35. Tabard-Cossa V, Trivedi D, Wiggin M, Jetha NN, Marziali A. 2007. Noise analysis and reduction in solid-state nanopores. *Nanotechnology* **18**:305505.
36. Fologea D, Gershow M, Ledden B, McNabb DS, Golovchenko JA, Li J. 2005. Detecting single stranded DNA with a solid state nanopore. *Nano Lett* **5**(10):1905–1909.
37. Fologea D, Brandin E, Uplinger J, Branton D, Li J. 2007. DNA conformation and base number simultaneously determined in a nanopore. *Electrophoresis* **28**:3186–3192.
38. Wanunu M, Meller A. 2007. Chemically modified solid-state nanopores. *Nano Lett* **7**(6):1580–1585.
39. Gershow M, Golovchenko JA. 2007. Recapturing and trapping single molecules with a solid-state nanopore. *Nat Nanotechnol* **2**:775–779.
40. Smeets RMM, Keyser UF, Krapf D, Wu M-Y, Dekker NH, Cees D. 2006. Salt dependence of ion transport and DNA translocation through solid-state nanopores. *Nano Lett* **6**(1):89–95.
41. Fologea D, Uplinger J, Thomas B, McNabb D, Li J. 2005. Slowing DNA translocation in a solid-state nanopore. *Nano Lett* **5**(9):1734–1737.

8

COMPLEX APPLICATIONS OF SIMPLE FRAP ON MEMBRANES

Minchul Kang[1] and Anne K. Kenworthy[1,2]
[1]Department of Molecular Physiology and Biophysics, [2]Department of Cell and Developmental Biology, Vanderbilt University School of Medicine, Nashville, Tennessee

8.1. INTRODUCTION

There is an old adage that says "To see is to believe," and it still seems to be true in many fields of biology. For an experimental validation of hypotheses, modern biology takes advantage of various fluorescence-based techniques (fluorescence microscopy, digital image analysis) for visualization and quantification. Fluorescence Recovery after Photobleaching (FRAP), a widely used fluorescence-based technique to visualize and quantify the diffusion of fluorescently tagged proteins, is one good example.

The first observation of fluorescence was made by Sir John Frederick William Herschel (1845) from a quinine solution, and the concept of fluorescence was first named after fluorite by Sir George G. Stokes (1857). On a microscopic scale, fluorescence was first observed by August Köhler (1904), who discovered that a biological tissue could autofluoresce under ultraviolet light irradiation. Later, it became popular in the biological field after M. Haitinger (1933) succeeded in staining histological specimens with fluorescent dyes for the first time, which is called the technique of secondary fluorescence, distinguishing it from autofluorescent tissue previously observed by M. Haitinger. Haitinger and others extended the application of the technique of secondary fluorescence to stain not only specific tissues but also bacteria, or other pathogens that are not autofluorescent [1]. Although the technique of secondary fluorescence demonstrated that nonfluorescent cells can be made fluorescent, it was a nonspecific

Address correspondence to Anne K. Kenworthy, Department of Molecular Physiology and Biophysics, Vanderbilt University School of Medicine, 718 Light Hall, Nashville, TN 37232, (615) 322-6615, (615) 322-7236 (fax), <anne.kenworthy@vanderbilt.edu>.

R. Faller et al. (eds.), *Biomembrane Frontiers: Nanostructures, Models, and the Design of Life*, Handbook of Modern Biophysics, DOI: 10.1007/978-1-60761-314-5_8,

staining technique. The breakthrough in a specific immunofluorescence staining technique was provided by Albert Coons in 1941 by attaching a fluorescent dye to an antibody [2]. Later, Coons and N.H. Kaplan developed the fluorescein isothiocyanate (FITC) immunofluorescence technique [3].

Although August Köhler invented the ultraviolet microscope to discover autofluorescent tissues for the first form of fluorescent microscope in 1904, it was not until the 1950s that a primitive form of confocal microscope was fashioned, in which light was collected from the same focal position of the specimen by spatial filtering by a pinhole, thereby eliminating out-of-focus light from specimens that are thicker than the plane of focus [4]. The discovery of green fluorescent protein (GFP) from the jellyfish *Aequorea victoria* in 1961 [5], which can be fused to virtually any protein of interest, added momentum to fluorescence microscopy after it was cloned [6] and shown that it can be expressed in a variety of cell types [7,8].

Although fluorescence microscopy is often used to collect static images of fixed cells, it is also capable of following dynamic events. In living cells, one of the most prevalent ways to transport molecules is by diffusion or Brownian motion. The rate of transport by diffusion is largely dependent on the diffusion coefficient of the molecule, which is a function of the size and shape of the molecule as well as the viscosity and temperature of the medium in which it is localized. FRAP experiments provide a way to measure diffusion of fluorescently labeled molecules by using a fluorescence microscope to monitor how rapidly the fluorescence intensity (FI) recovers in a small region after photobleaching by a high-intensity laser. FRAP was first developed in the 1970s to study the diffusion of proteins in cell membranes [9–13]. With the development of GFP fusion proteins and confocal microscopes with the capacity to perform controlled photobleaching, FRAP has become one of the most widely used techniques in fluorescence microscopy and is now used to study diffusion in a variety of intracellular compartments [14–18]. In this chapter we discuss the theoretical basis of FRAP, with an emphasis on the mathematical derivations of FRAP formulae, special considerations for confocal FRAP measurements, and the equations used to analyze normal diffusion, anomalous diffusion, and diffusion-binding kinetics in biological membranes.

8.2. PRINCIPLES OF FLUORESCENCE

8.2.1. Basic Concepts in Fluorescence

8.2.1.1. Electromagnetic Radiation

Electromagnetic radiation (or light) is a self-propagating energy wave in space with electric and magnetic field components, which are orthogonal to the direction of propagation and also orthogonal to each other. Paradoxically, electromagnetic radiation shows both wave and particle characteristics depending on how the radiation is observed, which is known as the wave–particle duality of electromagnetic radiation. Because of the wave-like property, each electromagnetic radiation has its own unique wavelength and frequency [19,20]. However, despite the differences in wavelength and frequency, all electromagnetic radiation travels at the constant speed of light $c = 2.99792 \times 10^8$ m/s in vacuum, which provides the relation between the speed of light (c, m/s), wavelength (λ, nm), and frequency (ν, Hz) as

$$c = \lambda \cdot \nu \,. \tag{8.1}$$

On the other hand, the particle characteristic of electromagnetic radiation implies light would then consist of particles (or quanta) called photons, the energy and momentum (yet, no mass) carrier of electromagnetic radiation of all wavelengths. It is not hard to imagine that rapidly oscillating photons have more energy $(E \propto \nu)$,

$$E = h \cdot \nu = h \cdot (c / \lambda) \tag{8.2}$$

for some constant h, which is now known as Planck's constant (h = 6.62618 × 10^{-34} J·s). The frequencies of electromagnetic radiation not only provide a way to measure the energy of photons in light, but also a convenient way to categorize electromagnetic radiation as radiowaves, microwaves, terahertz radiation, infrared radiation, visible light, ultraviolet radiation, X-rays, and gamma rays (Fig. 8.1).

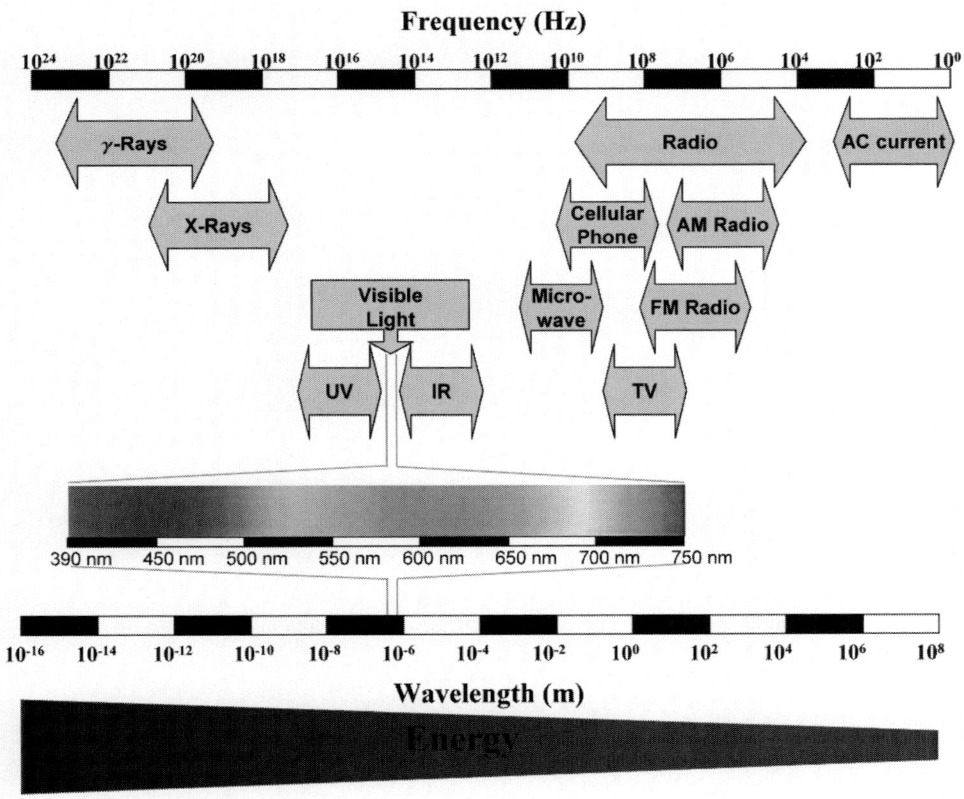

Figure 8.1. Electromagnetic radiation. The wavelength (λ, nm) and frequency (ν, Hz) are related by $c = \lambda \cdot \nu$, where c is the speed of light. From the equation $E = h \cdot \nu = h(c / \lambda)$, the energy level is higher for electromagnetic radiation with higher frequencies or shorter wavelengths. The range of electromagnetic radiation that is visible to the naked eye is called visible light. Please visit http://www.springer.com/series/7845 to view a high-resolution full-color version of this illustration.

8.2.1.2. Fluorescence, Fluorophores

Fluorescence is also a type of electromagnetic radiation from a certain type of substance called a fluorophore that is stimulated by absorption of light, mechanical friction, or a chemical reaction. Among these sources of stimuli, the absorption of light will be our main interest in fluorescence microscopy. Typically, a laser in the ultraviolet to blue-green wavelength is used for absorption radiation.

Origin of fluorescence, fluorescence lifetime, and the Jablonski energy diagram

In atomic physics, it is proposed that atoms have electron shells (main energy levels), a group of atomic orbitals around their nuclei. Electrons of an atom occupy these electron shells according to a certain probabilistic distribution, and this electron configuration of electrons and structure of electron shells defines an atom uniquely [19].

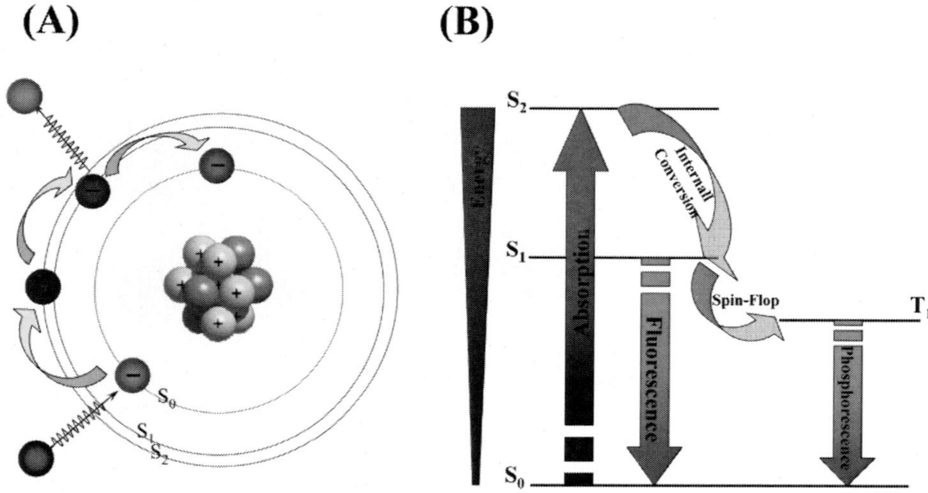

Figure 8.2. Origin of fluorescence and the Jablonski Energy Diagram. See text for details. Please visit http://www.springer.com/series/7845 to view a high-resolution full-color version of this illustration.

The fluorescence process is usually illustrated by a Jablonski energy diagram as the transitions of an electron between the different energy levels (Fig. 8.2). According to atomic physics theory, each electron of an atom is characterized by a unique set of quantum numbers — principal, angular momentum, spin, and magnetic. Any two electrons in the same orbital have the same principal, angular momentum, and magnetic quantum number but their spins have different signs (spin-paired). If all the electrons in an atom are spin-paired, the atoms are said to be in a singlet state, while they are referred to as residing in a triplet state if one set of electron spins is unpaired. Singlet states may have different energy levels, as do triplet states. In ascending order, singlet states of different energy level are termed ground state singlet level (S_0) and excited singlet level (S_i, $i = 1, 2, \ldots$). Similarly, excited triplet states are named as T_i, ($i = 1, 2, \ldots$). Note that the lowest energy levels of singlet and triplet states are S_1 and T_1, and that

Energy (S_0) < Energy (T_1) < Energy (S_1) < Energy (T_2) < Energy (S_2) holds between the energy levels of each state [19].

A simplified Jablonski energy diagram is shown in Figure 8.2. When the electrons in the ground singlet state of a fluorophore (S_0) absorb electromagnetic energy from incident photons (absorption), they are promoted to an inherently unstable higher energy level (S_1 or S_2) in approximately 10^{-15} s. A molecule in a high vibrational level of the excited state (S_1 or S_2) will quickly fall to the lowest excited singlet level (S_1) by losing energy to other molecules through collision, with a few exceptions. This process is called vibrational relaxation and occurs on the order of 10^{-12} s. Molecules in the S_1 state are eventually stabilized when the excited electrons return to their original energy level (S_0) with a certain probability based on the decay rates through a number of different decay pathways ($S_1 \rightarrow S_0$ or $S_1 \rightarrow T_1 \rightarrow S_0$) by emitting energy in the form of electromagnetic radiation (emission). When molecules undergo the $S_1 \rightarrow S_0$ decay pathway, the difference in energy between states S_1 and S_0 is released in the form of visible light (fluorescence). This occurs on the order of 10^{-8} s, which is known as a typical fluorescence lifetime.

In an ensemble description, fluorescence emission can be described as a simple exponential model with a lifetime typically of the order of 10^{-8} s, the fluorescence lifetime. In some cases, molecules in the lowest excited singlet state (S_1) can transit to the lowest excited triplet state (T_1) by a spin-flop, and eventually return the ground singlet state (S_0). Similar to fluorescence, the energy difference between T_1 and S_1 is released as visible light (phosphorescence). However, due to the delay in $S_1 \rightarrow T_1$ by a spin-flop, phosphorescence processes occur on the order of 10^{-3}~10^{2} s, which is long enough to generate emission after exposure to the excitation source is ceased. Different from phosphorescence, fluorescence stops immediately after the excitation source is removed.

Spectra, Stokes's law or Stokes shift, and quantum yield

As already mentioned, August Köhler discovered that a biological tissue could autofluoresce under ultraviolet light irradiation. This observation indicates two important points. First of all, autofluorescent biological tissues respond only to specific types of light (ultraviolet light), and, second, there exists an energy difference between the light autofluorescent tissue absorbs (ultraviolet) and emits (visible).

Fluorescence arises from the relaxation of excited electrons after molecules absorb light that excites the ground-state singlet level to a higher energy level. Because each fluorophore species has a unique quantized energy structure, each has unique and characteristic spectra for absorption. This characteristic absorption of light of a fluorophore species as a function of wavelength is called the absorption spectrum of the fluorophore (Fig. 8.3). Similarly, the characteristic emission of light (fluorescence) of a fluorophore species as a function of wavelength is called the emission spectrum.

As is common in many examples from physics, not all absorbed energy from light is converted into emitted light. There are many atomic-level processes that are responsible for this energy loss, including a loss of vibrational energy due to the vibrations of a chemical bond in a molecule. As a result, emission light has longer wavelengths than the absorption, and the emission spectrum is located to the right of the absorption spectrum. This phenomenon is known as Stokes's Law or a Stokes shift, the difference (in wavelength or frequency units) between positions of the band maxima of the absorption and emission spectra.

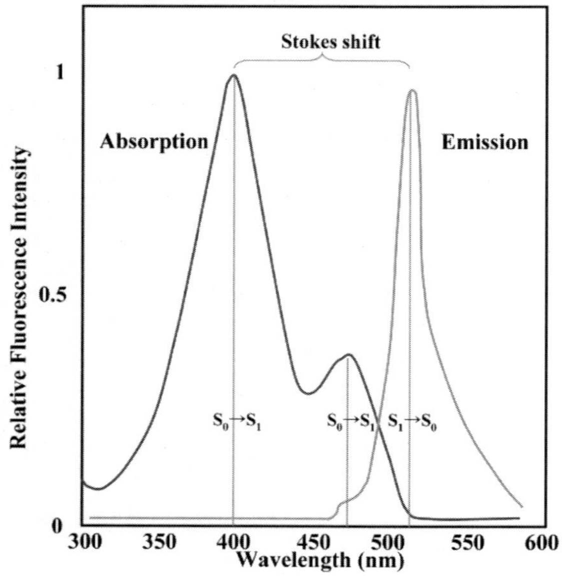

Figure 8.3. Absorption, emission spectra, and Stokes shift. See text for details. Please visit http://www.springer.com/series/7845 to view a high-resolution full-color version of this illustration.

When a fluorophore is excited by photons within its absorption spectrum, not all the photons are used to excite the electrons in the fluorophores. Some of them participate in non-radiative pathways such as heat or photochemical reactions. The ratio of the number of photons that follow a radiative pathway to the number of photons that entered the atomic system is different from one fluorophore to another, and this ratio is defined as the quantum yield of the fluorophore. Quantum yield typically ranges between 0.1 and 0.9.

8.2.1.3. Photobleaching (Fading)

Photobleaching is a phenomenon in which a fluorophore permanently loses its ability to undergo fluorescence emission due to photochemical destruction and covalent modification, usually when fluorophores interact with molecular oxygen before emission. Although the underlying mechanism of photobleaching is not yet fully understood, it is believed to occur when the excited singlet state of a fluorophore is transited to the excited triplet state that is relatively long lived and is chemically more reactive.

The ability of a fluorophore to undergo light absorption and fluorescence emission is not infinite. The characteristic number of excitation and emission cycles a fluorophore can undergo is fluorophore-specific and determines how long it can fluoresce under a given intensity of excitation light. Individual fluorophores undergo an abrupt transition from fluorescent to photobleached. However, prolonged illumination of a population of fluorophore molecules leads to progressive fading of the fluorescence intensity as a function of time. In general, the higher the excitation light intensity, the shorter time a given fluorophore emits before photobleaching. As a result, above a certain level of excitation laser intensity fluorophores are photobleached almost immediately, which is a major tool for fluorescence recovery after photobleaching (FRAP).

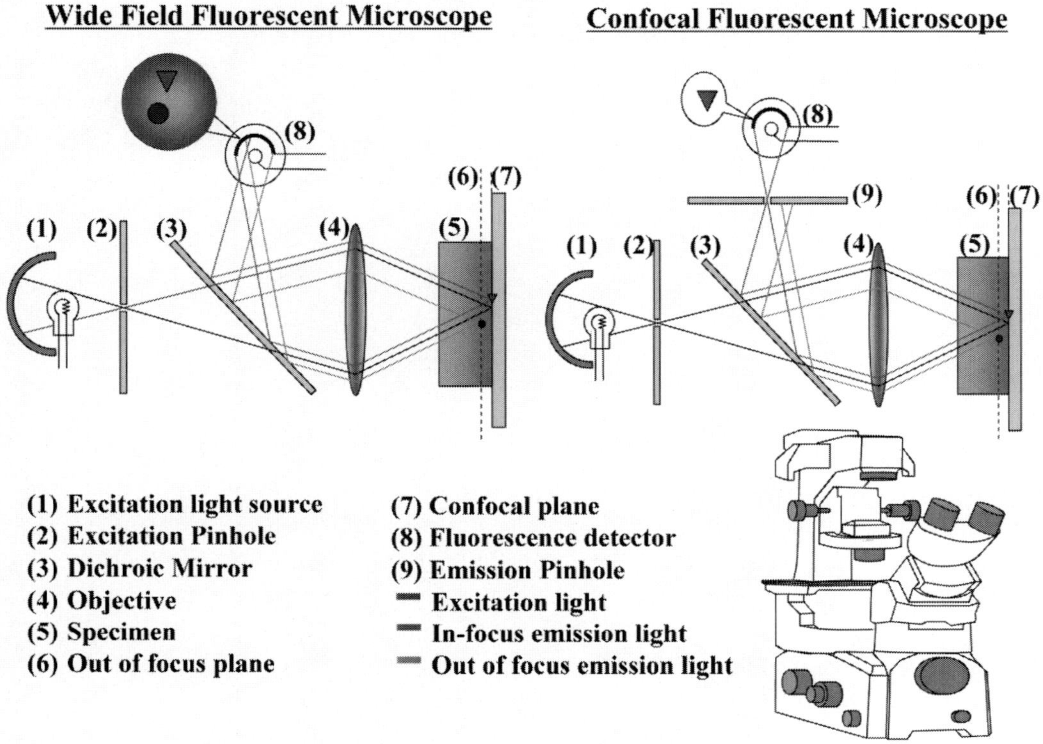

Figure 8.4. Wide-field fluorescence microscope vs. confocal fluorescence microscope. Redrawn from the patent by Marvin Minsky (1957). An example of a modern LSCM system is also shown (bottom right). Please visit http://www.springer.com/series/7845 to view a high-resolution full-color version of this illustration.

8.2.2. Laser Scanning Confocal Microscopes (LSCMs)

8.2.2.1. Comparison with a Wide-Field Fluorescence Microscope

Biological specimens or cells are three-dimensional objects, and often contain fluorescently labeled molecules distributed throughout their volume. When they are illuminated by excitation light, all the fluorescent molecules in the light path emit fluorescence. In a wide-field microscope, the fluorescence from the out-of-focus volume interferes with the fluorescence from the focal plane and the resulting fluorescent image is blurred, losing detailed information. However, a confocal microscope takes the thickness of specimens into consideration and provides a mechanism for rejection of out-of-focus fluorescence signals.

The small yet fundamental difference in optics that enables a confocal microscope but not a wide-field microscope to eliminate fluorescence from the out-of-focus volume is illustrated in Figure 8.4. When excitation light is turned on (Fig. 8.4(1)), the laser narrowed by the excitation pinhole (Fig. 8.4(2)) passes through the dichroic mirror (Fig. 8.4(3)). At the dichroic mirror, only the light with specific wavelengths (excitation light: blue line) can pass through, while light of other wavelengths (emission fluorescent light: green/gray lines) is reflected. Once the excitation light passes the dichroic mirror, it is refracted by an objective lens (Fig. 8.4(4)) and is focused in the specimen (Fig. 8.4(7)). Now, after absorbing the excitation light, the fluorophores emit fluorescence, which travels back through the objective lens (Fig. 8.4(4)). Due to the Stokes shift, the wavelength of emission fluorescence is longer than that of the excitation light and the fluorescence is reflected on the dichroic mirrors (Fig. 8.4(3)) in the direction of a detector device. Note that not only the fluorophores in the focal plane but also those in the out-of-focus volume emit fluorescence (Fig. 8.4(6)). In a wide-field fluorescence microscope, this out-of-focus fluorescence signal is collected by the detector (Fig. 8.4(8)). Because the detected fluorescence image in wide-field microscopy is due to all the photons emitted in the illuminated volume (wide field), not only the fluorescent molecules on the focal plane (∇ in Fig. 8.4) but also fluorescent molecules in the out-of-focus volume (o in Fig. 8.4) show up. Consequently, the images are bright due to the large number of photons reaching the detectors, but the images are also blurred.

However, in a confocal microscope, an emission pinhole only lets photons from the focal plane pass through it, blocking others from outside of the focal volume (Fig. 8.4(9)). By doing this, a confocal microscope has the "same focus" (confocal) as the focus that corresponds to the point on the focal plane. Because only the photons from one point (the focus) on the focal plane pass the emission pinhole, the image is restricted to a small region near the focus on the focal plane, but is also dim. To overcome this drawback, the confocal microscope adopts photon multiplier tubes (PMTs), which amplify the number of photons detected to obtain brighter and sharper images. In addition, in order for a detector to receive a satisfactory level of photons, powerful light sources with ample photons are required. Accordingly, lasers are widely used to study fluorescence as a source of excitation light. The laser scanning method collects images pointwise following each line scan over the regions of interest (ROIs), which are reassembled later to obtain the whole image. The ability to isolate a focal plane from other out-of-focus volumes puts an LSCM in a superior position over a wide-field microscope to image samples that contain fluorescent molecules throughout their volume.

8.2.2.2. Spinning-Disc Confocal Microscope

Although LSCMs provide high-definition images by removing out-of-focus fluorescent noise, it takes about one second for a conventional LSCM to produce a full-frame image. Thus. they are not optimal for imaging fast processes due to the relatively long time required to scan a sample and reconstruct the image from it. To address this problem, spinning-disk confocal microscopes (SDCMs) have been developed. Different from the LSCM's single confocal pinhole, SDCMs contain a rapidly spinning disc (1800~5000 rpm) on which an array of pinholes is located [21]. Excitation light passes through the pinholes on the disc, and as many confocal images as there are pinholes are obtained. As the disk rotates, the array of pinholes shifts and confocal images from other locations are also obtained. By utilizing the spinning disk, an SDCM can obtain as many as 1000 frames of images per second. In addition, because it uses low-intensity laser excitation, the damage to cells is minimal [21].

8.2.3. Mathematical Description of Laser Profiles and Photobleaching

8.2.3.1. Laser Profiles

As we will see below, FRAP on membranes is mathematically described by the diffusion equation in the entire plane ($\mathbb{R}^2$). A diffusion coefficient and the initial conditions completely determine the FRAP curves. From this point of view, the laser profile is important because it provides the initial data of the diffusion equation.

Although LSCM performs laser scanning over lines that fill an ROI, in most cases of classical photobleaching, modeling of the laser beam is treated as a single spot, not an integration of scanning lines. Generally, a laser beam profile is treated as either a Gaussian laser beam profile (Fig. 8.5Aa) or a uniform laser beam profile (or a circular disc profile; Fig. 8.5Ba). In the former case, a laser beam with maximum fluorescence intensity I_0 at the origin and radius ω (Fig. 8.5A1) is described by

$$I(x,y) = I_0 G_\omega \tag{8.3}$$

in terms of the Gaussian distribution

$$G_\omega = \frac{2}{\pi\omega^2}\exp\left(-\frac{2(x^2+y^2)}{\omega^2}\right)$$

with standard deviation, $\omega/2$ [13]. On the other hand, in the latter case, a laser beam with radius ω is described as (Fig. 8.5B1)

$$I(x,y) = \begin{cases} \dfrac{I_0}{\pi\omega^2} & x^2+y^2 \le \omega^2 \\ 0 & x^2+y^2 > \omega^2 \end{cases}. \tag{8.4}$$

In a uniform circle laser beam profile, the laser intensity within the beam is constant within the ROI. Note that in both cases $\iint_{\mathbb{R}^2} I(x,y)\,dx\,dy = I_0$.

8.2.3.2. Photobleaching

As mentioned above, illumination with high-intensity light within the absorption spectrum of a fluorophore induces an almost immediate irreversible photobleaching, while low-intensity illumination generates fluorescence. In FRAP experiments, the excitation laser intensity is switched between photobleaching mode and illumination mode. Under photobleaching conditions $I(x, y)$, the fluorescent molecule concentration (x, y) at time t decreases with time (i.e., $dC/dt < 0$). The rate of decrease in fluorescent molecule concentration increases with laser intensity, since more fluorescent molecules are excited under conditions of high laser intensity (high fluorescent molecule concentration or large $C(x,y,t)$). Therefore, we have $dC/dt \propto -I(x,y)C(x,y,t)$, and for some constant α

$$\frac{dC(x,y,t)}{dt} = -\alpha I(x,y)C(x,y,t). \tag{8.5}$$

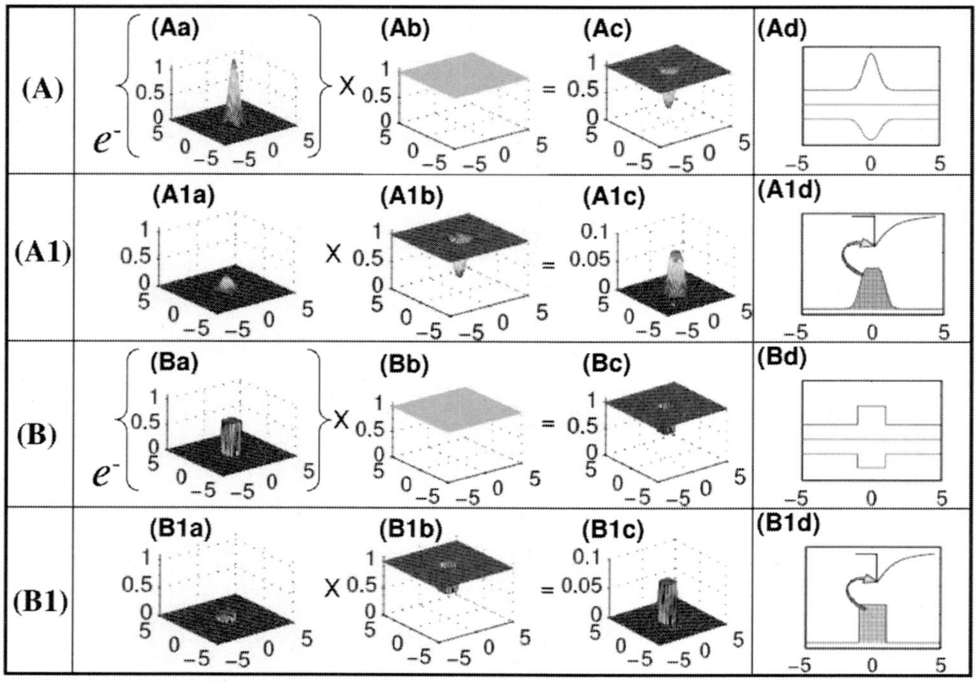

Figure 8.5. Laser profiles and post-bleach fluorescence distribution. (A) Graphs of each component in $C(x,y,0)=C_i\exp(-\alpha TI(x,y))$ for a Gaussian laser profile. (Aa) and (Ab) are graphs of $z=e^{-\alpha TI(x,y)}$ and $z=C_i$, and (Ac) represents $C(x,y,0)=C_i\exp(-\alpha TI(x,y))$, of which a cross-section is shown in (Ad). (A1) shows each component of $F(x,y,0)=qI_\varepsilon(x,y)C(x,y,0)$, where (A1a) and (A1b) are graphs of $z=qI_\varepsilon(x,y)$ and z = $C(x,y,0)$. (A1c) and (A1d) represent $F(x,y,0)=qI_\varepsilon(x,y)C(x,y,0)$ and the cross-section of $F(t)=q\iint_{\mathbb{R}^2}I_\varepsilon(x,y)C(x,y,t)\,dx\,dy$. Exactly the same is true for columns (B) and (B1) for the uniform circle profile laser. (Ba) $z=-\alpha TI(x,y)$. (Bb) z = C_i. (Bc) $C(x,y,0)=C_i\exp(-\alpha TI(x,y))$. (B1a) $z=qI_\varepsilon(x,y)$. (B1b) z = $C(x,y,0)$. (B1c) $F(t)=q\iint_{\mathbb{R}^2}I_\varepsilon(x,y)C(x,y,t)\,dx\,dy$. Finally, (Bd) is the cross-section of (Ba)–(Bc) and (B1d) is the cross-section of the integral $F(t)=q\iint_{\mathbb{R}^2}I_\varepsilon(x,y)C(x,y,t)\,dx\,dy$. Please visit http://www.springer.com/series/7845 to view a high-resolution full-color version of this illustration.

Setting the beginning of the recovery phase as $t = 0$, and assuming that photobleaching occurs for time $t = -T$ to $t = 0$ with the pre-bleach distribution of C at $t = -T$, $C(x,y,-T) = C_i$ (Fig. 8.5A2,B2), the solution is found as $C(x,y,t)=C_i\cdot e^{-\alpha I(x,y,t)}$. We assume that T is small compared with the characteristic times for molecular transport by diffusion. The fluorescent molecule concentration profile at the beginning of the recovery phase (initial data for the diffusion equation) is given (Fig. 8.5A3,B3) by

$$C(x,y,0)=C_i e^{-\alpha TI(x,y)}. \tag{8.6}$$

At the origin at t = 0, the minimal fluorescence concentration is given by $C(0,0,0) = C_i e^{-K}$, where $K = \alpha T\left(2I_0 / \pi\omega^2\right)$ [13].

Recall that low-intensity excitation light is still needed to record fluorescence from the specimen, while high-intensity excitation light photobleaches the fluorophores. For a small number, $\varepsilon \ll 1$, low-intensity laser excitation can be written as $I_\varepsilon = \varepsilon I(x,y)$ (Fig. 8.5A1a, B1a), and the fluorescence detected is

$$\begin{aligned} F(x,y,t) &= qI_\varepsilon(x,y)C(x,y,t) \\ &= q\varepsilon I(x,y)C(x,y,t). \end{aligned}$$

where q is the quantum yield. $F(x,y,0) = q\varepsilon I(x,y)C(x,y,0)$ is plotted in Figure 8,5(A1c), (B1c). Finally, the FI over the ROI can be represented (Fig. 8.5A1d, B1d) by

$$F(t) = q\varepsilon \iint_{\mathbb{R}^2} I(x,y)C(x,y,t)dx\,dy, \tag{8.7}$$

which gives us FRAP curves according to the evolution of $C(x,y,t)$ in time. Note that K can be computed from the initial fluorescence concentration (Eq. (8.6)) and by solving the following equation in K numerically:

$$F(0) = q\varepsilon \iint_{\mathbb{R}^2} I(x,y)C_i e^{-\alpha T I(x,y)}dx\,dy = \frac{F_i}{K}\left(1 - e^{-K}\right), \tag{8.8}$$

where $F_i = q\varepsilon I_0 C_i$ is the pre-bleach FI [13].

8.3. BASIC THEORY OF FRAP

8.3.1. Mathematical Model of FRAP

8.3.1.1. Fluorescence Recovery by Diffusion

Having derived a model for photobleaching for $-T \le t \le 0$ in the previous section, we now derive a mathematical model for fluorescence recovery for $t > 0$. Before photobleaching, fluorescent molecules are in a pre-bleach steady state and their concentration is uniform over the whole cell. After the fluorescent molecules in the ROI are converted into nonfluorescent molecules by photobleaching, a steep concentration difference exists between the ROI and its surrounding area. The concentration gradient introduced by the photobleaching process works as a driving force for fluorescent molecules to move until a new post-bleach steady state is reached by the process of diffusion. To describe the diffusion process mathematically, we make two assumptions. First of all, we assume that fluorescent molecules move along the concentration gradient, $\nabla C = \left(dC / dx, dC / dy\right)$, and, second, the net rate of change in the number of molecules in a unit area A surrounded by boundary S equals the net flux through S. The first assumption is called Fick's first law and the second assumption is called conservation of mass. The net rate of change in the total number of fluorescence molecules in A is computed as

$$\frac{\partial}{\partial t}\iint_A C(x,y,t)dx\,dy,$$

and for the total flux density **F**, the net flux through the boundary S is

$$-\int_S \mathbf{F}\cdot\mathbf{n}dS .$$

Where **n** is an outer normal vector on S. Therefore, the conservation of mass assumption can be rewritten as

$$\frac{\partial}{\partial t}\iint_A C(x,y,t)dx\,dy = -\int_S \mathbf{F}\cdot\mathbf{n}dS = -\iint_A \left(\frac{\partial \mathbf{F}}{\partial x}+\frac{\partial \mathbf{F}}{\partial y}\right)dx\,dy \qquad (8.9)$$

by applying the Gauss-Green theorem in the second equality. Sometimes, $\nabla\cdot\mathbf{F}$ is used for $\partial\mathbf{F}/\partial x + \partial\mathbf{F}/\partial y$. Because Fick's first law states that $\mathbf{F}\propto -\nabla C$, the following equality holds for some constant D:

$$\mathbf{F} = -D\nabla C . \qquad (8.10)$$

Now, combining the two assumptions together,

$$\frac{\partial}{\partial t}\iint_A C dx\,dy = -\iint_A \nabla\cdot\mathbf{F}dx\,dy = D\iint_A \nabla^2 C dx\,dy .$$

Since A is arbitrary, for

$$\iint_A \left(\frac{\partial C}{\partial t} - D\nabla^2 C\right)dx\,dy = 0$$

to be true, it follows that

$$\frac{\partial C(x,y,t)}{\partial t} = D\nabla^2 C(x,y,t) , \qquad (8.11)$$

which is also written as $\partial C/\partial t = D\Delta C$, where the Laplacian operator defines $\Delta = \nabla^2 = \partial^2/\partial x^2 + \partial^2/\partial x^2$. When the bleaching spot is circular, it is convenient to use polar coordinates rather than rectangular coordinates because C is symmetric in the radial direction [22]. By changing spatial variables $x = r \sin\theta$ and $y = r\cos\theta$, Eq. (8.11) can be converted to

$$\frac{\partial C}{\partial t} = \frac{1}{r}\frac{\partial C}{\partial r}\left(rD\frac{\partial C}{\partial r}\right) = D\left(\frac{\partial^2 C}{\partial r^2}+\frac{1}{r}\frac{\partial C}{\partial r}\right). \qquad (8.12)$$

Note that there is no term in θ because of the radial symmetry of C, and the spatial dimension of Eq. (8.12) is reduced from two to one.

D ($\mu m^2/s$) is called the diffusion coefficient, which provides information on how fast the molecules diffuse in a unit time. In general, D depends on both diffusing particles and the media in which diffusion occurs. The velocity of diffusing particles (or drift velocity), the geometry (shape and size) of particles, the density of particles, and the temperature and viscosity of the media are known parameters that a diffusion coefficient depends on. Let us assume that low-density spherical particles with small radius r are diffusing in a media with viscosity η. In the kinetic theory of physics, the Einstein relation states that the diffusion coefficient is proportional to the mobility of the particles (μ) and the absolute temperature (T), where the mobility is the ratio of the particle's terminal drift velocity (v_{drift}) to an applied force (F). With the proportionality constant κ_B, the Boltzmann constant, we have

$$D = \kappa_B T \mu = \kappa_B T \frac{v_{\text{drift}}}{F}.$$

On the other hand, if we assume that the ratio of inertial forces to viscous forces is small (low Reynolds number), the mobility of the particles (μ) is also given by the Stokes drag (γ), which is the limiting value of μ obtained by solving the Navier-Stokes equation in the limit of the Reynolds number approaching zero, where inertial forces are ignored. For a viscosity η, the Stokes drag (γ) is given by $\gamma = 6\pi\eta r$ and the mobility is the inverse of the drag, i.e., $\mu = 1/(6\pi\eta r)$, which leads us to the Einstein-Stokes equation:

$$D = \frac{\kappa_B T}{6\pi\eta r}. \tag{8.13}$$

Brownian motion in membranes is classically defined by the Saffman-Delbrück model [23].

8.3.1.2. Solution of the Diffusion Equation

Now, FRAP can be described from Eq. (8.7) while C can be found by solving Eq. (8.11). From PDE theory, it is known that the solution of Eq. (8.11), subject to an initial condition $C(x,y,0)$, can be found by

$$C(x,y,t) = \iint_{\mathbb{R}^2} C(x,y,0)\Phi(x-x',y-y',t)dx'\,dy', \tag{8.14}$$

where $\Phi(x,y,t)$ satisfies

$$\frac{\partial\Phi(x,y,t)}{\partial x} = D\Delta\Phi(x,y,t), \quad \Phi(x,y,0) = \delta_0,$$

where δ_0 is the Dirac delta funciton.

To find $\Phi(x,y,t)$, we apply the Fourier transform in the spatial variables x and y to obtain a differential equation involving only t, and then transform back by taking the inverse transform to get

$$\Phi(x,y,t) = \frac{1}{4\pi Dt} e^{-\left(\frac{x^2+y^2}{4Dt}\right)}, \tag{8.15}$$

which is called the fundamental solution of the diffusion equation. Notice that Eq. (8.15) can be regarded as a normal distribution density function with mean 0 and standard deviation $\sqrt{2Dt}$ and that $\iint\Phi(x,y,t)dx\,dy = 1$, which indicates that the shape of $\Phi(x,y,t)$ becomes very narrow when t is small, while when t is large it becomes very broad, corresponding to our experience that a diffusing substance tends to a spread out toward a uniform density. Equation (8.15) also gives the mean square displacement of diffusing molecules in $\mathbb{R}^2$ as

$$\langle r^2(t)\rangle = \iint_{\mathbb{R}^2} (x^2 + y^2)\Phi(x,y,t)dx\,dy = 4Dt. \tag{8.16}$$

In general, $<r^2(t)> = 2nDt$ in $\mathbb{R}^n$.

2D model with circular ROI on R_2

Gaussian laser beam. From Eqs. (8.3), (8.7), and (8.14), we have

$$F(t) = \frac{q\varepsilon}{4\pi Dt} \iint_{\mathbb{R}^2} I(x,y) \left(\iint_{\mathbb{R}^2} C(x',y',0) e^{-\left(\frac{(x-x')^2+(y-y')^2}{4Dt}\right)} dx'\,dy' \right) dx\,dy$$
$$= \frac{I_0 C_i q\varepsilon \nu}{K^\nu} \int_0^K \left(e^{-u} u^{\nu-1}\right) du$$
$$= Fi \frac{\nu}{K^\nu} \gamma(\nu, K),$$

where $\nu = (2t/\tau_D + 1)^{-1}$, $\tau_D = 4D/\omega^2$, $K = \alpha T\left(\frac{2I_0}{\pi\omega^2}\right)$, and $\gamma(\nu,K)$ is the lower incomplete gamma function [13].

For practical purposes, a series solution is often used. If we define

$$g_\omega = \exp\left(-2\left(x^2+y^2\right)/\omega^2\right),$$

then

$$I(x,y) = \frac{2I_0}{\pi\omega^2} g_\omega(x,y),$$

and $C(x,y,0) = C_i \exp(-Kg_\omega)$. From Eq. (8.7),

$$F(t) = \frac{q\varepsilon}{4\pi Dt} \iint_{\mathbb{R}^2} \frac{2I_0}{\pi\omega^2} g_\omega(x,y) \left[\iint_{\mathbb{R}^2} C_i e^{-Kg_\omega(x',y')} e^{-\frac{(x-x')^2+(y-y')^2}{4Dt}} dx'\,dy' \right] dx\,dy$$
$$= \frac{q\varepsilon C_i}{4\pi Dt} \frac{2I_0}{\pi\omega^2} \iint_{\mathbb{R}^2} \iint_{\mathbb{R}^2} g_\omega(x,y) \left[\sum_{n=0}^{\infty} \frac{(-K)^n}{n!} g_\omega^n(x',y') \right] e^{-\frac{(x-x')^2+(y-y')^2}{4Dt}} dx'\,dy'\,dx\,dy$$
$$= I_0 C_i q\varepsilon \sum_{n=0}^{\infty} \frac{(-K)^n}{n!} \frac{1}{1+n(1+2t/\tau_D)}$$
$$= F_i \sum_{n=0}^{\infty} \frac{(-K)^n}{n!\left[1+n(1+2t/\tau_D)\right]}, \qquad (8.17)$$

whose only parameter is τ_D [13].

Uniform laser beam: Similar to the Gaussian laser beam case, we first solve the differential equation involving only t after the Fourier transform, and take an inverse Fourier transform to get $C(x,y,t)$ with the uniform laser profile $I(x,y)$:

$$
\begin{aligned}
F(t) &= q\varepsilon \iint_{\mathbb{R}^2} I(x,y)\left[\frac{1}{2\pi}\iint_{\mathbb{R}^2} e^{-i(\xi x+\varsigma y)}\left(\left\{\frac{1}{2\pi}\iint_{\mathbb{R}^2} e^{i(\xi x'+\varsigma y')}C_i e^{-\alpha TI(x',y')}dx'\,dy'\right\}e^{-(\xi^2+\varsigma^2)Dt}\right)d\xi\,d\varsigma\right]dx\,dy \\
&= \frac{q\varepsilon}{4\pi^2}\iint_{\mathbb{R}^2}\left[\left(\iint_{\mathbb{R}^2} I(x,y)e^{-i(\xi x+\varsigma y)}dxdy\right)\left(\iint_{\mathbb{R}^2} e^{[i(\xi x'+\varsigma y')-\alpha TI(x',y')]}dx'\,dy'\right)\right]e^{-(\xi^2+\varsigma^2)Dt}d\xi\,d\varsigma \\
&= F_i\left[1-2(1-e^{-K})\int_0^{\infty}\left[\frac{\mathbf{J}_1(s\omega)}{s}\right]^2 e^{-s^2Dt}s ds\right], \\
&= \left(1-\frac{\tau_D\exp(-2\tau_D/t)}{t}\left[\mathbf{I}_0(2\tau_D/t)+\mathbf{I}_2(2\tau_D/t)\right]+2\sum_{n=0}^{\infty}\frac{(-1)^n(2n+2)!(n+1)!(\tau_D/t)^{n+2}}{(n!)^2[(n+2)!]^2}\right)(F_i-F_0)+F_0
\end{aligned}
\tag{8.18}
$$

where $\mathbf{I}_0$ and $\mathbf{I}_2$ are modified Bessel functions, and δ is the Dirac delta function. Note that $F_0 = F_i e^{-K}$. In scaled form,

$$
f(t)=\left(1-\frac{\tau_D\exp(-2\tau_D/t)}{t}\left[\mathbf{I}_0(2\tau_D/t)+\mathbf{I}_2(2\tau_D/t)\right]+2\sum_{n=0}^{\infty}\frac{(-1)^n(2n+2)!(n+1)!(\tau_D/t)^{n+2}}{(n!)^2[(n+2)!]^2}\right), \tag{8.19}
$$

where $f(t) = (F(t) - F_0)/(F_i - F_0)$, which gives $f(0) = 0$ and $f(\infty) = 1$, because 100% recovery is assumed in $\mathbb{R}^2$, i.e., $F_i = F(\infty)$ [13]. Due to the singularity at $t = 0$ and the infinite series, Eq. (8.19) is inconvenient for numerical work. Rather than applying $2\mathbf{J}_1(\omega s) = \omega s\,(\mathbf{J}_0(\omega s) + \mathbf{J}_2(\omega s))$ in Eq. (8.18), which gives the infinite series form, Eq. (8.18) can be converted into a differential equation by differentiating both sides of Eq. (8.18) with respect to t. The solution of the differential equation can be found explicitly as a closed form:

$$
f(t)=\exp\left(-\frac{2\tau_D}{t}\right)\left[\mathbf{I}_0\left(\frac{2\tau_D}{t}\right)+\mathbf{I}_1\left(\frac{2\tau_D}{t}\right)\right], \tag{8.20}
$$

where $\mathbf{I}_0$ and $\mathbf{I}_1$ are modified Bessel functions [24]. Note also that the singularity at $t = 0$ in Eq. (8.20) is removable, i.e., $\lim_{t\to 0} f(t)=0$.

8.3.2. FRAP Data

A typical FRAP curve is depicted in Figure 8.6. At steady state, fluorescent molecules on membranes maintain their steady states by a two-dimensional random walk. We call the steady-state FI before photobleaching the pre-bleach FI (F_i: Fig. 8.6A). When a high enough intensity laser excitation is applied to an ROI to photobleach fluorescent molecules in a short time, a fraction of fluorescent molecules lose their fluorescence and the FI in the ROI drops to F_0, the initial FI of an FRAP curve. In LSCM, multiple iterations of a series of laser lines are scanned over the ROI to obtain the desired bleaching depth ($F_i - F_0$). Due to the time required for multiple iterations of the bleaching scans, confocal FRAP curves are divided into three different regimes: pre-bleach mode, photobleaching mode, and FRAP mode (Fig. 8.6A). However, immediate photobleaching is assumed in practice (inset of Fig. 8.6A). After photobleaching, both bleached and fluorescent molecules coexist in the ROI and both diffuse by random walk. When the bleached molecules are replaced by the fluorescence molecules from outside of the ROI, fluorescence recovery occurs, and when bleached molecules become uniformly distributed over the

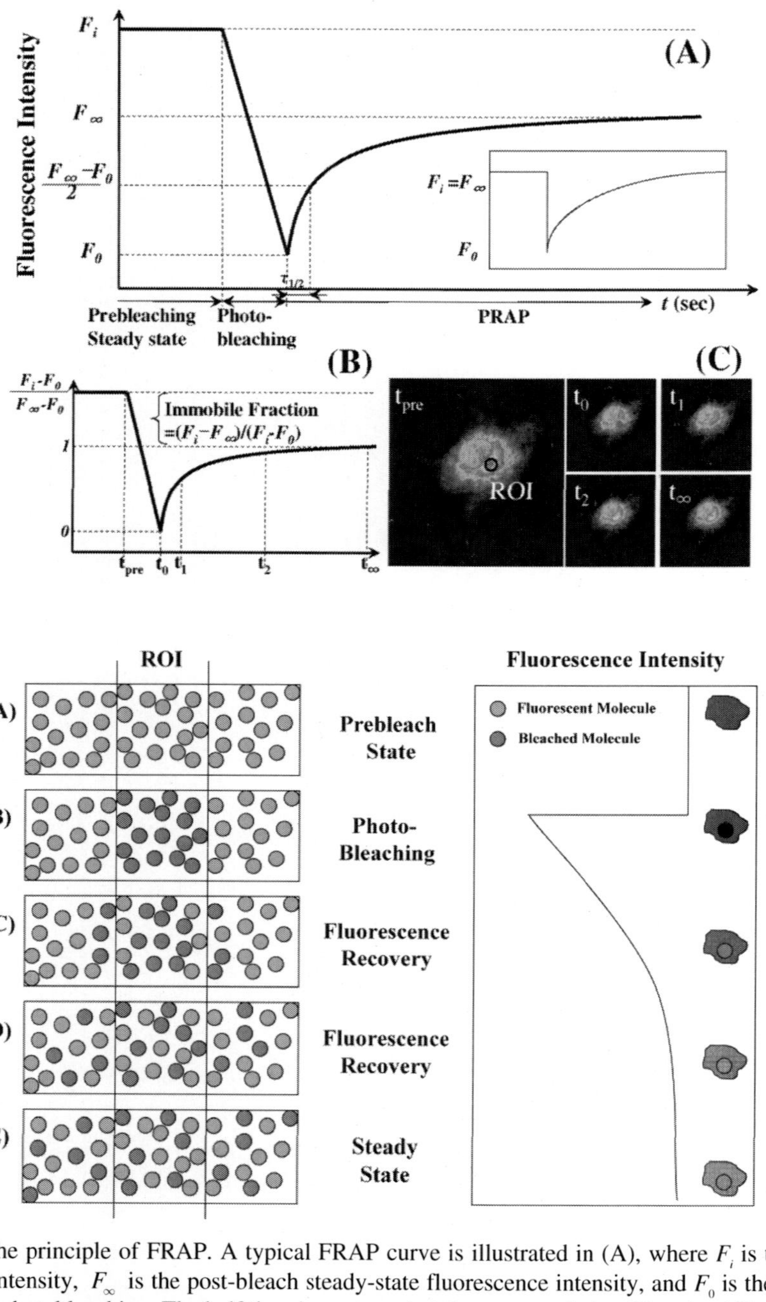

Figure 8.6. The principle of FRAP. A typical FRAP curve is illustrated in (A), where F_i is the pre-bleach fluorescence intensity, F_∞ is the post-bleach steady-state fluorescence intensity, and F_0 is the fluorescence intensity after photobleaching. The half time is measured at time $\tau_{1/2}$ such that $F(\tau_{1/2}) = (F_\infty - F_0)/2 + F_0 = (F_\infty + F_0)/2$. The FRAP curve is usually normalized as $f(t) = [F(t) - F_0]/[F_\infty - F_0]$, which sets the post-bleach fluorescence intensity and post-bleach steady-state fluorescence intensity as 0 and 1, respectively. The fraction difference between pre- and post-bleach steady states is defined as the immobile fraction, which is computed as $([F_i - F_\infty]/[F_i - F_0])$. When there are no immobile fluorescent molecules in the ROI and the size of the ROI is small, full recovery occurs (inset of (A)). (C) FRAP data for a palmitoylation mutant of HRas tagged with GFP. The pool of fluorescent protein associated with the Golgi complex was bleached in this experiment. (D) Schematic depiction of an FRAP experiment. See text for further details. Please visit http://www.springer.com/series/7845 to view a high-resolution full-color version of this illustration.

whole membrane, the FI reaches post-bleach steady state (F_∞: Fig. 8.6A), which is usually lower than F_i. Assuming the ROI size is small, the difference $F_i - F_\infty$ is from an immobile pool of non-diffusing fluorescence molecules, which does not contribute to FRAP. If the ROI size is not small, the FI loss in the ROI decreases the average FI over the entire membrane, which also influences the pre- and post-bleaching FI differences ($F_i - F_\infty$). In practice, in the absence of an immobile pool of fluorescent molecules in the ROI, 100% fluorescence recovery is assumed ($F_i = F_\infty$: inset of Fig. 8.6A).

FI is usually rescaled so that data can be compared across cells with different FIs. A commonly used scaled form of FI is given as

$$f(t) = \frac{\left[F(t) - F_0\right]}{\left[F_\infty - F_0\right]}, \tag{8.21}$$

where $f(0) = f_0 = 0$ and $f_\infty = 1$(Fig. 8.6B) [13,24]. Under this scaling, the mobile fraction is defined as $1/f_i = (F_\infty - F_0)/(F_i - F_0)$ while the immobile fraction is defined as $1 - (1/f_i)$ = (pre- post-bleaching FI difference)/(Photobleach depth) = $(F_i - F_\infty)/(F_i - F_0)$. Also, the half time of fluorescence recovery can be found as $\tau_{1/2} = f^{-1}(1/2)$, where f^{-1} is the inverse function of f in $t \geq 0$.

8.3.3. Finding a Diffusion Coefficient from FRAP Curves

8.3.3.1. Data Fitting

We have studied four FRAP formulae in a closed form or a series form depending on the laser profiles (see Table 8.1). Each of them has two unknown parameters, K and τ_D. Once K is determined numerically from the ratio F_0/F_i by solving

$$\frac{F_i}{F_0} = \frac{\left(1 - e^{-K}\right)}{K}, \tag{8.22}$$

the only free parameter is τ_D [13]. This can be determined from FRAP data by data fitting to minimize the mean square error:

$$E = \int_0^\infty \left|f(t) - f^{\text{data}}(t)\right|^2 dt = \sum_{n=0}^{N} \left|f(t_n) - f^{\text{data}}(t_n)\right|^2$$

where $f^{\text{data}}(t) = [F^{\text{data}}(t) - F_0^{\text{data}}]/[F_\infty^{\text{data}} - F_0^{\text{data}}]$ is the scaled FRAP data and $f(t) = [F(t) - F_0]/[F_\infty - F_0]$ is a scaled FRAP curve from the formulae in Table 8.1. Because FRAP data are given in discrete times in finite time as $\{F(t_n) : n = 1, 2, 3, \ldots, N\}$, we use the summation form of the mean square error.

8.3.3.2. The Half Time of Recovery and the Size of ROI

From Eq. (8.16) we understand that the mean-square displacement is linearly proportional to the time of observation. From this observation, we can expect a linear relation between the half time of recovery and ROI size to exist. For a uniform circle laser beam profile, by solving $\frac{1}{2} = \exp(-2\tau_D/\tau_{1/2})\,[\mathbf{I}_0(2\tau_D/\tau_{1/2}) + \mathbf{I}_1(2\tau_D/\tau_{1/2})]$ numerically, one gets

$$D = \gamma_D \frac{\omega^2}{4\tau_{1/2}} = 0.22\frac{\omega^2}{\tau_{1/2}}, \tag{8.23}$$

where $\gamma_D \approx 0.88$ [13,24].

Table 8.1. FRAP Formulae

Laser	Closed form	Series form
Gaussian	$F(t)=F_i \frac{\nu}{K^{\nu}}\gamma(\nu,K)$	$F(t)=F_i\sum_{n=0}^{\infty}\frac{(-K)^n}{n![1+n(1+2t/\tau_D)]}$
Uniform	$F(t)=\exp\left(-\frac{2\tau_D}{t}\right) \times\left[I_0\left(\frac{2\tau_D}{t}\right)+I_1\left(\frac{2\tau_D}{t}\right)\right]\left(F_i-F_0\right)+F_0$	$F(t)=\left(1-\frac{\tau_D}{t}\exp\left(-\frac{2\tau_D}{t}\right)\left[\mathbf{I}_0\left(\frac{2\tau_D}{t}\right)+\mathbf{I}_2\left(\frac{2\tau_D}{t}\right)\right] +2\sum_{n=0}^{\infty}\frac{(-1)^n(2n+2)!(n+1)!}{(n!)^2[(n+1)!]^2}\left(\frac{\tau_D}{t}\right)^{n+2}\right)(F_i-F_0)+F_0$

$\nu=(2t/\tau_D+1)^{-1}$, $\tau_D=4D/\omega^2$, $K=\alpha T\frac{2I_0}{\pi\omega^2}$, $\gamma(\nu,K)=\int_0^K\left(e^{-u}u^{\nu-1}\right)du$, and I_k are modified Bessel functions (k = 0, 1, 2).

8.3.4. Summary

In living cells, one of the most prevalent ways to transport molecules is diffusion, which results from either random motion of molecules or a concentration gradient. The rate of transport by diffusion largely depends on the diffusion coefficient of a molecule, which is a function of the size of the molecule and the viscosity and temperature of the medium. FRAP experiments provide a way to measure mobility or a diffusion coefficient by monitoring how fast the fluorescence intensity recovers in a small area after photobleaching by a high-intensity laser.

Figure 8.6 explains how FRAP works in general. In the pre-bleach state (Fig. 8.6A), fluorescent molecules are uniformly distribution in the region including the region of interest (ROI). These molecules maintain their equilibrium by constant random motion. Due to the nature of the equilibrium maintained by random motion, the local fluorescence intensity tends to fluctuate. However, when the ROI size is large enough, the average fluorescence intensity is essentially constant, which corresponds to the pre-bleach state FI on the FRAP curve.

When a high-intensity laser scans the ROI, the major part of fluorescent molecules are irreversibly photobleached and lose their ability to fluoresce. Consequently, the FI of the FRAP curve takes its minimum value at this point (Fig. 8.6B). Even though fluorescent molecules have lost their fluorescence, they still undergo constant random motion. Notice that the constant random motion guarantees that the total number of molecules at given time in the ROI remains constant whatever types of molecules they are (bleached or fluorescent).

After photobleaching of the ROI, bleached molecules mix with fluorescent molecules in such a way that part of the bleached molecules diffuse out from the ROI while their counterpart fluorescence molecules diffuse into the ROI (Fig. 8.6C–D). This mixing of fluorescence molecules gives rise to fluorescence recovery. The higher the gradient of fluorescent to nonfluorescent molecules, the faster the rate of fluorescence recovery. As photobleached molecules globally distribute and uniformly mix with fluorescent molecules, the FRAP curve reaches a constant level (Fig. 8.6E), the post-bleach steady state.

Finally, the experimentally determined FRAP curve is compared with theoretically computed FRAP curves with given diffusion coefficients. The formulae for theoretical FRAP curves depend on the laser profiles and can be represented as either a closed form or an infinite series (Table 8.1). The major difference between the FRAP formulae of two different laser profiles is their dependence on bleaching depth. FRAP formulae for a Gaussian laser have two parameters, K and τ_D, where K is associated with the initial FI as in Eq. (8.22). For this reason, FRAP fitting of a Gaussian laser profile is done in two steps: finding K (Eq. (8.22)) and finding a τ_D that minimizes the mean-square error between the FRAP data and theoretical FRAP curve given either in closed form or infinite series, as shown in Table 8.1.

On the other hand, FRAP formulae for a uniform circular laser beam are independent of the bleaching parameter K. Therefore, in the case of FRAP with a uniform circular laser beam, direct fitting to determine the τ_D that generates the minimal mean-square error is enough. Even better, since the half time and the area of the bleach ROIs are linearly related in the case of a uniform circle laser beam, Eq. (8.23) can be used to determine diffusion coefficients. However, for a Gaussian laser beam, D in Eq. (8.23) is also a function of K, and additional steps are required to use Eq. (8.23) to determine D.

8.4. FURTHER APPLICATIONS OF FRAP

8.4.1. Binding-Diffusion Model

In a cell, many proteins can exist in both a soluble form and bound to membranes. The reversible attachment of proteins to membranes can regulate many important cellular events, including cellular signaling. Most peripheral membrane proteins such as G proteins, certain types of protein kinases, polypeptide hormones, and possibly others are included in this class [25]. The binding-diffusion model is a major tool that can be used to analyze the mobility of such proteins by FRAP. This model is also used to study the reversible binding of nuclear proteins to DNA [17,26–28]. Here, we describe the theoretical basis for this model.

Throughout this section we assume a uniform laser profile. Also, a small ROI size is assumed to guarantee that full recovery occurs. For a binding site S, unbound molecule U, and $U \cdot S$ complex C, a binding reaction can be described as

$$U + S \underset{k_{\mathrm{off}}}{\overset{K_{\mathrm{on}}}{\rightleftharpoons}} C.$$

If we represent the concentrations of molecules as $u = [U]$, $s = [S]$, and $c = [C]$, a binding-diffusion coupled equation can be written as

$$\begin{cases} \dfrac{\partial u}{\partial t} = D_1 \nabla^2 u - k_{\mathrm{on}} u + k_{\mathrm{off}} c \\ \dfrac{\partial c}{\partial t} = D_2 \nabla^2 c + k_{\mathrm{on}} u - k_{\mathrm{off}} c \end{cases}, \tag{8.24}$$

where D_1 and D_2 are diffusion coefficients in the soluble and bound states.

Here, we make the assumption that the binding site is immobile and exists as a large population so that $k_{\mathrm{on}} = K_{\mathrm{on}} S = K_{\mathrm{on}} S_{\infty}$, i.e., S is constant in its steady state [24]. From Eq. (8.7), we may identify u and c as corresponding FIs because an FI in the ROI is a constant multiplied by the concentration of fluorescent molecules. We also assume that u and c are in scaled form similar to Eq. (8.21), so that

$$\begin{cases} u(x,y,0) = u_i e^{-\alpha TI(x,y)} \\ c(x,y,0) = c_i e^{-\alpha TI(x,y)} \end{cases} \tag{8.25}$$

where α and T are as in Eq. (8.6), and the pre-bleach steady state u_i and c_i satisfy $u_i = k_{\text{off}}/(k_{\text{on}}+k_{\text{off}})$ and $c_i = k_{\text{on}}/(k_{\text{on}}+k_{\text{off}})$. This can be derived from the definition of steady-state $\partial u/\partial t = \partial c/\partial t = 0$ (i.e., $k_{\text{on}}u_\infty = k_{\text{off}}c_\infty$) and the assumption of a small ROI (i.e., full recovery: $u_i + c_i = u_\infty + c_\infty = 1$). The solution to Eq. (8.24) was reported [29] as

$$\begin{aligned} u(\mathbf{x},t) &= e^{-k_{\text{on}}t}\Phi(\mathbf{x},D_1t) * u(\mathbf{x},0) \\ &+ \frac{e^{\lambda t}\sqrt{k_{\text{off}}}}{D_1 - D_2}\int_{D_2t}^{D_1t} e^{-\mu s}\left(\sqrt{k_{on}}\xi\sum\mathbf{I}_1(\eta)\Phi(\mathbf{x},s) * u(\mathbf{x},0)\right. \\ &\left. + \sqrt{k_{\text{off}}}\mathbf{I}_0(\eta)\Phi(\mathbf{x},s) * b(\mathbf{x},0)\right)ds, \end{aligned} \tag{8.26a}$$

$$\begin{aligned} b(\mathbf{x},t) &= e^{-k_{\text{off}}t}\Phi(\mathbf{x},D_2t) * b(\mathbf{x},0) \\ &+ \frac{e^{\lambda t}\sqrt{k_{\text{on}}}}{D_1 - D_2}\int_{D_2t}^{D_1t} e^{-\mu s}\left(\sqrt{k_{\text{off}}}/\xi\mathbf{I}_1(\eta)\Phi(\mathbf{x},s) * b(\mathbf{x},0)\right. \\ &\left. + \sqrt{k_{\text{on}}}\mathbf{I}_0(\eta)\Phi(\mathbf{x},s) * u(\mathbf{x},0)\right)ds, \end{aligned} \tag{8.26b}$$

where

$$\lambda = \frac{k_{\text{on}}D_2 - k_{\text{off}}D_1}{D_1 - D_2},$$

$$\mu = \frac{k_{\text{on}} - k_{\text{off}}}{D_1 - D_2},$$

$$\xi = \sqrt{\frac{s - D_2t}{D_1t - s}},$$

$$\eta = \frac{2\sqrt{k_{\text{on}}k_{\text{off}}(D_1t - s)(s - D_2t)}}{D_1 - D_2},$$

$$\Phi(x,Dt) = \frac{1}{4\pi Dt}e^{-\frac{x^2+y^2}{4Dt}}.$$

In Eq. (8.26), $\mathbf{I}_0$ and $\mathbf{I}_1$ are modified Bessel functions, and $*$ represents convolution. Recall that Φ is the fundamental solution of the diffusion equation introduced in Eq. (8.15).

8.4.1.1. *Simplifications of the Binding-Diffusion Model*

Although the solution of Eq. (8.24) is complicated, it can be reduced to a simpler form depending on the size of the parameters. Throughout this section, we use $A \ll B$ when the size of B is much larger than that of A, and $A \sim B$ if A and B are similar in size. Notice that the timescale of diffusion is determined by the characteristic diffusion time, $\tau_D = \omega^2/(4D)$ and that the timescale of binding kinetics is governed by $\tau_w = 1/k_{on}$ and $\tau_r = 1/k_{off}$. Since k_{on} is the binding rate per unit time, τ_w measures the wandering time between binding events. Similarly, τ_r measures the residence time of bound molecules. For example, when most molecules exist in the unbound state, binding kinetics can be ignored and Eq. (8.24) can be approximated by a pure diffusion equation. On the other hand, when the binding kinetics are fast compared to diffusion, i.e., molecules undergo constant rapid binding and release during diffusion, Eq. (8.24) can be reduced to a pure diffusion equation but with slower diffusion coefficients. We formally derive the reduced model by considering the timescale differences in Eq. (8.24). Additionally, we assume that $D_2 = 0$ in Eq. (8.24).

Diffusion-dominant model

Consider the case when molecules rarely bind to their binding sites (i.e., their wandering time is long, $\tau_w \gg 1$ or $k_{on} \ll 1$), and once they bind, they stay for a short time (i.e., their residence time is of the order of the characteristic diffusion time, $\tau_r \sim \tau_D$, $k_{off} \sim D$), so that most of molecules exist in an unbound state. This condition guarantees that the fraction of unbound to bound pools is large at steady state ($u_\infty / c_\infty = k_{off}/k_{on} \gg 1$). If we write $k_{on} = \varepsilon\kappa_{on}$ for some $\kappa_{on} \sim 1$, the following holds: $\varepsilon\kappa_{on} \ll k_{off} \sim D \sim 1$.

Under this nomenclature, Eq. (8.24) becomes

$$\begin{cases} \dfrac{\partial u}{\partial t} = D\nabla^2 u - \varepsilon\kappa_{on} u + k_{off} c \\ \dfrac{\partial c}{\partial t} = \varepsilon\kappa_{on} u - k_{off} c \end{cases} \tag{8.27}$$

with initial conditions

$$u(0) = \begin{cases} 0 & x^2 + y^2 \le \omega^2 \\ \dfrac{k_{off}}{\varepsilon\kappa_{on} + k_{off}} & x^2 + y^2 > \omega^2 \end{cases}, \quad c(0) = \begin{cases} 0 & x^2 + y^2 \le \omega^2 \\ \dfrac{\varepsilon\kappa_{on}}{\varepsilon\kappa_{on} + k_{off}} & x^2 + y^2 > \omega^2 \end{cases}. \tag{8.28}$$

Now, we expand the solution of Eq. (8.24) in the small parameter ε so that

$$\begin{aligned} u &= \sum_{n=0}^{\infty} \varepsilon^n u_n = u_0 + \varepsilon u_1 + \varepsilon^2 u_2 + \cdots, \\ c &= \sum_{n=0}^{\infty} \varepsilon^n c_n = c_0 + \varepsilon c_1 + \varepsilon^2 c_2 + \cdots, \end{aligned} \tag{8.29}$$

and plug this series expansion into Eqs. (8.27)–(8.28). Then the leading-order problem not involving ε is obtained as

$$\begin{cases} \dfrac{\partial u_0}{\partial t} = D\nabla^2 u_0 + k_{\text{off}} c_0 \\ \dfrac{\partial c_0}{\partial t} = -k_{\text{off}} c_0 \end{cases}$$

with initial condition

$$u_0(0) = \begin{cases} 0 & x^2 + y^2 \le \omega^2 \\ 1 & x^2 + y^2 > \omega^2 \end{cases}, \quad c_0(0) = 0,$$

which can be reduced to $\partial u_0 / \partial t = D\nabla^2 u_0$ [26]. This is exactly the same as Eq. (8.11) with a uniform circle laser profile, and the solution is therefore given by Eq. (8.20) (Fig. 8.7A) [26].

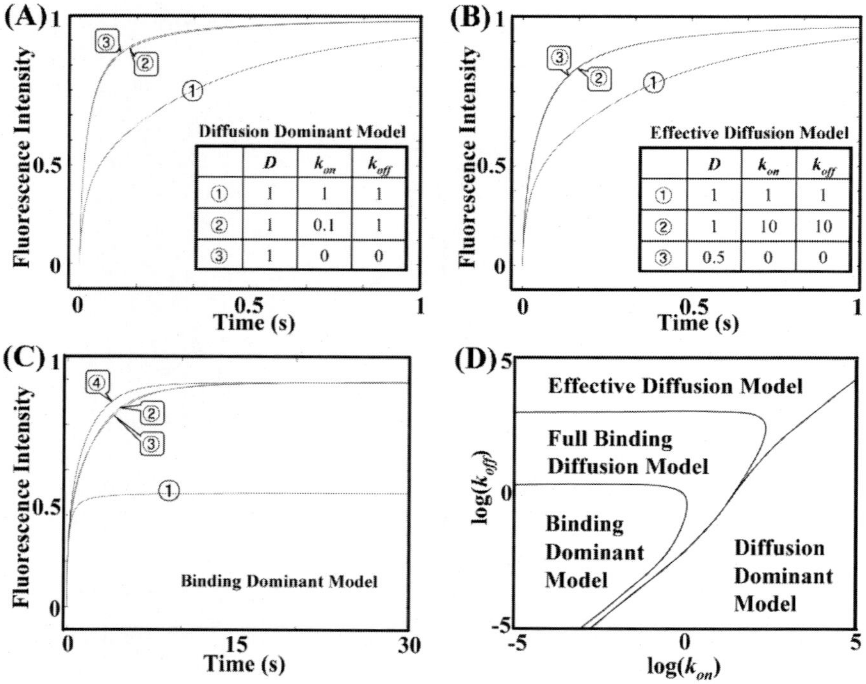

Figure 8.7. Binding diffusion model and its simplification. (A) Diffusion dominant model. Predicted FRAP curves were calculated for the parameters shown in the inset table. (B) Effective diffusion model. Predicted FRAP curves were calculated for the parameters shown in the inset table. (C) Binding-dominant model. FRAP curves were calculated as follows: (1) $f(t) = \exp\left(-\frac{2\tau_D}{t}\right)\left[I_0\left(\frac{2\tau_D}{t}\right) + I_1\left(\frac{2\tau_D}{t}\right)\right]$ with τ_D = 0.0625; (3) $f(t) = 0.5\exp\left(-\frac{2\tau_D}{t}\right)\left[\mathbf{I}_0\left(\frac{2\tau_D}{t}\right) + \mathbf{I}_1\left(\frac{2\tau_D}{t}\right)\right] + 0.5\left(1 - e^{-k_{\text{off}}t}\right)$ with τ_D = 0.0625 (D = 1, ω = 0.5) and k_{off} = 0.01; and (4) $f(t) = 0.5\exp\left(-\frac{2\tau_D}{t}\right)\left[\mathbf{I}_0\left(\frac{2\tau_D}{t}\right) + \mathbf{I}_1\left(\frac{2\tau_D}{t}\right)\right] + 0.5\left(1 - e^{-k_{\text{off}}t}\right)$ with τ_D = 0.0625 and k_{off} = 1 are used. (D) Range of parameter space covered by each of the regimes of the binding-diffusion model as re-drawn from [27]. Please visit http://www.springer.com/series/7845 to view a high-resolution full-color version of this illustration.

Effective diffusion model

Consider the case of rapid binding and release ($\tau_w \sim \tau_r \ll \tau_D$, or $k_{off} \sim k_{off} \gg D$). In this case, diffusing molecules are slowed down by fast binding kinetics and overall appear to undergo slow diffusion without any binding. Under these assumptions, if we write $k_{on} = \kappa_{on}/\varepsilon$ and $k_{off} = \kappa_{off}/\varepsilon$, where $\kappa_{on} \sim \kappa_{off} \sim D \sim 1$ and $\varepsilon \ll 1$, then Eq. (8.24) becomes

$$\begin{cases} \dfrac{\partial}{\partial t}(u+c) = D\nabla^2 u \\ \varepsilon \dfrac{\partial c}{\partial t} = k_{on}u - k_{off}c \end{cases},$$

with the initial condition given by Eq. (8.25). By replacing u and c with the series expansion in ε as in Eq. (8.29), we get the leading-order approximation as

$$\frac{\partial u_0}{\partial t} = \left(\frac{D}{1+k_{on}/k_{off}}\right)\nabla^2 u_0, \quad \frac{\partial c_0}{\partial t} = \left(\frac{D}{1+k_{on}/k_{off}}\right)\nabla^2 c_0. \qquad (8.30)$$

One can imagine this as the movement of a ball passing an alley between two walls by bouncing off the walls in a zigzag manner. Although the ball moves forward, the velocity is slower than when it passes the alley without touching the walls. Also, if the track of the ball is traced on both walls, they will be identical, as we can imagine from Eq. (8.30).

Because the total FI is given by $u + c$, their leading-order approximation is given by

$$\frac{\partial}{\partial t}(u_0 + c_0) = \left(\frac{D}{1+k_{on}/k_{off}}\right)\nabla^2 (u_0 + c_0), \quad (u_0 + c_0)(0) = \begin{cases} 0 & x^2+y^2 \le \omega^2 \\ 1 & x^2+y^2 > \omega^2 \end{cases},$$

where the solution is given by Eq. (8.20) (Fig. 8.7B). We define the effective diffusion coefficient as

$$D_{eff} = \frac{D}{1+k_{on}/k_{off}}$$

which is smaller than the original diffusion coefficient ($D > D_{eff}$) [26,27].

Binding-dominant model (biphasic model)

Opposite to the effective diffusion model, when binding kinetics are much slower than diffusion ($\tau_w \sim \tau_r \gg \tau_D$, or $k_{off} \sim k_{off} \ll D$), the FRAP curve becomes biphasic: the early rapid recovery phase is dominated by fast diffusion, and slower first-order exponential recovery due to binding occurs in the later phase. Because the time interval of diffusion driving the recovery phase is much shorter than that of binding, it is also called the binding-dominant model [27]. If we write $k_{on} = \varepsilon\kappa_{on}$ and $k_{off} = \varepsilon\kappa_{off}$, then Eq. (8.24) can be rewritten as

$$\begin{cases} \dfrac{\partial u}{\partial t} = D\nabla^2 u - \varepsilon\kappa_{on}u + \varepsilon\kappa_{off}c \\ \dfrac{\partial c}{\partial t} = \varepsilon\kappa_{on}u - \varepsilon\kappa_{off}c \end{cases},$$

with the initial conditions as in Eq. (8.25). After expanding u and c in ε as in Eq. (8.29), we find the leading-order approximation,

$$\begin{cases} \dfrac{\partial u_0}{\partial t} = D\nabla^2 u_0 \\ \dfrac{\partial c_0}{\partial t} = 0 \end{cases}$$

with initial condition

$$u_0(0) = \begin{cases} 0 & x^2+y^2 \le \omega^2 \\ \dfrac{k_{\text{off}}}{k_{\text{on}}+k_{\text{off}}} & x^2+y^2 > \omega^2 \end{cases}, \quad c_0(0) = \begin{cases} 0 & x^2+y^2 \le \omega^2 \\ \dfrac{k_{\text{on}}}{k_{\text{on}}+k_{\text{off}}} & x^2+y^2 > \omega^2 \end{cases}.$$

Therefore, the solution for u is given by Eq. (8.20) and $c_0 = c_0(0)$, which describes the FRAP due to diffusion in the early phase.

On the other hand, to study long-timescale behavior of the solution, we introduce a slow time variable, $t' = \varepsilon t$, which shrinks the time interval over which diffusion occurs. Under this timescale, since we have $\partial/\partial t' = (\partial/\partial t)(\partial t/\partial t') = (\partial/\partial t)/\varepsilon$, with the initial condition given by Eq. (8.25),

$$\begin{cases} \varepsilon\dfrac{\partial u}{\partial t} = D\nabla^2 u - \varepsilon\kappa_{\text{on}}u + \varepsilon\kappa_{\text{off}}c \\ \varepsilon\dfrac{\partial c}{\partial t} = \varepsilon\kappa_{\text{on}}u - \varepsilon\kappa_{\text{off}}c \end{cases}.$$

From the series expansions of u and c in ε as in Eq. (8.29), the leading-order approximations are found as

$$\begin{cases} 0 = D\nabla^2 u_0 \\ \dfrac{\partial c_0}{\partial t} = \kappa_{\text{on}}u_0 - \kappa_{\text{off}}c_0 \end{cases},$$

$$u_0(0) = \begin{cases} 0 & x^2+y^2 \le \omega^2 \\ \dfrac{k_{\text{off}}}{k_{\text{on}}+k_{\text{off}}} & x^2+y^2 > \omega^2 \end{cases}, \quad c_0(0) = \begin{cases} 0 & x^2+y^2 \le \omega^2 \\ \dfrac{k_{\text{on}}}{k_{\text{on}}+k_{\text{off}}} & x^2+y^2 > \omega^2 \end{cases}.$$

Because the steady-state solution of the heat equation satisfies $0 = D\nabla^2 u_0$, we have $u_0 = u_i = k_{\text{off}}/(k_{\text{on}}+k_{\text{off}})$. The equation in c_0 satisfies

$$\varepsilon\frac{\partial c_0}{\partial t}e^{\kappa_{\text{off}}t'} + \varepsilon\kappa_{\text{off}}c_0e^{\kappa_{\text{off}}t'} = \varepsilon\kappa_{\text{on}}u_ie^{\kappa_{\text{off}}t'},$$

or

$$\frac{d}{dt'}\left(c_0e^{\kappa_{\text{off}}t'}\right) = \kappa_{\text{on}}u_ie^{\kappa_{\text{off}}t'},$$

which can be directly integrated as

$$c_0 = \frac{\kappa_{\text{on}} u_i}{\kappa_{\text{off}}}\left(1 - e^{\kappa_{\text{off}} t'}\right) = c_i\left(1 - e^{\kappa_{\text{off}} t'}\right).$$

This result indicates that under conditions where that binding kinetics are much slower than diffusion, after a while the FRAP curve looks like a simple exponential curve (Fig. 8.7C) [26].

Full binding-diffusion model

We have seen that Eq. (8.24) can be simplified depending on the size of the parameters or timescales of binding and diffusion. However, not all cases can be reduced to simple forms. There exist cases with parameter sets that have to be analyzed by the full binding-diffusion model.

One may ask if there is a more simplified case than the diffusion-dominant model, effective-diffusion model, and binding-dominant model (biphasic model) since we have considered only a few possible cases. To the contrary, it has been demonstrated [27] that Eq. (8.24) can be simplified into the three cases we studied in the previous sections. This region can be found in the $k_{\text{on}} \times k_{\text{off}}$ parameter space shown in Figure 8.7D.

8.4.2. Anomalous Diffusion

8.4.2.1. Pure Diffusion vs. Anomalous Diffusion

So far we have studied molecular diffusion in a homogeneous medium, which is characterized by a simple linear relationship between the mean-square displacement and time as $\langle r^2(t)\rangle = 4Dt$ on a membrane ($\mathbb{R}^2$). However, cell membranes are not necessarily homogeneous. Inhomogeneity of cell membranes usually slows down diffusion, which is commonly referred to as anomalous subdiffusion. The underlying mechanisms of anomalous subdiffusion on a cell membrane are not fully understood, but asymptotic behaviors of anomalously diffusing molecules can be described as

$$\left\langle r^2(t)\right\rangle = 4Dt^{\alpha} \tag{8.31}$$

for $0 < \alpha < 1$ (Fig. 8.8) [30]. In cell membranes, anomalous subdiffusion is most likely the result of both obstacles to diffusion and traps with a distribution of binding energies or escape times. Possible mechanisms are obstruction by mobile or immobile proteins, transient binding to mobile or immobile membrane structures, binding by extracellular matrix, confinement by membrane skeletal corrals, hydrodynamic interaction, and possibly more [30–39].

In mathematics theory, diffusing molecules are often described as a continuous-time random walk model, in which a particle jumps from one location to another and waits a certain amount of time before the next jump. If the jump length $\mathbf{r} = (x,y)$ and waiting time t follow certain probability rules, i.e., random variables, the probability that a molecule jumps the distance $\mathbf{r}$ after waiting t is given by the joint probability $\Psi(\mathbf{r},t)$. From the joint probability $\Psi(\mathbf{r},t)$, we compute the jump length probability density function $\psi(\mathbf{r})$ and the waiting time probability density function $\phi(t)$ as

$$\psi(\mathbf{r}) = \int_0^{\infty} \Psi(\mathbf{r},t)\,dt, \quad \phi(t) = \iint_{\mathbb{R}^2} \Psi(\mathbf{r},t)\,dx\,dy,$$

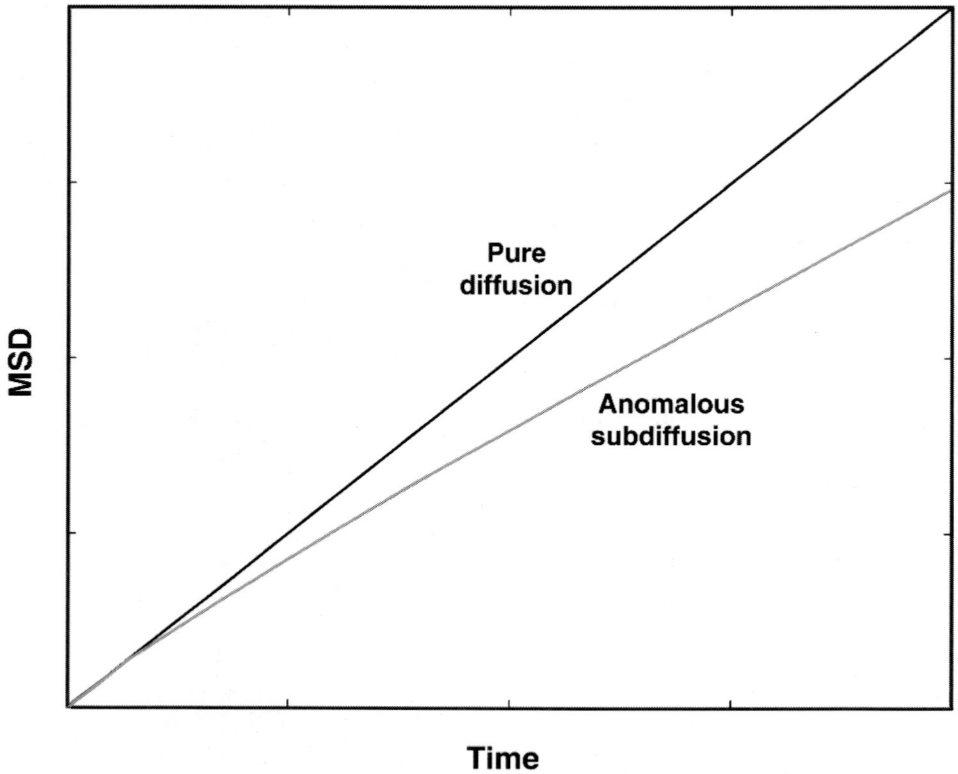

Figure 8.8. Mean-square displacement versus time for normal diffusion and anomalous subdiffusion. Please visit http://www.springer.com/series/7845 to view a high-resolution full-color version of this illustration.

The central limit theorem states that if the mean waiting time $\overline{t} = \int_0^\infty t\phi(t)\,dt$ and the jump-length variance $\sigma_r = \iint_{\mathbb{R}^2} (x^2 + y^2)\,dx\,dy$ are finite numbers, then the movement of the molecule is Brownian motion, or pure diffusion [40].

Anomalous subdiffusion occurs when the central limit theorem fails: the mean waiting time is infinite while the jump length variance is finite. The most studied case is when the probability density function of the waiting time has the asymptote

$$\phi(t) \sim \frac{1}{t^{\alpha+1}},$$

which is called a heavy tail distribution or a fat tail distribution [41]. In the case of pure diffusion, the waiting time probability density function decays exponentially ($\phi(t) \sim e^{-t}$).

8.4.2.2. Anomalous Diffusion FRAP Models

Feder's model

The simplest FRAP model for anomalous subdiffusion is based on Eq. (8.31), which can be re-written as

$$\langle r^2(t)\rangle = 4\left[Dt^{\alpha-1}\right]t = 4\Gamma(t)t$$

for $0 < \alpha < 1$ and solves

$$\frac{\partial u}{\partial t} = \frac{\Gamma(t)}{r}\frac{\partial}{\partial r}\left(r\frac{\partial u}{\partial r}\right).$$

In this case, by $t' = \int_0^t \Gamma(s)\,ds = Dt^\alpha/\alpha$ we can convert the above equation to the pure diffusion equation

$$\frac{\partial u}{\partial t'} = \frac{1}{r}\frac{\partial}{\partial r}\left(r\frac{\partial u}{\partial r}\right),$$

which was already solved in a previous section. Therefore, the anomalous FRAP formula is obtained by simply switching t' by Dt^α/α. For example, for FRAP formulae for the Gaussian laser profile similar to Eq. (8.17), the FRAP formula for anomalous diffusion can be written as

$$F(t) = F_i\left(\sum_{n=0}^{\infty}\frac{(-K)^n}{n!\left[1+n\left(1+2\dfrac{t^\alpha}{\alpha\tau_D}\right)\right]}\right)R + (1-R)F_0 \tag{8.32}$$

for mobile fraction $R = (F_\infty - F_0)/(F_i - F_0)$ [42]. Other formulae similar to those listed in Table 8.1 are also derived in a similar way. Notice that the FRAP formula includes a mobile fraction term. Using this formula, it was speculated that the immobile fraction is due to particles undergoing anomalous subdiffusion associated with restricted lateral mobility [42]. Although this approach provides a very simple anomalous diffusion model satisfying $< r^2(t) >= 4Dt^\alpha$, $0 < \alpha < 1$, it does not describe any underlying physics causing anomalous diffusion.

Fractional-diffusion model

Another approach to analyze anomalous diffusion is based on a continuous-time random walk model. As briefly mentioned, in this approach, it is assumed that the joint probability for a molecule to jump length **r** after waiting t is given by $\Psi(\mathbf{r},t)$ and the waiting time probability density function $\phi(t)$ satisfies asymptotically a fat tail distribution or molecules undergo long rest kinetics. Now, we define the arrival probability density function $\varphi(\mathbf{r},t)$ to give the p-robability for a molecule to arrive at a location **r** at t. Suppose that a molecule that arrived at a location $\mathbf{r}'$ at time t' eventually arrived at location **r** at time t. In the simplest case, it may jump from $\mathbf{r}'$ to **r** (the distance will be $\mathbf{r}-\mathbf{r}'$) after waiting $t-t'$, and this satisfies

$\varphi(\mathbf{r},t)=\Psi(\mathbf{r}-\mathbf{r}',t-t')\varphi(\mathbf{r}',t')$ [42]. However, because the molecule could arrive at $(\mathbf{r};t)$ after multiple jumps from $(\mathbf{r}',t')$, if all possible independent paths are included it will satisfy

$$\varphi(\mathrm{r},t)=\int_0^\infty\iint_{\mathbb{R}^2}\Psi(\mathrm{r}-\mathbf{r}',t-t')\,\varphi(\mathbf{r}',t')d\mathbf{r}'dt+\delta(\mathrm{r})\delta(t).$$

Finally, we define the probability density function for a molecule to be found at a certain location $\mathbf{r}$ at a certain time t as $\Phi(\mathbf{r},t)$. For a molecule at $\mathbf{r}$ at time t' (with probability $\varphi(\mathbf{r},t')$) to be still found at $\mathbf{r}$ at time t, it should not move between t' and t, i.e., the waiting time for the molecule should be longer than $t-t'$. The probability for a molecule to wait longer than $t-t'$ is given by $\phi^c(t-t')=1-\int_0^{t-t'}\phi(s)\,ds$ [43]. In this case, the waiting time until time t can be any arbitrary number less than t ($t'<t$), and

$$\Phi(\mathbf{r},t)=\int_0^t\varphi(\mathbf{r},t')\,\phi^c(t-t')\,dt'.$$

In the pure diffusion case, $\Phi(\mathbf{r},t)$ is given by Eq. (8.15). The Fourier-Laplace transform (the Fourier transform in spatial variable and the Laplace transform on time variable) satisfies

$$\hat{\tilde{\Phi}}(\xi,s)=\left(\frac{1-\hat{\phi}(s)}{s}\right)\left(\frac{\tilde{\Phi}_0(\xi)}{1-\hat{\tilde{\Psi}}(\xi,s)}\right),$$

where $\tilde{\Phi}_0(\xi)$ is the Fourier transform of the initial distribution of molecules. For a fat tail waiting time distribution $\phi(t)\sim t^{-\alpha-1}$, the Fourier-Laplace transform becomes

$$\hat{\tilde{\Phi}}(\xi,s)=\frac{1}{s}\left(\frac{\tilde{\Phi}_0(\xi)}{1-D\dfrac{|\xi|^2}{s^\alpha}}\right),$$

whose inverse transform satisfies

$$\Phi(\mathbf{r},t)-\Phi_0(\mathbf{r})={}_0D_t^{-\alpha}\left(D\nabla^2\Phi(\mathbf{r},t)\right),$$

where the Riemann-Liouville operator ${}_0D_t^{-\alpha}$ is defined as

$${}_0D_t^{-\alpha}\Phi(\mathbf{r},t)=\frac{1}{\Gamma(\alpha)}\frac{\partial}{\partial t}\int_0^t\frac{\Phi(r,t)}{(t-t')^{1-\alpha}}\,dt'.$$

Finally, by differentiating on both sides with respect to t, we get a fractional diffusion equation:

$$\frac{\partial\Phi(\mathbf{r},t)}{\partial t}={}_0D_t^{1-\alpha}\left(D\nabla^2\Phi(\mathbf{r},t)\right), \tag{8.33}$$

of which the solution can be found in terms of Fox functions, which also satisfies anomalous diffusion, $<r^2(t)>=4Dt^\alpha$ [43,44].

Recently, the FRAP formula based on Eq. (8.33) has been reported as [44]

$$F_{\alpha}(t)=\int_{0}^{\infty}\left[A(s,t)F(s)\right]ds\,,$$

$$A(s,t)=\frac{1}{s}\sum_{n=0}^{\infty}\frac{(-1)^{n}}{\Gamma(1-\alpha-\alpha n)\Gamma(1+n)}\left(\frac{s}{t^{\alpha}}\right)^{n},$$

$$F(s)=F_{i}\left(\sum_{n=0}^{\infty}\frac{(-K)^{n}}{n!\left[1+n(1+2t/\tau_{D})\right]}\right)R+(1-R)F_{0}\,.$$

8.5. CHALLENGES AND ARTIFACTS IN CONFOCAL FRAP

FRAP has been one of the most successful approaches to measure protein and lipid diffusion coefficients in cells. However, it is important to understand the limitations of FRAP in order to avoid reaching erroneous conclusions. In this section, we will review limitations of confocal FRAP. As we saw in previous sections, FRAP theory is based on many simplifying assumptions, which are not always fulfilled. In general, these assumptions fall into categories related to LSCM, fluorescence molecules, and cells.

8.5.1. Limitations Due to LSCM

For convenience, we assume that photobleaching follows Eq. (8.5), with laser profiles given by either a uniform circle or a Gaussian profile. This also implies that we have assumed all the molecules within the ROI to be photobleached at the same time and that no diffusion occurs during photobleaching. However, neither of these is necessarily true when an LSCM is used to photobleach or excite fluorescent molecules [45,46].

8.5.1.1. Diffusion during Photobleaching

Commonly used LSCMs can bleach a circular ROI with diameter 1 μm in approximately 1/100 s, which can be regarded as a very short time. However, in practice, in order to obtain well-defined FRAP curves, more than one cycle of photobleaching scans has to be performed. In this way, one can obtain deeper bleaching depth and a higher gradient of concentration of fluorescent and nonfluorescent molecules between the bleach ROI and surrounding area. However, it has been reported that as more photobleaching scans (iterations) are performed, diffusion coefficients are progressively underestimated [45]. This difference is much more significant for fast diffusing molecules. Mathematically, this has been modeled as

$$\frac{\partial C(x,y,t)}{\partial t}=D\nabla^{2}C(x,y,t)-\alpha I(x,y)C(x,y,t)\,. \tag{8.34}$$

This effect is thought to occur as the result of diffusion of molecules into and out of the bleach ROI during photobleaching. When diffusion occurs during photobleaching, the initial state of FRAP is not correctly represented by Eq. (8.6), but instead yields wider yet shallower bleaching profiles. When taking a shallower FI in the ROI as an initial value, it takes a longer time for fluorescent molecules to fill the bleached area. As a result, diffusion during iterative photobleaching yields a smaller diffusion coefficient than predicted by Eq. (8.13). To overcome this, one can compute the initial FI distribution from Eq. (8.34) rather than using Eq. (8.6) [46].

8.5.1.2. Size of the Bleach ROI

In FRAP, the bleach ROI size is important for several reasons. First of all, the characteristic time used to compute diffusion coefficients depends on the square of the ROI radius (Eq. (8.16)). Second, most theoretical FRAP curve-fitting formulae assume the ROI size is much smaller than the object under study to guarantee full recovery. However, too small an ROI size results in very noisy data.

Similar to when different numbers of photobleaching scan iterations are employed, the use of different-sized ROIs can yield different values of diffusion coefficients [46]. A smaller ROI size tends to give underestimated diffusion coefficients. This occurs when there is a mismatch between the area of the experimentally obtained bleach ROI and the ROI size set by a user. In one example, the bleach ROI had an almost 50% larger radius than the ROI set by the LSCM system in the case of an ROI radius of 1 μm [47]. The difference between the radius set by the user and the radius determined immediately post-bleach decreased as ROI size was increased and the values were almost identical when ROI size was set to 5 μm [47].

This mismatch is likely also due to diffusion of molecules during the photobleach scans. To overcome this problem, it is possible to do image analysis of the first post-bleach image to determine the "effective radius" of the photobleached area in the case of uniform circle laser profiles. In the case of Gaussian laser profiles, fitting to Eq. (8.3) can be used to find an appropriate effective radius (ω_{eff}) for the laser profile [46].

8.5.2. Artifacts Associated with Photobleaching

Another hidden assumption Eq. (8.5) makes is that fluorescence molecules are irreversibly photobleached following first-order kinetics. However, it is known that high photobleaching laser intensity can cause excitation saturation in the triplet state (Fig. 8.2) rather than bleaching to ground level. This is called reversible photobleaching. Once reversible photobleaching happens, FRAP may occur even without diffusion via relaxation from the triplet state to the ground state at a timescale of the order of milliseconds to several seconds. While reversible photobleaching is insensitive to fluorescent molecule concentration, it increases when the photobleaching time and/or bleaching laser beam intensity increase as well as the viscosity of the matrix where fluorescence molecules are embedded.

The phenomenon of reversible photobleaching has been documented for fluorescent proteins [48]. To check if such reversible photobleaching interferes with FRAP by diffusion, FRAP experiments can be done in fixed samples or over different-sized ROIs. In the absence of reversible photobleaching, fluorescence recovery should not be seen in fixed samples.

8.5.3. Assumptions about Biological Membrane Structure

In analysis of FRAP data, a pure diffusion model can be employed when molecules move randomly over certain distances with an equal probability (Eq. (8.11) or (8.12)). Contrary to this assumption, membranes of cells are heterogeneous, containing both mobile and immobile proteins as well as membrane skeletal corrals [49]. As a result, in cells the diffusion of most membrane proteins is slower than predicted by the Saffman-Delbrück equation [23]. These features can also give rise to anomalous subdiffusion, which is covered in a previous section.

To deal with the problem of anomalous diffusion, mathematical models in which the mean-square displacement is related to time to a certain exponent less than one have to be used. One approach employs a diffusion equation with a time-dependent diffusion coefficient. A fractional diffusion equation also shows this nonlinear relationship between mean-square displacement and time.

8.6. SUMMARY

Fluorescence is a type of electromagnetic radiation from a fluorophore stimulated by a certain wavelength of light. Fluorescence can be well described by the Jablonski energy diagram, which describes the transitions of an electron between different energy levels. When a fluorophore absorbs light within its excitation spectrum, electrons in the ground singlet state of a fluorophore (S_0) become excited and promoted to an inherently unstable higher energy level (S_1 or S_2) in approximately 10^{-15} s, and then quickly fall to the lowest excited singlet level (S_1) in the next 10^{-12} s. Molecules in the S state are eventually stabilized when the excited electrons return to their original energy level (S_0) by releasing the energy difference in a form of visible light, which is called fluorescence. The event occurs on the order of 10^{-8} s, which is known as the fluorescence lifetime. Before re-excitation, the stabilized molecules undergo a short refractory period, during which they are insensitive to absorption light. In addition, the number of excitation and emission cycles of a given fluorophore undergoes is finite and strongly fluorophore-specific. This characteristic number of excitation and emission cycles of a fluorophore determines how long it can fluoresce under excitation light. In general, the higher the excitation light intensity, the shorter time a fluorophore will emit before photobleaching. As a result, above a certain level of excitation laser intensity, fluorophores are photobleached almost immediately, which is a major advantage of FRAP.

In FRAP experiments, a series of fluorescence intensities in a region of interest (ROI) are recorded as a function of time under low laser intensity, after the fluorescent molecules within the ROI are irreversibly photobleached by a pulsed laser with high intensity. The FRAP curves may have different characteristics, depending on the kinetics (binding/diffusion) of the fluorescence-tagged molecules. Kinetic constants can be found by fitting FRAP curves to a mathematical model.

Mathematical models can be incorporated into FRAP analysis by an FRAP formula. In general, FRAP formulae have the form

$$F(t) = q\iint_{\mathbb{R}^2} [\varepsilon I(x,y)] C(x,y,t)\,dx\,dy\,,$$

where q is quantum yield, $\varepsilon I(x,y)$ is attenuated excitation laser, and $C(x,y,t)$ is a solution of mathematical models that describe the evolution of fluorophore concentration in time and space. For $C(x,y,t)$, the pure diffusion equation, a reaction-diffusion equation, and a fractional diffusion equation were considered in this chapter, which allows us to study diffusion of molecules in membranes as well as the structure of membranes.

For confocal FRAP by LSCMs, the finite scanning time and diffusion during photobleaching should be taken into account. These artifacts usually lead to underestimated diffusion coefficients in the pure diffusion case.

ACKNOWLEDGMENTS

Supported by NIH R01 GM 073846 to AK.

PROBLEMS

8.1. In an FRAP experiment, the scaled fluorescence intensities are $F_i = 1$, $F_0 = 0.4$, and $F_\infty = 0.8$. Assume the bleaching spot is a disk with radius 1 μm and the half time of recovery (τ_D) is 0.2 s.

a. Find the diffusion coefficient for this protein. (Assume a uniform circular laser profile.)

b. Find the mobile and immobile fractions of this protein.

8.2. Two FRAP experiments were performed for the same protein using circular ROIs of different size. The half times of recovery (τ_D) were 0.5 and 1 s. What is the ratio of the ROI radii in two experiments? (Assume a uniform circular laser profile.)

8.3. If the diffusion coefficient of a protein X is approximately 40 μm^2/s, what is the expected half time of recovery for a circular ROI with a 10-pixel diameter? (Assume a uniform circular laser profile and 1 pixel = 0.11 μm).

8.4. For two soluble proteins X and Y with molecular weights W_x and W_y and diffusion coefficients D_x and D_y, respectively, derive the following relation from the Stokes-Einstein equation. (Assume the shapes of these proteins are spherical.)

a. $D_y = D_x \sqrt[3]{\dfrac{W_x}{W_y}}$.

b. If the molecular weight of X is 27 kDa, and the diffusion coefficient of X is 10 μm^2/s, what is the expected diffusion coefficient of a protein with a molecular weight of 40 kDa?

8.5. For Eq. (8.24), assume $D_2 \neq 0$ and $k_{on} \gg D_1/(4\omega^2),\ D_2/(4\omega^2)$. D_2 may not necessarily be smaller than D_1.

a. Show that Eq. (8.24) can be approximated by the effective diffusion model with an effective diffusion coefficient

$$D_{eff} = \frac{k_{off}D_1 + k_{on}D_2}{k_{on} + k_{off}}.$$

b. From this show that the partition coefficient ($k_p = k_{on}/k_{off}$) is given by $k_p = \dfrac{D_1 - D_{\text{eff}}}{D_{\text{eff}} - D_2}$.

8.6. If Eq. (8.24) is in the reaction-dominant regime, the concentration of fluorescence intensity is represented in terms of binding complex concentration, $c(x,y,t) \simeq c_i \left(1 - e^{\kappa_{\text{off}} t'}\right)$. From this and employing $F(t) = q\varepsilon \iint_{\mathbb{R}^2} I(x,y) C(x,y,t)\, dx\, dy$, derive the corresponding FRAP formula.

FURTHER STUDY

Lakowicz JR. 2006. *Principles of fluorescence spectroscopy*. New York: Springer.

Weiss TF. 1996. *Cellular biophysics: transport*. Cambridge: MIT Press.

Rogers LCG, Williams D. 2000. *Diffusions, markov processes and martingales*, Vol. 1: *Foundations*. Cambridge: Cambridge UP.

Keener JP. 2000. *Principles of applied mathematics*. Boulder, CO: Westview Press.

Crank J. 1980. *The mathematics of diffusion*. New York: Oxford University Press.

Carslaw HS, Jaeger JC. 1986. *Conduction of heat in solids*. Oxford University Press.

Podlubny I. 1999. *Fractional differential equations*. New York: Academic Press.

REFERENCES

1. Valeur B. 2001. On the origin of the terms fluorescence, phosphorescence and luminescence. In *New trends in fluorescence spectroscopy: applications to chemical and life sciences*, pp. 3–6. Ed B Valeur, J Brochon. Berlin: Springer-Verlag.
2. Coons A, Creech H, Jones R. 1941. Immunological properties of an antibody containing a fluorescent group. *Proc Soc Exp Biol Med* **47**:200.
3. Coons AH, Kaplan MH. 1950. Localization of antigen in tissue cells: improvements in a method for the detection of antigen by means of fluorescent antibody. *J Exp Med* **91**:1–13.
4. Minsky M. 1988. Memoir on inventing the confocal scanning microscope. *Scanning* **10**:128–138.
5. Shimomura O. 1995. A short story of aequorin. *Biol Bull* **189**:1–5.
6. Prasher DC, Eckenrode VK, Ward WW, Prendergast FG, Cormier MJ. 1992. Primary structure of the *Aequorea victoria* green-fluorescent protein. *Gene* **111**:229–233.
7. Chalfie M, Tu Y, Euskirchen G, Ward WW, Prasher DC. 1994. Green fluorescent protein as a marker for gene expression. *Science* **263**:802–805.
8. Inouye S, Tsuji FI. 1994. Aequorea green fluorescent protein: expression of the gene and fluorescence characteristics of the recombinant protein. *FEBS Lett* **341**:277–280.
9. Liebman PA, Entine G. 1974. Lateral diffusion of visual pigment in photoreceptor disk membranes. *Science* **185**:457–459.
10. Poo M, Cone RA. 1974. Lateral diffusion of rhodopsin in the photoreceptor membrane. *Nature* **247**:438–441.
11. Edidin M, Zagyansky Y, Lardner TJ. 1976. Measurement of membrane protein lateral diffusion in single cells. *Science* **191**:466–468.
12. Schlessinger J, Webb WW, Elson EL, Metzger H. 1976. Lateral motion and valence of Fc receptors on rat peritoneal mast cells. *Nature* **264**:550–552.
13. Axelrod D, Koppel DE, Schlessinger J, Elson E, Webb WW. 1976. Mobility measurement by analysis of fluorescence photobleaching recovery kinetics. *Biophys J* **16**:1055–1069.
14. Lippincott-Schwartz J, Snapp E, Kenworthy AK. 2001. Studying protein dynamics in living cells. *Nat Rev Mol Cell Biol* **2**:444–456.
15. Lippincott-Schwartz J, Altan-Bonnet N, Patterson G. 2003. Photobleaching and photoactivation: following protein dynamics in living cells. *Nat Cell Biol* (Suppl):S7–S14.

16. Reits EAJ, Neefjes JJ. 2001. From fixed to FRAP: measuring protein mobility and activity in living cells. *Nat Cell Biol* **3**:E145–E147.
17. Sprague BL, McNally JG. 2005. FRAP analysis of binding: proper and fitting. *Trends Cell Biol* **15**:84–91.
18. Stavreva DA, McNally JG. 2004. Fluorescence recovery after photobleaching (FRAP) methods for visualizing protein dynamics in living mammalian cell nuclei. *Methods Enzymol* **375**:443–455.
19. Lakowicz JR. 1999. *Principles of fluorescence spectroscopy*, 2nd ed. New York: Springer-Verlag.
20. Feynman R, Leighton R, Sands M. 2005. *The Feynman lectures on physics*. Reading, MA: Addison Wesley.
21. Nakano A. 2002. Spinning-disk confocal microscopy—a cutting-edge tool for imaging of membrane traffic. *Cell Struct Funct* **27**:349–355.
22. Evans L. 1998. *Partial differential equations*. Providence, RI: American Mathematical Society.
23. Saffman PG, Delbrück M. 1975. Brownian motion in biological membranes. *Proc Natl Acad Sci USA* **72**:3111–3113.
24. Soumpasis DM. 1983. Theoretical analysis of fluorescence photobleaching recovery experiments. *Biophys J* **41**:95–97.
25. Cafiso D. 2005. Structure and interactions of C2 domains at membrane surfaces. In *Protein–lipid interactions: from membrane domains to cellular networks*, pp. 403–422. Ed L Tamm. Weinheim: Wiley-VCH.
26. Carrero G, Crawford E, Hendzel MJ, de Vries G. 2004. Characterizing fluorescence recovery curves for nuclear proteins undergoing binding events. *Bull Math Biol* **66**:1515–1545.
27. Sprague BL, Pego RL, Stavreva DA, McNally JG. 2004. Analysis of binding reactions by fluorescence recovery after photobleaching. *Biophys J* **86**:3473–3495.
28. McNally JG. 2008. Quantitative FRAP in analysis of molecular binding dynamics in vivo. *Methods Cell Biol* **85**:329–351.
29. Kang M, Kenworthy AK. 2008. A closed-form analytic expression for FRAP formula for the binding diffusion model. *Biophys J* **95**(2):L13–L15.
30. Saxton MJ, Jacobson K. 1997. Single-particle tracking: applications to membrane dynamics. *Annu Rev Biophys Biomol Struct* **26**:373–399.
31. Saxton MJ. 2001. Anomalous subdiffusion in fluorescence photobleaching recovery: a Monte Carlo study. *Biophys J* **81**:2226–2240.
32. Saxton MJ. 1996. Anomalous diffusion due to binding: a Monte Carlo study. *Biophys J* **70**:1250–1262.
33. Saxton MJ. 1994. Anomalous diffusion due to obstacles: a Monte Carlo study. *Biophys J* **66**:394–401.
34. Saxton MJ. 2007. A biological interpretation of transient anomalous subdiffusion, I: qualitative model. *Biophys J* **92**:1178–1191.
35. Saxton MJ. 2008. A biological interpretation of transient anomalous subdiffusion, II: reaction kinetics. *Biophys J* **94**:760–771.
36. Nicolau DV Jr., Hancock JF, Burrage K. 2007. Sources of anomalous diffusion on cell membranes: a Monte Carlo study. *Biophys J* **92**:1975–1987.
37. Ritchie K, Shan XY, Kondo J, Iwasawa K, Fujiwara T, Kusumi A. 2005. Detection of non-Brownian diffusion in the cell membrane in single molecule tracking. *Biophys J* **88**:2266–2277.
38. Jin S, Verkman AS. 2007. Single particle tracking of complex diffusion in membranes: simulation and detection of barrier, raft, and interaction phenomena. *J Phys Chem B* **111**:3625–3632.
39. Weiss M, Hashimoto H, Nilsson T. 2003. Anomalous protein diffusion in living cells as seen by fluorescence correlation spectroscopy. *Biophys J* **84**:4043–4052.
40. Hughes B. 1995. *Random walks and random environments*. New York: Oxford University Press.
41. Nagle JF. 1992. Long tail kinetics in biophysics? *Biophys J* **63**:366–370.
42. Feder TJ, Brust-Mascher I, Slattery JP, Baird B, Webb WW. 1996. Constrained diffusion or immobile fraction on cell surfaces: a new interpretation. *Biophys J* **70**:2767–2773.
43. Metzler R, Klafter J. 2000. The random walk's guide to anomalous diffusion: a fractional dynamics approach. *Phys Rep* **339**:1–77.
44. Lubelski A, Klafter J. 2008. Fluorescence recovery after photobleaching: the case of anomalous diffusion. *Biophys J* **94**:4646–4653.
45. Weiss M. 2004. Challenges and artifacts in quantitative photobleaching experiments. *Traffic* **5**:662–671.
46. Braga J, Desterro JM, Carmo-Fonseca M. 2004. Intracellular macromolecular mobility measured by fluorescence recovery after photobleaching with confocal laser scanning microscopes. *Mol Biol Cell* **15**:4749–4760.

47. Pucadyil TJ, Chattopadhyay A. 2006. Confocal fluorescence recovery after photobleaching of green fluorescent protein in solution. *J Fluoresc* **16**:87–94.
48. Sinnecker D, Voigt P, Hellwig N, Schaefer M. 2005. Reversible photobleaching of enhanced green fluorescent proteins. *Biochemistry* **44**:7085–7094.
49. Jacobson K, Sheets ED, Simson R. 1995. Revisiting the fluid mosaic model of membranes. *Science* **268**:1441–1442.

9

Punching Holes in Membranes: How Oligomeric Pore-Forming Proteins and Lipids Cooperate to Form Aqueous Channels in Membranes

Cécile Fradin,[1,2] Dmitri Satsoura,[1] and David W. Andrews[1]
[1]*Department of Biochemistry and Biomedical Sciences,*
[2]*Department of Physics and Astronomy, McMaster University, Hamilton, Ontario, L8S4M1, Canada*

9.1. INTRODUCTION

Many important biological processes are carried out by a small number of proteins working together as a team to accomplish a specific task. Cooperation between the different proteins is often accomplished through the formation of a supramolecular complex, comprised of either identical or different subunits. Although the formation of protein assemblies is a favored mechanism throughout the cell, it becomes especially important in lipid membranes, as evidenced by the numerous cellular events that are either triggered by or result in the formation of protein complexes in membranes. However, due to the difficulties associated with the study of membrane proteins, the formation of oligomers in lipid membranes is perhaps one of the least understood cellular processes.

In this chapter we focus our attention on a subset of membrane complexes — namely, those formed by proteins that are able to pass from a water-soluble to a transmembrane form in order to create a water-filled channel through the lipid membrane. These pore-forming proteins (PFPs) are found in many organisms throughout different kingdoms of life, from bacteria to human. They are often involved in cell death mechanisms through their capacity to break membrane permeability barriers, which can lead to dissipation of the membrane potential as well as

Address correspondence to Cécile Fradin, Department of Physics and Astronomy, McMaster University, 1280 Main Street W, Hamilton, Ontario, L8S4M1, Canada, 905 525-9140, 905 546-1252 (fax), <fradin@physics.mcmaster.ca>.

R. Faller et al. (eds.), *Biomembrane Frontiers: Nanostructures, Models, and the Design of Life*, Handbook of Modern Biophysics, DOI: 10.1007/978-1-60761-314-5_9,

introduction or leakage of enzymatic proteins. In fact, a large subset of the PFPs are toxins, and referred to in the literature as pore-forming toxins (PFTs). The association of several monomers into an oligomer is almost always an important aspect of the modus operandi of these proteins. Oligomerization can be useful in several ways: it results in structures large enough to delineate nanometer-size water-filled channels in lipid bilayers, it ensures the presence of large hydrophobic surfaces that can support insertion in the membrane, and it permits cooperative formation and insertion mechanisms.

The PFPs are fascinating for a number of reasons. From a purely structural point of view, they raise the question of how to design proteins that can exist both in soluble and membrane form, and how this information can be simply and entirely encoded in the protein sequence. From a molecular machine point of view, the very simple pore-forming system consisting of the activated protein, the lipid bilayer, and the aqueous solution constitutes an ideal system for structural investigations or studies of pore formation mechanisms. Some toxins also achieve translocation of a subunit across the membrane, and this is again done in a self-contained manner. From the pharmaceutical and pest-fighting points of view, they have an enormous potential as toxic molecules that are able to target specific organisms and not others. Finally, from a practical point of view, they are very attractive systems for biotechnology applications.

This chapter is organized as follows. We first give an overview of the prominent PFP families studied to date (§9.2). This is followed by a discussion of the different biophysical methods available for membrane oligomer structure determination and how they have been applied to the study of PFPs (§9.3). In the next section (§9.4), we consider the role played by the lipids in the formation and structure of membrane pores. We next detail the different steps often involved in the transition from a soluble monomeric protein to an oligomeric membrane pore (§9.5). Finally, the problems included at the end of the chapter invite the reader to take a brief look at the energetic aspects involved in membrane pore formation.

9.2. PORE-FORMING PROTEINS

PFPs are often categorized according to the classification proposed by Gouaux a decade ago into α-helix-PFPs and β-barrel-PFPs, depending on the structure adopted by the pore-forming region of the protein in the membrane [1]. Gouaux noted at the time that his classification "may, of course, require revision as more structures are determined." Yet, although many more PFPs have indeed since been discovered and characterized, their division into these two clearly separate classes still holds. Furthermore, antimicrobial peptides, which also cause membrane permeabilization starting from a water-soluble form, also tend to adopt either an α-helical or a β-hairpin structure in lipid bilayers [2]. It therefore seems that the α-helical and β-barrel structures represent two different rational answers to the membrane pore formation problem. Indeed, both allow the membrane-inserted protein chains to form peptide hydrogen bonds through intra- or interchain interactions without the help of any water molecules. Moreover, as will be discussed below, both are commonly found in the design of membrane proteins. It is interesting to note that these two different pore structures are formed by fundamentally different types of water-soluble proteins. The α-PFPs can be seen as essentially membrane proteins forced to adopt an inside–out configuration in solution in order to bury their large pore-forming hydrophobic region (usually a hydrophobic α-helix hairpin) in their core. The β-PFPs, on the other hand, can be considered as inherently soluble proteins, with only a small hydrophobic content, which therefore absolutely need to join forces (oligomerize) in order to form a hydrophobic surface large enough to insert in the membrane [3].

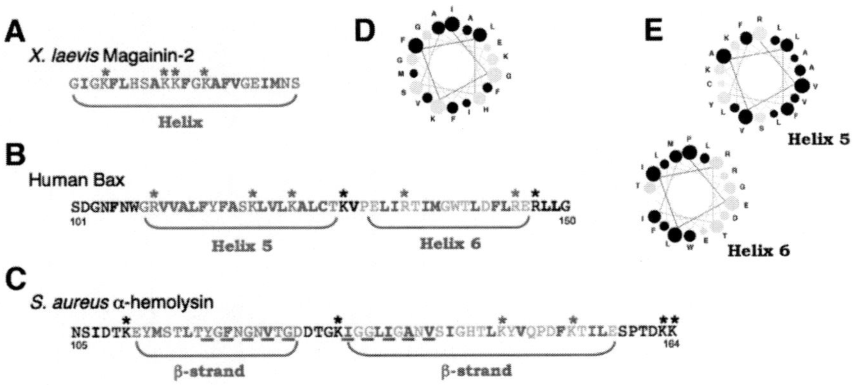

Figure 9.1. (A) Primary sequence of the bacterial peptide magainin-2. Hydrophobic amino acids are in bold font, and positively charged amino acids are marked by a star. (B) Part of the primary sequence of the α-PFP Bax containing the pore-forming domain. The indicated α-helices are those observed in the protein solution structure. (C) Part of the primary sequence of the β-PFP α-hemolysin containing the pore-forming domain. The indicated β-strands are those observed in the oligomer formed in the presence of detergent. Residues with sidechains orientated toward the hydrophobic part of the membrane are underlined. (D) Helical wheel representation of the magainin-2 α-helix. (E) Helical wheel representation of the two pore-forming helices of Bax. Helical wheels were plotted with http://cti.itc.Virginia.EDU/~cmg/Demo/wheel/wheelApp.html. Hydrophobic residues are indicated by a black disc, and polar and charged residues are indicated by a grey disc.

Not surprisingly, PFPs exhibit very diverse primary sequences. A common theme is the presence of an amphipathic region, which is the part of the protein interacting with the bilayer, and in the case of antimicrobial peptides often covers the entire length of the chain. Figure 9.1 shows several examples of amphipathic sequences found in PFPs, as compared to the sequence of an antibacterial peptide. Because they are neither strongly hydrophilic nor hydrophobic, it is difficult to recognize these sequences as corresponding to a pore-forming region without prior independent knowledge of the pore-forming capacity of the polypeptide chain. In other words, it is very difficult to predict the pore-forming capacity of a polypeptide chain based on its sequence. Furthermore, despite this common requirement for amphipathicity, pore-forming regions of PFPs do form different types of secondary, tertiary, and quaternary structures in the soluble form of the protein.

We describe here a few of the better-studied families of PFPs, starting with α-PFPs (§9.2.1) and followed by β-PFPs (§9.2.2). The prominent α-PFPs and β-PFPs highlighted in this review are listed in Tables 9.1 and 9.2, respectively.

Table 9.1. Characteristics of a Few α-PFPs

Family	Name	Organism of origin	Function	Activation mechanism	Membrane targeting	Pore structure	References
Pore-forming colicins	Colicin E1, Ia, Ib, A, B, N and K	Bacteria, including *E. coli*	Bacterial toxicity, via membrane permeabilization	Conformational change due to acidic conditions in the periplasm	Receptor binding (outer membrane) and affinity for negatively charged membranes	Variable size (internal diam. 0.6 nm and up) and stoichiometry (possibly monomeric)	[8,11,23,230]
	Diphtheria toxin	*Corynebacterium diphtheriae* bacterium	Mammalian cell toxicity, via translocation of the A chain	Conformational change due to acidic conditions encountered in the endosome	Binding to a glycoprotein receptor and affinity for negatively charged membranes	Variable size and stoichiometry (possibly monomeric)	[23,27,30,32,33]
Cry δ-endotoxins	Cry3A, Cry1Aa, Cry1Ac, Cry3Bb1, Cry2Aa...	*Bacillus thuringiensis* bacterium	Insecticide, via membrane permeabilization	C-terminal cleavage	Binding to a glycoprotein receptor		[39–41]
Pore-forming Bcl-2 family proteins	Bcl-XL Bax, Bid, etc.	Mammals	Regulators of outer mitochondrial membrane permeabilization	Cleavage (Bid) or interaction with a BH3-only Bcl-2 family member (Bax)	Affinity for neg. charged membranes (truncated Bid), interaction with truncated Bid (Bax)	Variable size (Bax: internal diam. 1 nm and up) and stoichiometry (Bax: dimeric and up)	[45,47,57, 172,196,198, 208, 221]
	α-synuclein	Vertebrates	Unknown	Denaturing conditions	Affinity for neg. charged and curved membranes	2–5 nm internal diam., probably pentameric or hexameric; also forms larger oligomers and fibrils	[62,63,65,66,69, 131,216]
Actinoporins	Equinatoxin I & II, Sticholysin I & II	Sea anemones (*Actinia equinam, Stichodactyla helianthus*)	Widespread cytotoxicity to animal cells		Affinity for sphingomyelin-containing membranes		[80,81,129,231]

2.1. α-PFPs

The formation of α-helices is a general mechanism for insertion of polypeptide chains into lipid membranes [4,5]. For instance, a transmembrane α-helical pore design is often observed for receptors (e.g., the light receptor bacteriorhodopsin [6]), and for ion channels (e.g., the KcsA potassium channel [7]). Similarly, for a large number of PFPs, the interaction with the membrane is mediated by a hydrophobic or amphipathic hairpin formed by two α-helices (as opposed to generally one α-helix for antimicrobial peptides). Since no structure of an α-PFP pore is available yet at atomic resolution, this had to be inferred by indirect methods, as will be discussed in Section 9.3.

The α-PFPs come from a wide variety of protein families. The better characterized are the flagship pore-forming colicins, followed by several others, which share structural (but not sequence) homologies with the colicins: the diphtheria toxin, the Cry δ-endotoxins and pore-

Table 9.2. Characteristics of a Few β-PFPs

Family	Name	Organism of origin	Function	Activation mechanism	Membrane targeting	Pore structure	References
Toxins of the α-hemolysin family	α-hemolysin, γ-hemolysins, leukocidins	*Staphylococcus aureus* bacterium	Animal cell toxicity, via membrane permeabilization			1.4 nm internal diam., 10 nm external diam. (α-hemolysin), generally heptameric	[85,88,89,91,92, 122,232]
Cholesterol-dependent cytolysins	Pneumolysin, streptolysin O, perfringolysin O	Different types of Gram-negative bacteria	Animal cell toxicity, via membrane permeabilization		Affinity for sterol-rich membranes	From 15 to 30 nm internal diameter, formed with up to 50 submits	[93,97,99, 100,140]
	Protective antigen of anthrax toxin	Bacterium *Bacillus anthracis*	Animal cell toxicity, via delivery of a catalytic subunit of the anthrax	Proteolysis	Receptor binding		[87,105,106, 108,109,201]
Cyt δ-endotoxins	CytB	Bacterium *Bacillus thuringiensis*	Insecticide	Proteolysis leading to dimer dissociation	Affinity for unsaturated phospholipids		[41,110,200]

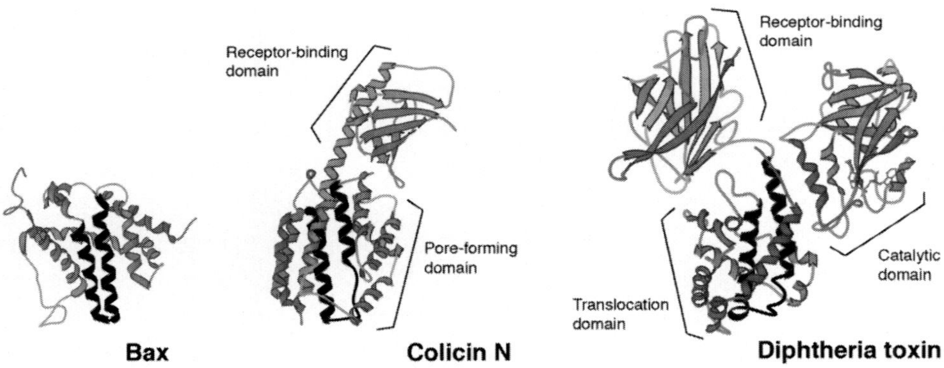

Figure 9.2. Water-soluble form of different α-PFPs. The structures of Bax (PDB: 1f16), Colicin N (PDB: 1a87) and diphtheria toxin (PDB: 1ddt) are presented, with the hydrophobic α-helical hairpin involved in pore formation highlighted in black. The structures have been drawn using KiNG software.

forming Bcl-2 family proteins. Figure 9.2 shows the structure of the water-soluble form of several α-PFPs, with the hydrophobic hairpin highlighted in black. These proteins tend to have a large α-helical content and typically adopt a structure that is sometimes described as a three-layer or sandwich structure, where the hydrophobic hairpin is part of the inside layer, that is, buried in the hydrophobic core of the protein. There are exceptions to this rule, such as the protein α-synuclein, which has a disordered structure in solution.

In membrane form, the hydrophobic α-hairpin is inserted in the lipid membrane, and brings about membrane permeabilization by forming a structure that is still ill-defined. Since an α-helix satisfies all the peptide hydrogen bonds through intrachain interactions, hydrophobic or amphipathic α-helices can exist within the lipid bilayer without the need to interact with other proteins. This might explain why α-PFPs sometimes seem to be able to form monomeric channels, and why the size of the pores formed by α-PFPs is often variable. This latter observation points to the possibility of a loose structure with strong involvement of the lipids (a lipidic pore), a supposition supported by the fact that α-PFP pores are rather unstable structures that often allow the quick exchange of lipid molecules between the membrane leaflets.

9.2.1.1. Pore-Forming Colicins

Colicins are highly toxic proteins expressed by different species of bacteria, including *E. coli* [8–10]. They target other related bacteria through a mechanism involving binding to a specific receptor on the outer membrane of the cell, energy-dependent translocation into the periplasm, and insertion into the plasma membrane. Accordingly, colicins are usually made up of three distinct domains: a receptor-binding domain, a translocation domain, and a "killing" domain. A subset of these proteins (colicins E1, Ia, Ib, A, B, N, and K) kill their target cells by forming small pores in the plasma membrane, in which case the "killing" domain is a pore-forming domain. The pore-forming domains of colicins for which a solution structure is available (full-length colicins Ia [11] and N [12], and pore-forming domain of colicins A [13] and E1 [14]) all have an α-helical globular structure with a hydrophobic hairpin buried in the interior of the domain. As an illustration, the water-soluble structure of colicin N is shown in Figure 9.2. Note that in this case the translocation domain is unstructured and does not appear in the X-ray structure of the protein. The pore-forming domain of colicins is thought to undergo a conformational change due to the acidic environment in the periplasm [15], which triggers insertion of the hydrophobic hairpin in the plasma membrane [16,17]. This step occurs without any loss of α-helical structure [18,19]. Colicins interact preferentially with negatively charged membranes [20]. Channel formation is thought to occur through further rearrangement of the protein, and the insertion of additional α-helices in the membrane [21,22]. Although it seems that monomeric colicins can form nonspecific ~1 nm ion channels [8,23], the fact that their pore-forming domain can translocate large hydrophilic segments of the protein [21,24] or even foreign epitopes [25] across lipid membranes could suggest the involvement of a larger oligomeric pore structure. Pore formation by a given colicin can be inhibited through direct binding to its corresponding immunity protein [26], which is found in the plasma membrane of the producing bacteria, and confers protection to the cell against its own toxin.

9.2.1.2. Diphtheria Toxin

Diphtheria toxin is the protein responsible for the serious upper respiratory tract disease diphtheria [27]. It is secreted by some strains of the bacterium *Corynebacterium diphtheriae*, and targets mammalian cells. It is synthesized as a single polypeptide chain, and later cleaved into two fragments [28]. The structure of the soluble cleaved form of the protein is shown in Figure 9.2. The N-terminal A chain contains the catalytic domain (C domain), which is an enzyme that interferes with protein synthesis and kills cells very efficiently [29]. The C-terminal B chain contains both the receptor-binding domain that recognizes an epidermal growth factor–like precursor and triggers endocytosis [30], and the translocation domain (T domain) that forms membrane pores [31] and allows translocation of the A chain to the cytoplasm by a mechanism that is still controversial. Although some experiments indicate that the channels formed by the diphtheria toxin are monomeric, their size has also been shown to vary with protein concentration, pointing to an oligomerization process [32]. The structure of the T domain in the water-soluble form of the protein is purely α-helical, with a central hydrophobic hairpin (helices 8 and 9) buried amongst the other seven helices [33,34]. The two hydrophobic helices are thought to insert in the membrane when encountering low-pH conditions in the endosome [35], which may cause the T domain to go to a molten globule state [36]. These helices are inserted perpendicular to the lipid bilayer and situated at the surface of the water-filled pore [37].

9.2.1.3. Proteins of the Cry δ-Endotoxin Family

The Cry δ-endotoxins are insecticide proteins produced by the bacterium *Bacillus thuringiensis* [38]. After synthesis, these toxins are stored in crystallized form in cellular inclusions, and released only when bacteria find themselves inside the insect midgut. The toxins are then activated through C-terminal protease cleavage. The Cry δ-endotoxins act through receptor-mediated binding, and each of them therefore targets specific insect species. This feature makes them very attractive as potential biopesticides. Their toxic activity is thought to come from their pore-forming ability, as these proteins form pores 1 to 2 nm in diameter, resulting in osmotic swelling and cell lysis. Although the amount of sequence homology between different Cry δ-endotoxins varies, those for which the structure of the water-soluble form has been solved (Cry3A [39], Cry1Aa [40], Cry1Ac [41], Cry3Bb1 [42], Cry2Aa [43]) show strong structural homologies between themselves. In addition, their pore-forming domain has the same overall structure as that of the colicins.

9.2.1.4. Pore-Forming Proteins of the Bcl-2 Family

The Bcl-2 family of proteins plays a crucial role in multicellular organisms, by regulating the activation of the caspases responsible for the final cell dismantlement in apoptosis [44–46]. This family is constituted of proteins having up to four homology regions, and has been traditionally divided into two groups: pro-apoptotic members, which promote caspase activation, and anti-apoptotic members, which tend to prevent it. When the first structure became available for a Bcl-2 family protein, Bcl-XL, its structural homology with the translocation domains of diphtheria toxin and the colicins was immediately noticed [47]. It is now suspected that many proteins in this family adopt a structure homologous to that of colicins [48]. This is certainly true of the following family members, for which the water-soluble monomeric structure is available: anti-apoptotic Bcl-XL [47], Bcl-2 [49], Bcl-w [50], CED-9 [51], and pro-apoptotic Bax [52] and Bid [53,54]. All these proteins possess the tell-tale mainly hydrophobic α-hairpin buried amongst six or seven amphipathic α-helices, and interact with membranes either constitutively or in response to apoptotic stimuli [55]. It is very well established that their antagonistic action regulates the permeability of the outer mitochondrial membrane [56]. Like the bacterial toxins, several of them were shown to form ion channels, and in some cases pores, in artificial membranes [57,58]. Of particular physiological relevance are the pores formed by the protein Bax, which are thought to be the cause for the release of apoptotic factors from the mitochondrial intermembrane space into the cytoplasm [59]. Although clearly oligomeric, the structure of these pores has remained elusive.

9.2.1.5. *α–Synuclein*

Although the function of α-synuclein is for the time being unknown, this synaptic protein has been under intense scrutiny since its aggregation into large fibrils was linked to the pathogenesis of Parkinson's disease and Lewy body dementia [60]. As for other amyloid diseases, the neuronal death observed in these diseases is thought to be due to protofibrils, which are the precursors of protein fibrils [61]. In its soluble cytosolic form, α-synuclein is a disordered protein. However, the N-terminal domain of the protein can interact with negatively charged, phosphatidylethanolamine (PE)-containing and highly curved lipid membranes, resulting in the formation of an α-helical secondary structure [62–65]. Interaction with membranes was shown to result in the formation of low-molecular-weight oligomers acting as ion channels [66], as well as pores with a ~2.0 nm inside diameter [67–69]. Pore formation can be inhibited by the related protein β-synuclein, through an unknown mechanism [70,71]. The pore-formation activity of α-synuclein has led to the hypothesis that its cytotoxic effects could be due to membrane permeabilization [72], possibly a general mechanism in amyloid diseases [73,74]. However, it is also likely that interaction with lipid membranes is part of the protein's normal function [75]. The structure adopted by α-synuclein when part of a membrane pore is still unclear. The soluble α-synuclein protofibrils are rich in β-sheets, as are all amyloid aggregates, and thus α-synuclein pores were initially thought to have a β-barrel structure [76]. However, the α-helical structure observed in the presence of detergents and lipids [62,64,65] makes α-synuclein a likely α-PFP.

9.2.1.6. *Others*

Many other proteins have been proposed to be α-PFPs but are much less well characterized than those described above. Among these are, for example, the *Pseudomonas aeruginosa* exotoxin A [77], the protein tryalisin produced by the salivary glands of the insect *Triatoma infestans* [78,79], and the sea anemone's actinoporins [80], including *Stichodactyla helianthus* Sticholysin II [81,82] and *Actinia equina* equinatoxin II [83]. In addition, it is likely that many α-PFPs remain to be discovered, so that the current list of α-helical PFPs is far from complete.

9.2.2. β-PFPs

For β-PFPs, the interaction with the membrane is mediated by a β-sheet, and probably consists in all cases of a transmembrane β-barrel [1,84]. This structural design was observed in atomic detail for the β-PFP α-hemolysin [85], which forms a heptameric membrane pore, as shown in Figure 9.3. All β-PFPs are thought to form pores involving a transmembrane β-barrel. This structure is also observed for different membrane proteins found in the outer membrane of Gram-negative bacteria or mitochondria (e.g., the bacterial outer membrane proteins or Omps [86]).

The universality of the transmembrane β-barrel design can be explained by the fact that, for a β-sheet to satisfy all of its peptide hydrogen bonds through interchain interactions and insert into a lipid membrane, it needs to be folded into a cylindrical structure. This often requires oligomerization, as each protein in the pore usually contributes only one or two β-hairpins (this part of the protein is called the insertion peptide) to form the β-barrel. The sidechains facing the acyl chains in the lipid bilayer then tend to be hydrophobic, while those facing the water-filled channel tend to be charged or polar. As an example, the alternating pattern of hydrophobic and hydrophilic residues in the insertion peptide of α-hemolysin is shown in Figure 9.1C. Discrete fixed stoichiometries are often (but not always) observed for β-PFP pores, which also have a propensity to be more stable than α-PFP pores. This is in keeping with the idea of a well-defined proteinaceous pore structure such as that of the α-hemolysin pore.

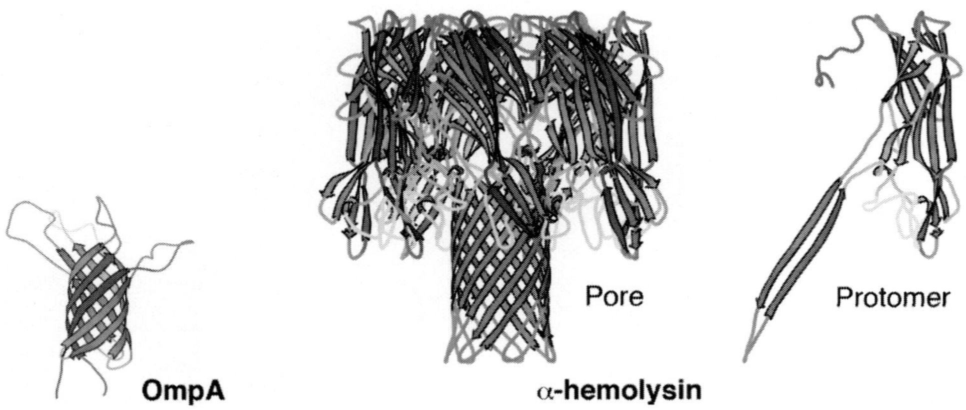

Figure 9.3. Transmembrane β-barrel pores. Structure of the pore formed by *Staphylococcus aureus* α-hemolysin (PDB: 7ahl), compared to the transmembrane domain of the outer membrane protein A (PDB: 2ge4).

In the soluble form of the β-PFPs, the insertion peptide can assume very different structures: a β-sheet for α-hemolysin [85], a disordered loop for the anthrax protective antigen [87], α-helices for the cytolysins. Contrary to the hydrophobic hairpin of the α-PFPs, the insertion peptide is only marginally hydrophobic and is usually solvent exposed in the soluble form of β-PFPs. This also explains why oligomerization is necessary in order to generate a large enough hydrophobic surface to drive insertion into the lipid bilayer. Figure 9.4 shows the solution structures of different β-PFPs, with the putative pore-forming domain (insertion peptide) indicated in black.

9.2.2.1. Toxins of the Staphylococcal α-Hemolysin Family

The α-hemolysin family is a family of toxins secreted by bacteria of the *Staphylococcal* species [88,89]. They are toxic to animals, in whose cell membranes they form lytic pores. The prototypical member of this family, α-hemolysin, is produced by *Staphylococcus aureus*. Although a solution structure of α-hemolysin is not available, it is the only PFP for which an atomic resolution structure of the membrane pore has been determined [85]. The α-hemolysin pore has a mushroom shape, and its stem is a transmembrane β-barrel with an internal diameter of 1.4 nm (Fig. 9.3). It is a heptameric structure, with each monomer contributing two antiparallel β-strands to the barrel, orientated at 38° from the lipid bilayer normal in a right-handed manner.

Other well-studied proteins in the family are γ-hemolysins and leukocidins, which share ~20–30% sequence identity with α-hemolysin [90]. These proteins are bicomponent toxins that fall into two classes (S and F) and form heptameric or octameric heterooligomers, with class S and class F proteins in a roughly 1:1 ratio [91,92]. When going from their water-soluble conformation to their pore-forming conformation, staphylococcal hemolysins and leukocidins transform their three β-strand insertion peptides into an extended β-hairpin. This conformational change can be visualized by comparing the structure of water-soluble γ-hemolysin shown in Figure 9.4 with that of the membrane inserted α-hemolysin protomer shown in Figure 9.3.

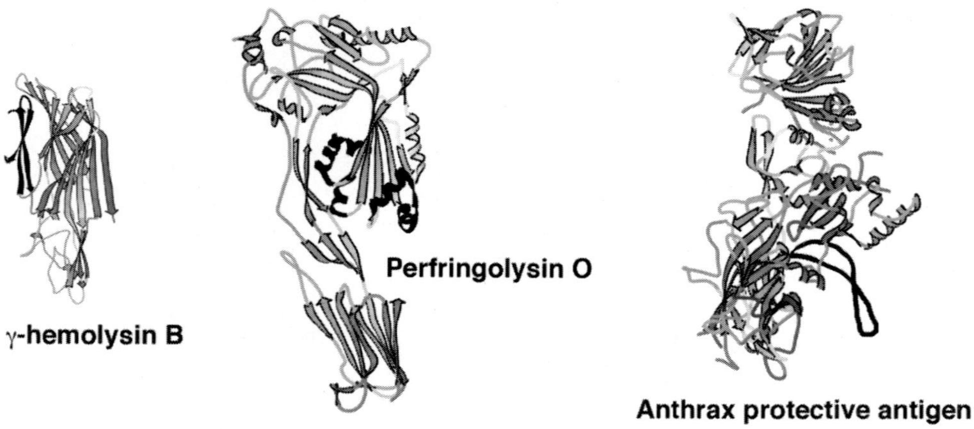

Figure 9.4. Water-soluble form of different β-PFPs. The structures of γ-hemolysin B (PDB: 1lkf), anthrax protective antigen (PDB: 1acc), and perfringolysin O (PDB: 1m3i) are presented. In each case, the insertion peptide is highlighted in black. In the case of the anthrax protective antigen, the insertion peptide, represented here by a black line, is unstructured and therefore did not appear in the crystal structure of the protein.

9.2.2.2. Cholesterol-Dependent Cytolysins

The cholesterol-dependent cytolysins are toxins secreted by several different types of Gram-positive bacteria (*Pneumoccoci, Streptococci, Listeriae, Bacillus, Clostridium*) that target animal cell membranes containing cholesterol or related sterols [93–96]. They have been classified as a single family based on their high level of sequence identity (between 40 and 80%), with the conserved sequence principally corresponding to that of pneumolysin, which is the smallest known cholesterol-dependent cytolysin. The prototypical member of the family, perfringolysin O, is secreted by *Clostridium perfringens*, and presents an elongated four-domain structure in its soluble form [97,98], which is shown in Figure 9.4. Proteins of this family can form very large pores, allowing the passage of folded proteins [99–101]. Streptolysin O, for example, can form pores up to 35 nm in diameter, which are probably constituted of about 45 to 50 subunits [99]. Cholesterol-dependent cytolysins are secreted along with enzymes such as proteases or nucleases, which can then penetrate and kill the cells permeabilized by the toxins. Proteins of this family are also known to cause membrane disruption, fusion, and lipid extraction [102]. The pore-forming domain is folded into several α-helices in the soluble form of the protein (as can be seen in Fig. 9.4), and transforms into two amphipathic transmembrane β-hairpins after the protein has oligomerized to form a transmembrane β-barrel [103].

9.2.2.3. Protective Antigen of Anthrax Toxin

The anthrax toxin is at the root of the symptoms of the lethal anthrax disease that can afflict animals, including humans [104]. It is constituted of three separate proteins secreted by the bacterium *Bacillus anthracis*. Like the diphtheria toxin, it is an example of A–B toxin. The B subunit (the protective antigen) forms pores in order to deliver a catalytic A subunit (either the lethal factor or the oedema factor). The protective antigen is targeted to host cells through receptor binding [105], and subsequently activated by protease cleavage of a 20-kDa N-terminal fragment [106]. The cleavage exposes binding sites for the catalytic proteins and also makes the protective antigen capable of oligomerization by removing steric interactions. Low-pH conditions trigger the insertion of the cleaved protein into the membrane, maybe through the protonation of histidine residues [107,108], and the formation of stable heptamers [109]. Those correspond to cation-selective pores 0.9 nm in diameter [107]. The structure of the water-soluble monomer has been solved (and is shown in Fig. 9.4), as well as the structure of a water-soluble homo-heptamer [87]. This water-soluble oligomer has a ring-shaped structure, with a 2.0–3.5 nm negatively charged inside diameter, explaining the cation selectivity but not the passage of a large A subunit. A heptameric membrane pore model based on this heptameric structure, with a β-barrel formed by 14 β-strands, was consequently proposed. A considerable conformational change would have to take place to go from the water-soluble to the membrane-inserted heptamer.

9.2.2.4. Others

Just as for α-PFPs, many more β-PFPs are known and many more remain to be discovered. Among the known β-PFPs that have not been discussed in detail here are the *Bacillus thuringiensis* Cyt δ-endotoxins [110], the *Aeromonas hydrophila* aerolysin [111–113], the *Clostridium septicum* α-toxin [114,115], the *Clostridium perfringens* iota toxin [116], and the *Pseudomonas aeruginosa* cytotoxin [117,118].

9.3. PORE STRUCTURES

When trying to understand the biological role of the proteins implicated in the formation of a particular complex, the complex structure, including its stoichiometry, is always an important clue, as are the interactions between the different complex subunits. In the case of membrane complexes, the embedding of the proteins in a lipid bilayer puts important constraints on the possible structures adopted by the complex.

A biological membrane constitutes a unique environment for proteins to be in. It is mainly composed of glycerophospholipids arranged into a bilayer, where the lipid hydrocarbon chains form a hydrophobic core lined on either side by the polar or charged lipid headgroups (Fig. 9.5). The organization of the lipids into this two-dimensional structure is the result of the excessively strong energy penalty associated with contacts between the apolar lipid tails and the polar water molecules. It results in a very large change in polarity over the roughly 3-nm distance that separates the two sides of the membrane. These changes are reflected in the very abrupt variations in dielectric permittivity that are encountered across the bilayer [119].

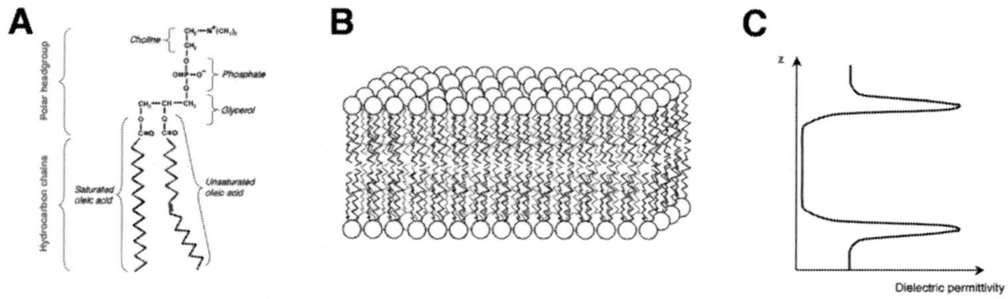

Figure 9.5. Structure of a lipid membrane. (A) The structure of a particular glycerophospholipid (dioleoylphosphatidylcholine, or DOPC). (B) Schematic arrangement of the lipids in a bilayer. For the sake of clarity, lipid chains have been represented as more ordered than they would be in a real lipid membrane. (C) Shape of the dielectric permittivity profile across a lipid bilayer, showing the large difference between the apolar membrane core and the very polar headgroup regions.

From structural studies of the pores formed by PFPs, it seems that the constraints provided by the lipid bilayer for the creation of defined water-filled channels can be fulfilled only in two different ways. Either the proteins can provide a continuous interface between the core of the bilayer and the water channel (protein-lined channel), or they can help the lipids bend into a toroidal shape (lipid-lined channel). Here we describe these two types of pores in detail (§9.3.1) before discussing different biophysical methods that have been employed to observe these structures (§9.3.2).

9.3.1. Pore Categories

9.3.1.1. Proteinaceous Pores

Initially, a model called the barrel-stave model was introduced to describe the pores formed by the α-helical antimicrobial peptide alamethicin [120]. This model predicts that in their pore-forming configuration the peptides insert in the membrane perpendicular to the plane of the lipid bilayer, and that they form a ring delineating a water-filled channel, each α-helix being like the stave of a barrel. An extension of this model to PFPs is the so-called proteinaceous or protein-lined pore model, of which a schematic representation is shown in Figure 9.6A. Proteinaceous pores are formed by a ring of contiguous pore-forming protein domains (not necessarily α-helices), defining a channel in the lipid membrane.

This type of pore is characterized by a very-well-defined annular protein pore structure from which the lipids are excluded. This has a few consequences in terms of observable properties. First, because the pore-forming domains are in direct contact with one another, one can expect this type of pore to be stable (meaning that they have a long lifetime and that they can be expected to be extracted from the membrane using detergents without significant changes to their structure), to have well-defined fixed stoichiometries and diameters, and to exhibit cooperativity during the formation process. Second, because the lipids are excluded from the pore structure itself and are not in contact with the water-filled channel, this type of pore should not aid transbilayer lipid motions.

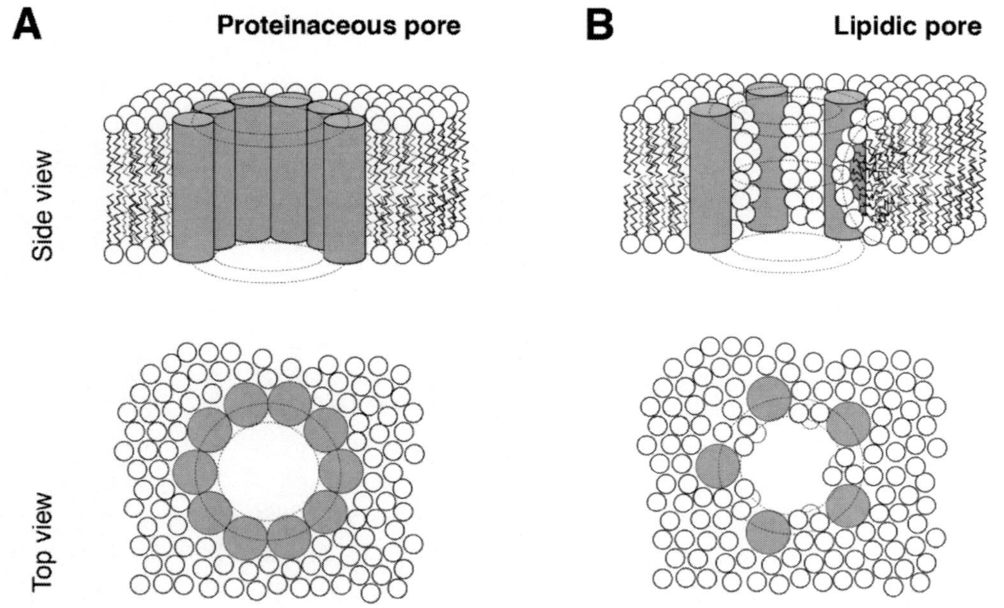

Figure 9.6. Schematic structure of two different types of membrane pores: (A) proteinaceous or barrel-stave pores, and (B) lipidic or toroidal pores. Each cylinder represents the pore-forming domain of an individual protein.

For proteinaceous pores, the transmembrane domain of the proteins forming the pore should have a large hydrophilic surface containing mainly polar residues (that can be characterized in the case of α-helices by the so-called "polar angle"), which will face the water-filled channel. In contrast, it should have a relatively small hydrophobic surface containing uncharged amino acids with aliphatic sidechains, which will face the hydrophobic core of the lipid bilayer. In addition, the two extremities of the pore-forming domain are close to the lipid headgroups. They can therefore be expected to contain basic residues like arginine (R) and lysine (K), which have a marked affinity for the phosphate group of phospholipids, as well as residues like tryptophan (W) or tyrosine (Y), which are both polar and aromatic, and which therefore have a preference for interfaces between polar and apolar media [121]. Indeed, polar and aromatic residues are often found on either side of the transmembrane region of membrane proteins, forming the so-called "aromatic belt."

A very clear example of a proteinaceous pore is that of α-hemolysin (Fig. 9.3). The seven β-hairpins contributed by the individual protein monomers to the stem region of the pore are in direct contact with each other through hydrogen bonds, and the β-barrel that they form constitutes a clear delimitation between the lipid bilayer on one side and the water-filled pore on the other (Fig. 9.3). As expected, residues on the inside of the β-barrel are hydrophilic, while on the outside they form a 2.8-nm thick apolar band [85]. The structure of the α-hemolysin pore bound to different glycerophospholipids has been resolved, and it was shown that each protomer in the pore possesses a lipid-binding pocket situated between the stem and the rim of the pore [122].

The headgroup of the phospholipid interacts with several residues in the binding pocket, including Y118 (at the top of the apolar band on the stem) and W179 and R200 (situated opposite on the inner surface of the rim). The hydrocarbon chains of the lipids interact with the apolar band of the stem. As expected, the α-hemolysin pore has a fixed stoichiometry (heptameric, although hexamers are also sometime detected) and is stable.

9.3.1.2. Lipidic Pores

A second pore model, known as the toroidal or wormhole model, was introduced to describe the pores formed by the antimicrobial peptide magainin [123,124]. The special feature of this model is that, with the help of the peptides, the membrane curves to form a torus-like channel, so that the water channel is in contact with both the peptides and the lipid headgroups. By extension, all membrane pores in which the membrane is curved into the shape of a torus are called lipidic, lipid-lined, toroidal, or wormhole pores. This type of structure is also sometimes referred to as a non-bilayer structure. A schematic representation of a lipidic pore is shown in Figure 9.6B.

For lipidic pores, the pore-forming domains of the proteins stabilizing the pores are not necessarily in contact with each other; instead, they are separated by lipids. Charged proteins are therefore more likely to form lipid-lined pores. In addition, a protein domain inserted into the membrane may be in contact with the lipid headgroups over its complete length, as shown in Figure 9.6B, so that instead of an "aromatic belt" one can expect to find interfacial residues (W, Y, R, and K) across the whole length of the pore-forming domain. In this type of pore, the role of the protein is not so much to create an interfacial area between the hydrophobic core of the membrane and the water-filled channel as to help reduce the stress caused by membrane deformation. Contrary to proteinaceous pores, the stoichiometry of a lipidic pore is not necessarily well defined, and the pores may be less stable. Also, compared to a protein-lined pore of the same size, fewer proteins are necessary to form a lipid-lined pore, and lipid-lined pores could have in principle very low stoichiometries. The formation of this type of pore will be facilitated by the presence of lipids promoting curvature of the lipid layer, and will clearly allow the transfer of lipids from one bilayer to the other.

Most antimicrobial peptides, which overwhelmingly adopt α-helical structures once inserted in lipid membranes, form pores whose properties are consistent with that of lipidic pores [125]. Similarly, most α-PFPs probably form lipidic pores, as proved by their capacity to promote transbilayer lipid transfer. This was shown, for example, for colicin E1 [126], for the Bcl-2 protein Bax [127], and for sticholysins [128].

9.3.2. Methods for Pore Structure Determination

There are several important difficulties associated with determination of the atomic structure of the pores formed by PFPs. A first difficulty is that pore formation requires the presence of a membrane, thus the pore structure must be solved in the presence of lipids or detergents. This usually precludes the use of X-ray scattering methods, which require the structural unit, in this case the pore, to be arranged into a crystalline superstructure. And this complicates the preparation of samples for nuclear magnetic resonance (NMR). A second difficulty is that the pores tend to be oligomeric structures with a large total molecular weight. This rules out the use of

NMR, which at the moment can only be applied to small proteins. A third difficulty is that a considerable number of PFPs seem to form pores with either disorganized, unstable, or polydisperse structures, which defeats the purpose of obtaining an atomic resolution structure in the first place. Although high-resolution structural methods are thus mostly not applicable to determination of the structures of the pores formed by PFPs, different low-resolution methods can provide complementary clues about these structures. We discuss here both high-resolution structure methods and alternative methods that can provide either low-resolution or partial structures for the pores, with an emphasis on those that give information about stoichiometry, and a particular focus on recently developed fluorescence methods.

9.3.2.1. Nuclear Magnetic Resonance

Solution NMR spectroscopy has proven useful for solving the structure of the water-soluble form of a number of small PFPs, for example, Bax [52] or Equinatoxin II [129]. It has also proven useful for studying the structure of antimicrobial peptides or small PFPs in their monomeric form in lipid micelles. For example, solution NMR was used to show that the 23-residue antimicrobial peptide magainin had an α-helical structure in several types of lipid micelles [130]. It was also used to produce a structure of monomeric α-synuclein in a lipid micelle [65], and this monomeric structure was later exploited in membrane-docking computer simulations to show that in a lipid membrane α-synuclein should form pentamers and hexamers with a pore-like structure [131]. However, the structure resolution of large oligomeric pores, whose slow tumbling motions result in broadened spectroscopic lines, still represent a challenge for solution NMR.

Unlike solution NMR, solid-state NMR is not size limited, and appears to be an ideally suited method for solving the structure of membrane proteins directly in orientated lipid bilayers [132,133]. In practice, however, the difficulty of correctly assigning each detected chemical shift to the correct atom still prevents the use of this technique for the structure determination of large oligomeric membrane structures such as the PFP pores. Nevertheless, solid-state NMR is being used successfully for determination of the membrane structure of antimicrobial peptides, and, as a definite advantage over solution NMR, this can be done in actual lipid bilayers. For example, the α-helical structure of gramicidin A [134] and magainin [135] in membranes was confirmed by solid-state NMR. Another advantage of solid-state NMR is that the orientation of the molecule with respect to the membrane can be detected. For example, it was shown that magainin is orientated parallel to the membrane, while gramicidin is perpendicular [136].

9.3.2.2. X-Ray Crystallography

X-ray diffraction has been very widely used for determination of the structure of the water-soluble form of many PFPs, for example, exotoxin A [77], colicin A [13], and aerolysin [113] among many others (see [23] for a recent review of the crystallographic structures of soluble PFPs). In the case of the anthrax protective antigen, the crystallographic structure of a soluble prepore was also obtained, where the protein is already assembled into a circular heptamer, but not yet inserted into a lipid membrane [87]. Still, as mentioned in the introduction, the requirement for the production of a crystal mostly eliminates X-ray diffraction as a useful method for PFP membrane pore structure determination. There is, however, a very important counterexample to this rule, which is the determination of the structure of the *S. aureus* α-hemolysin pore [85,122]. In that particular case, the pores were solubilized in detergent before crystallization was successfully attempted. As the α-hemolysin pore is very stable, this could be done without significant modification of the pore structure. The pore turned out to be a heptameric structure with a mushroom conformation, about 10 nm in height, defining a water-channel with a diameter ranging from 1.4 to 4.6 nm (Fig. 9.3). This structure confirmed the β-barrel nature of the α-hemolysin pore transmembrane region and established α-hemolysin as the prototypical β-PFP. It yielded numerous insights about the structural design of the pore and how this design relates to its function and promotes interaction with the membrane. To this day, α-hemolysin remains the only PFP for which the structure of the membrane pore is available in atomic details.

9.3.2.3. Electron Microscopy

The electron microscope produces two-dimensional projections of the object being imaged, and three-dimensional information is normally recovered by combining such projections, taken over a sufficient angular range [137]. Proteins or protein assemblies are observed as individual particles. It is then possible to precisely align the individual images of each single particle for the purpose of forming an average or, alternatively, the origin and relative orientation of each particle can be determined to combine all the images into a 3D map of the object. Electron microscopy (EM) combined with single-particle averaging methods enables the direct imaging of macromolecules with a resolution that can in theory be as good as ~1 Å [138]. In practice, because of instrumental limitations and because of beam-induced motion of the specimen during data acquisition, the achieved resolution is usually limited to ~10 Å. The value of the obtained three-dimensional maps, even with such a "low" resolution, cannot be overestimated. Besides providing information on quaternary structure, the EM map of a macromolecular assembly can be combined with the high-resolution model of its components to yield atomic resolution. To enable samples to sustain the high vacuum found inside the electron microscope and to achieve the highest possible resolution of proteins in their fully hydrated native state, samples are quickly frozen in vitreous ice, a method known as cryo-EM. With this method, molecules are not packed into a crystal and physiological buffers can be used; therefore, the obtained structures represent the physiologically relevant and fully functional form of the proteins, which is not always the case with X-ray crystallography. Although it is possible to maintain a protein in its native membrane environment using cryo-EM, a simpler and often-used procedure is to solubilize the membrane protein in detergent before sample preparation [139].

Because there is no restriction on the size of the molecular complexes that can be investigated with EM, it is not surprising that this technique has been extensively used for studies of protein assemblies. Maybe the most striking application of EM to the study of the oligomeric pores formed by PFPs so far has been the elucidation of the structure of the pores formed by the cholesterol-dependent PFP pneumolysin [140]. Cryo-EM images of pneumolysin oligomers formed in the presence of lipid membranes were averaged to obtain ~30-Å resolution structures of both the oligomeric prepore formed at the surface of the lipid membrane before full insertion of the proteins has taken place, and the fully membrane-inserted lipidic pores. Using atomic structure fitting of the perfringolysin O monomer (which is highly homologous to pneumolysin) into the cryo-EM map, it was possible to obtain atomic resolution structure for both the pore and the prepore. This study confirmed that the pneumolysin pores have a variable stoichiometry with an average of ~40 subunits and that the protomers extend into the membrane by collectively forming a β-barrel. More importantly, it allowed the visualization of the conformational changes that take place during membrane insertion of the prepore. A similar approach, combining cryo-EM imaging of 2D crystals formed in lipid bilayers or at the surface of lipid monolayers and atomic fitting of the monomer structure in the EM density map, had been used previously to reconstruct the structure of smaller pores — for example, the heptameric pores formed by *Aeromonas hydrophila* aerolysin [113,141] and the tetrameric pores formed by *Stichodactyla helianthus* Sticholysin II [81].

Even when atomic precision information is not obtained, the stoichiometry of a molecular complex can often be deduced from either the symmetry of the rotationally averaged EM image of the complex or complementary mass measurements done with scanning transmission electron microscopy (STEM) [142]. These methods were used in several instances to establish the stoichiometry of the pores formed by PFPs — for example, the pores formed by *E. coli* hemolysin E in detergent micelles were shown to be octameric [143], and the pores formed by *E. coli* cytolysin A in the presence of detergent were shown to be formed of 13 subunits [144].

9.3.2.4. Atomic Force Microscopy

Atomic force microscopy (AFM) is a technique that can provide high-resolution topographical images of biological samples, without the need for labeling or staining [145]. For structural studies of membrane proteins, it presents many of the same advantages as EM in terms of the possibility to image single molecular complexes at near-atomic resolution, while allowing imaging to take place in situ, with the protein inserted in a hydrated supported planar lipid bilayer. AFM has been successfully used to observe the stoichiometry of the pores formed by several PFPs, for example, the heptameric structure of the α-hemolysin pore and prepore [146,147]. Because the structure obtained does not need to be averaged, it has allowed detection of heterogeneities in pore structures, for example, the presence of hexameric α-hemolysin pores [148] or the size distribution of α-synuclein membrane oligomers [149]. Because it is nondestructive, AFM can also capture dynamical events, such as the collapse of the perfringolysin O prepore into a pore [150].

9.3.2.5. Fluorescence

The broad availability of fluorescent probes and intrinsically fluorescent proteins combined with the relative simplicity of site-directed mutagenesis or construction of fusion proteins have revolutionized modern biochemistry by making it possible to apply the very large toolbox of fluorescence-based techniques to the study of the structure and dynamics of biomolecules. This has proven especially useful in the case of α-PFPs, which tend to form molecular assemblies closely involving the participation of lipids, and which are very dynamic in nature, in which case the methods described above cannot be applied. Fluorescence resonance energy transfer (FRET), for example, has allowed the measurement of the distance between different helices of colicin A, an experiment which has led to the formulation of the umbrella model described in Section 9.5.3, and greatly contributed to our understanding of the membrane structure of colicins [151,152]. FRET also allows the assessment of the positioning of specific residues with respect to the lipid bilayer, as was done for example in the case of α-hemolysin [153]. Site-specific labeling with environment-sensitive probes has also proven to be a very useful method for identification of membrane-exposed residues and transmembrane domain structure [154]. A recent review has been devoted to the application of fluorescent techniques to the resolution of the structure of PFPs [155]. We depict here two recently developed fluorescence-based techniques — stepwise photobleaching and fluorescence fluctuation analysis — that specifically address the question of stoichiometry and have been applied to the study of PFP oligomers.

9.3.2.5.1. Stepwise photobleaching

Fluorophore photobleaching is an irreversible loss of fluorescence due to a light-induced change in the structure of the molecule. As photobleaching is a discrete process, the total fluorescence emission of a complex containing fluorescently labeled molecules will drop in a stepwise fashion if these molecules sequentially photobleach. It is thus possible to infer the number of fluorescently labeled molecules present in a complex if the individual photobleaching steps leading to the disappearance of the fluorescence emission of a single molecular complex can be resolved. This approach toward stoichiometry determination can be implemented with proteins labeled either with a covalently attached organic fluorophore or with a fused fluorescent protein. The latter approach is especially attractive, as it enables studying the complex stoichiometry in live cells. One limitation of this approach lies in the increasing difficulty of assessing the distribution of bleaching steps for complexes with a large number of subunits. Also, it is important to keep in mind the possibility of fluorophore self-quenching following complex assembly, which might lead to ambiguous results in interpretation of the stepwise loss of fluorescence.

In a pioneering study, the stepwise photobleaching method was tested on a number of complexes of known composition as well as being applied to determine the composition of multimeric receptors with as-yet-unresolved stoichiometries [156]. In particular, the stoichiometry of several membrane proteins was determined directly in live cells. Proteins of interest were fused to the green fluorescent protein (GFP) and expressed at very low levels in the absence of the endogenous protein. Imaging and photobleaching of the complexes formed by these proteins were done using total internal reflection fluorescence microscopy in order to eliminate the contribution of autofluorescence from the cytoplasm of the cells to the detected fluorescence signal. Stepwise photobleaching also was successfully used to compare the stoichiometries of the pores formed by the archetypal β-PFP, α-hemolysin, to that of a bicomponent β-PFP, leukocidin [157]. In contrast to the previous study, this was done by covalently labeling the proteins of interest with fluorescent probes. Moreover, oligomers of the two complexes in question were formed in vitro on rabbit red blood cell membranes.

9.3.2.5.2. Fluorescence fluctuation analysis

Fluorescence fluctuation methods are based on an analysis of the temporal fluctuations of the fluorescence signal obtained from a small, usually diffraction-limited, confocal detection volume. Different types of analyses can be carried out, resulting in different molecular information. Of interest for the resolution of molecular complex stoichiometries are the fluorescence fluctuation methods based on analysis of the photon counting histogram, which is the probability of recording a certain number of photons during a fixed time interval [158,159]. Because the photon counting histogram depends on the absolute concentration and the specific brightness of the diffusing fluorescent species present in solution, both these quantities can be retrieved using analysis methods such as fluorescence intensity distribution analysis (FIDA) [159]. The specific brightness is the number of fluorescence photons detected per particle per second. It is an important parameter in all types of fluorescence fluctuation experiments because it corresponds to the basic fluctuation in fluorescence intensity corresponding to the passage of one fluorescent particle through the detection volume. In the context of the study of molecular complexes, which diffuse as one single particle, it can be used as a measure of the stoichiometry if the specific brightness of the monomer is known, and if all the monomers in the complex are fluorescently labeled.

The first demonstration of fluorescence fluctuation analysis as a method to detect the stoichiometry of a complex was a proof-of-principle study of the dimerization of nuclear receptors in live cells [160]. Fluorescence fluctuation analysis was later applied to study the stoichiometry of a membrane protein oligomer, where analysis of the photon counting histogram using FIDA was used to monitor the oligomerization of EGF receptors in the plasma membrane of live cells under a variety of conditions such as cholesterol loading or depletion of the membrane [161]. Recently, we have taken advantage of this same method to investigate the stoichiometry of the membrane oligomers formed by an α-PFP, namely, the pro-apoptotic Bcl-2 family protein Bax [162]. We have used a reconstituted in-vitro system where insertion of a recombinant EGFP–Bax fusion protein into the membrane of diffusing large unilamellar vesicles is triggered by the action of a second pro-apoptotic Bcl-2 family protein, tBid. The membrane insertion of EGFP–Bax results in oligomerization of the protein and in formation of membrane pores. By first measuring the specific brightness of soluble EGFP–Bax, which is monomeric, and by placing ourselves in conditions where there is one or less membrane pore per vesicle, we were able to relate the specific brightness of a fluorescent vesicle, which we measured by FIDA, to the stoichiometry of the Bax membrane pore. We found that the pores varied in size in a concentration-dependent manner.

9.4. INFLUENCE OF THE LIPIDS ON PORE FORMATION AND STABILITY

The contribution of the lipids to pore formation is clearly very important, as they affect every stage of this process. Here we describe their influence on the binding (§9.4.1) and insertion (§9.4.2) of proteins in the membrane, and on the stability of the final pore structure (§9.4.3).

9.4.1. Membrane Recognition

Each type of PFP has a marked preference for a specific type of membrane, and this strong selectivity is essential to their function. For example, many bacterial pore-forming toxins preferentially target the plasma membrane of eukaryotic cells, in sharp contrast with antimicrobial peptides that selectively bind bacterial membranes. In several cases (e.g., colicins, δ-endotoxins), specific receptors present in the membrane of target cells explain how the PFP recognizes its target membrane, even if the receptor has not been identified yet. But in many other cases, selectivity is achieved uniquely by recognizing specific lipid compositions.

Cells of different type have membranes with different lipid compositions, which confer upon them very different physicochemical properties. For example, the bacterial surface tends to be negatively charged, with a high membrane content of anionic lipids such as phosphatidylglycerol (PG) and cardiolipin (CL), and with the presence of a layer of polyanionic lipopolysaccharides (LPSs) on the outer membrane of Gram-negative bacteria [163]. On the other hand, the plasma membrane of mammalian cells is characterized by a high cholesterol content and therefore a high rigidity, and has an overall neutral electric charge, containing large amounts of the neutral zwitterionic lipids phosphatidylcholine (PC) and sphingomyelin [164]. Also, different organelles can be distinguished according to their lipid composition. Mitochondria, for example, have high levels of PC and phosphatidylethanolamine (PE), and very low levels of cholesterol, phosphatidylserine (PS), and sphingomyelin, compared to the plasma membrane. The inner mitochondrial membrane is characterized by a high level of CL, while the outer mitochondrial membrane is characterized by a relatively high level of phosphatidylinositol (PI) [165].

Membrane selectivity based on lipid composition is an efficient solution for proteins involved in host defense or toxicity, because it allows adequate specificity combined with the possibility of a broad range of action against a series of cells of the same type with similar membrane properties. But how is this specificity achieved? Below we consider several mechanisms by which a PFP might be able to recognize a specific type of lipid membrane.

9.4.1.1. Membrane Composition

In several cases, the specificity exhibited by a PFP for a given type of membrane has been narrowed down to the presence of a specific lipid in the membrane. The most obvious case is that of cholesterol-dependent cytolysins, whose affinity for a membrane depends on the presence of sterols in the lipid bilayer [94,166–169]. These proteins thus target cholesterol-rich animal cells rather than the membrane of the producing bacteria. The affinity for membranes with high cholesterol content is due to a direct interaction between a tryptophan-rich motif in the protein and the sterol [170]. For example, perfringolysin O interacts with cholesterol via a tryptophan-rich motif on its C terminal [171]. Other examples of specific lipids that have been proposed to act as a receptor for a family of PFPs are sphingomyelin, which is required for the activity of actinoporins [80], and cardiolipin, which helps target the Bcl-2 family protein Bid to lipid membranes [172]. Also, Cyt δ-endotoxins bind to unsaturated lipids, which are more abundant in the cell membranes of dipteran species [173].

9.4.1.2. Electrostatic Properties of the Membrane

In the absence of specific protein/lipid interactions, the simplest mechanism by which a protein and a membrane might preferentially interact is through electrostatic interactions, with a positively charged protein attracted to the usually negatively charged lipid membrane (since charged phospholipids are usually anionic because of their phosphate group). Indeed, this is how antimicrobial peptides, which are expressed by animals and plants, achieve their strong selectivity for bacterial cells over eukaryotic cells. These peptides are typically cationic, with a high lysine and arginine content (e.g., Magainin 2, whose sequence is shown in Fig. 9.1, has an overall charge between +3 and +4 at physiological pH, taking into account the partial charge of the N-terminal amino group). Their positive charge gives them selectivity for the negatively charged outer leaflets of the membrane of Gram-negative bacteria. Clusters of positive charges are often observed for PFPs (e.g., see the sequence of the pore-forming domain of Bax in Fig. 9.1), and preference for negatively charged membranes is observed for many PFPs, including α-synuclein [63], Bcl-2 family member Bid [174,175], diphtheria toxin [176], and colicins [20]. Even though in the latter two cases recognition by a receptor is also involved, a negatively charged membrane accelerates membrane binding and insertion.

The main effect of an attractive electrostatic interaction between a protein and a membrane is to increase the local concentration of the protein just above the membrane surface [177]. This effect can be quantified by considering the membrane surface potential, ψ_0, and the net charge of the protein, q. The protein interfacial concentration (c_I) is related to its bulk concentration (c_B) by

$$c_I = c_B e^{-q\psi_0 / kT},$$

where k is Boltzmann's constant and T is the absolute temperature. The membrane surface potential can be calculated from the membrane surface charge density (σ) using Gouy-Chapman theory [178]. For a small surface potential, one simply has

$$\psi_0 = \frac{\sigma}{\varepsilon_0 \varepsilon_r \kappa}.$$

In this expression, ε_0 is the permittivity of free space, ε_r is the dielectric constant of the bulk, and $1/\kappa$ is the Debye length, defined as

$$\frac{1}{\kappa} = \sqrt{\frac{\varepsilon_0 \varepsilon_r kT}{2 N_A e^2 I}},$$

where N_A is Avogadro's number, e is the charge of the electron, and I is the ionic strength of the bulk aqueous solution.

Given a certain interfacial concentration c_I, the concentration of the protein in the membrane, c_M, can then be calculated using the intrinsic partition coefficient K:

$$c_M = K c_I.$$

The intrinsic partition coefficient is directly related to the free energy of binding, and in general does not depend on electrostatic interactions. Selectivity for the charged membrane can be achieved simply because the protein interfacial concentration is high in comparison to the bulk protein concentration. This selectivity has been studied in detail for several antimicrobial peptides [179–181].

9.4.1.3. Elastic Properties of the Membrane

The elastic properties of a lipid bilayer vary greatly with lipid composition. For example, cholesterol promotes the creation of lipid domains with increased order and rigidity, also known as lipid rafts. On the contrary, unsaturated lipids tend to partition to more disordered lipid phases. And lipids such as PE, which have relatively small headgroups, tend to destabilize lipid bilayers by forcing them to adopt high curvatures and therefore promote the formation of non-bilayer structures [182].

In theory, it is therefore possible that the bilayer elastic properties could play a role in the targeting of PFPs to membranes with specific compositions. Indeed, the binding of some amphipathic α-helices to lipid membranes was proposed to depend on curvature [183]. And α-synuclein was found to preferentially bind to membranes with high curvature, probably because of the presence of defects in headgroup packing [184]. This finding is consistent with the activity of the protein at neuronal membranes, which are rich in PE.

9.4.2. Pore Formation

9.4.2.1. Role of Electrostatic Interactions

Electrostatic interactions can also affect the insertion of PFPs in a lipid membrane. This influence of electrostatic interactions has been studied in detail for colicin A, because a pattern of negatively charged residues is conserved among different colicins, meaning that electrostatic interactions must play an important role [185]. A large conformational change is necessary to go from the water-soluble form of a PFP to its membrane-inserted form. It is believed that low pH conditions such as those encountered in the endosome (for diphtheria toxin) or the periplasm (for colicins) help bring about this conformational change. Because a negatively charged membrane also attracts positively charged ions, including protons, the local pH close to the membrane can be significantly lower than that in the bulk. Thus, negatively charged membranes might help trigger membrane insertion by providing low-pH conditions [186]. In addition, it was proposed that the strong electric potential gradient close to the charged membrane might help orientate proteins in order to promote correct insertion [185].

9.4.2.2. Membrane Thickness and the Aromatic Belt

The hydrophobic portion of the membrane-spanning domains of PFPs, like that of many membrane proteins, is often bordered by an "aromatic belt" on either side, with a high occurrence of the aromatic or heteroaromatic residues tryptophan, tyrosine, and histidine [187]. Those residues have a high affinity for the complex interfacial regions of the lipid membrane. They have both aromatic rings that can interact with the hydrocarbon chains through hydrophobic interactions and polar groups that can form hydrogen bonds with the polar lipid headgroups [121,188]. While the hydrophobic stretch of a PFP membrane-spanning domain is what provides anchorage of the protein in the membrane, the eventual presence of aromatic belts allows a stronger level of control of the exact positioning of the protein with respect to the membrane, and can stabilize specific membrane conformations of the protein [189,190].

The presence of a hydrophobic mismatch, that is, a discrepancy between the length of the hydrophobic stretch of the PFP and the thickness of the hydrophobic part of the membrane, can promote transmembrane domain tilt and oligomerization [191]. These effects will be even more dramatic in the presence of aromatic belts [192].

9.4.3. Pore Structure

9.4.3.1. Proteinaceous Pore

The atomic resolution structure of the proteinaceous α-hemolysin pore in the presence of different lipid headgroups suggested that, as for other membrane oligomers, the lipids were mediating interactions between protomers, with several of them occupying well-defined positions at the surface of contact between two protomers [122]. In the case of the α-hemolysin pore, phosphocholine headgroups serve as a direct link between the stem and the rim domains of the pore. Therefore, even in a proteinaceous pore, lipids are an inherent part of the pore structure. It is therefore not surprising that pore formation might be influenced by the membrane lipid composition, since binding of some lipids to the pore structure can be favored through specific interactions.

9.4.3.2. Lipidic Pore

The formation of a lipidic pore involves the bending of a usually planar lipid monolayer into a half torus (cf. Fig. 9.6). In this configuration, the membrane needs to adopt both a positive curvature (in the direction parallel to the pore axis) and a negative curvature (in the direction perpendicular to the pore axis). The positive curvature requirement is especially costly, as the membrane must bend 180° over a very short distance corresponding to the membrane thickness. Because of this, the presence of lipids promoting a spontaneous curvature of the membrane is expected to stabilize lipidic pore structures. Such lipids are those for which the cross-sectional area of the lipid headgroup is not matched to that of the acyl chain, creating a preference for a conical geometry (as opposed to a cylindrical geometry for lipids like PC for which both cross-sectional areas match). Lipids with relatively large headgroups (such as lysophosphatidylcholine, lysolecithin, or lysophosphatidic acid [193]) promote positive lipid monolayer curvature, while lipids with relatively small headgroups (such as PE or cardiolipin) promote negative lipid monolayer curvature. Both types of lipids promote the formation of non-bilayer structures. Both are also expected to help the formation of lipidic pores, but the former are expected to be especially helpful. An influence on pore formation of lipids inducing spontaneous curvature has been observed for several antimicrobial peptides — for example, magainin [194] and syringomycin [195] — as well as for different α-PFPs, in particular sticholysin [128], Bax [58,196], and colicin E1 [197]. This effect is often taken as an indication that the pores formed are lipid-lined, as protein-lined pores should not be affected by the intrinsic curvature of the membrane.

9.5. PORE FORMATION MECHANISM

The assembly of a protein membrane pore can generally be subdivided into the following steps (although not necessarily in this exact order): (1) activation, (2) diffusion to the membrane and binding, (3) conformational change and partial membrane insertion, (4) oligomerization (or further activation), and (5) insertion of transmembrane domains leading to formation of an aqueous channel. These different steps are illustrated in Figure 9.7 and detailed below.

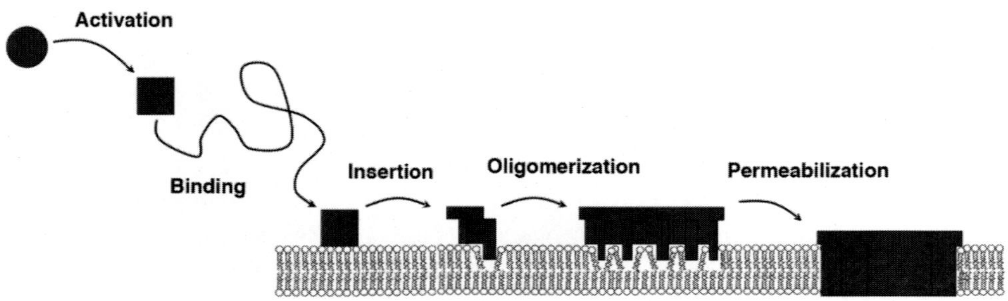

Figure 9.7. General pore formation mechanism for PFPs. The five different steps illustrated here do not necessarily occur in this exact order for all PFPs.

9.5.1. Activation

All the PFPs considered here start off as soluble proteins, and in the majority of cases their membrane pore-forming activity is activated as a response to a trigger. The result is usually exposure of one or several hydrophobic patches, which makes the protein competent for membrane insertion.

One common trigger is the enzymatic cleavage of a peptide (the activation peptide) at either the C or N terminal of the full-length water-soluble PFP, as is for example the case for most δ-endotoxins [41], for pro-apoptotic Bid [198], and for aerolysin [199]. Sometimes, as is the case for *B. thuringiensis* δ-endotoxin CytB, both the C and N terminal of the protein need to be cleaved for activation [200]. In addition, the activation peptide sometimes prevents oligomerization because of steric hindrance. This is the case for the anthrax protective antigen, whose N terminal must be cleaved off before oligomerization can occur [201].

Another common way in which the exposure of hydrophobic patches can be achieved is though a conformational change of the protein. The trigger for the conformational change can either be a change in the protein's environment, such as the acidic conditions encountered by colicins in the bacterial periplasm [202,203], or interaction with another protein, such as the interaction of Bax and cleaved Bid [204].

9.5.2. Binding to the Membrane

Efficient targeting to host membranes, which means that the targeting is both specific and fast, is an essential functional requirement for all PFPs. It ensures that the PFP is present in the host membrane at a high enough concentration in order to permit oligomerization and pore formation. It is, however, especially important for those PFPs that are secreted at low concentration in a rapidly clearing environment such as the intestinal or urinary tracts.

Membrane specificity can be achieved either through direct interactions with lipids in the membrane (including, for example, electrostatic interactions, as detailed in §9.4.1), or through protein receptor binding. Sometimes the same PFP uses both strategies; for example, colicins use receptor binding to target the outer membrane of bacteria, and electrostatic interactions to bind to the plasma membrane. Whereas interaction with the lipids allows targeting a broad range of host membranes, receptor binding permits restricting the targeting to a narrow range of host cells. For example, each δ-endotoxin targets the right insect species thanks to the use of specific membrane receptors. In addition, when the receptors are clustered in membrane domains (as is the case for the GPI-anchored proteins, which cluster in cholesterol-rich domains), receptor binding may allow further concentrating the protein locally on the membrane. Although some PFP receptors have been characterized (see Tables 9.1 and 9.2), many remain to be identified.

The binding of a PFP to a host membrane is in general fast and reversible. For cholesterol-dependent cytolysins, for which the kinetics of binding has been studied in detail, it was found that PFPs bind membranes through a first-order reaction, which means that they probably bind as monomers [205]. The rate at which a PFP may bind to a membrane is limited by the diffusion-limited flux of the protein on the membrane, F, which depends on the PFP concentration, c, as well as on the characteristic size of the target cells (or organelles), R: $F \approx 4\pi RDc$ [206]. If each collision results in binding of the PFP to the membrane, then the rate of binding events (per unit volume of solution) is $k_B \approx 4\pi RDCc$, where C is the concentration of target cells. Although the assumption that every collision results in binding might be reasonable when host membrane recognition occurs via protein-lipid interactions, it becomes unreasonable when the recognition occurs via receptor interaction. Therefore, increased specificity through receptor interactions is obtained at the cost of a decreased rate of binding.

Membrane binding of PFPs often leads to an important conformational change of the proteins characterized by a loosening or even disappearance of the tertiary structure, which helps subsequent membrane insertion [202,207]. For example, the binding of colicin A to negatively charged membranes, which places the protein at an effectively low pH, results in the formation of a molten globule, making the protein competent for membrane insertion [186]. For Cry δ-endotoxins, the observed conformational change is thought to be due to binding of the protein to its receptor [41]. The interaction of Bax with membranes has also been shown to result in a detectable conformational change before membrane insertion, although the origin of this change is unknown [208].

9.5.3. Insertion into the Membrane

Contrarily to membrane binding, the insertion step is usually slow and irreversible. This irreversibility is sometimes engineered in the design of the PFP. For example, in the case of diphtheria toxin, the decrease in pH experienced by the protein in the endosome neutralizes two acidic residues found at the tip of the hydrophobic α-helical hairpin, facilitating the membrane insertion of this domain [209,210]. However, once these residues reach the cytoplasmic side of the membrane, they become charged again, and this renders the membrane insertion process irreversible.

Insertion of the PFP in the membrane leads to yet another conformational change, which can make the protein competent for oligomerization, sometimes through allosteric effects [211]. The conformation adopted by PFPs after membrane insertion often remains uncertain, because of the difficulty of resolving membrane structures. However, several possible prototypical conformations have been proposed. Colicins, and in general α-PFPs, do not lose their high helical content upon membrane insertion [19,151,208]. Therefore, the membrane form of these proteins likely simply corresponds to a rearrangement of the helices present in the soluble form. Originally, a model called the "umbrella" model was proposed for colicins, where the hydrophobic hairpin is inserted perpendicular to the plane of the membrane, while the other helices are arranged like an umbrella above and parallel to the membrane [212]. This model was later challenged, as different measurements seemed to indicate that the hydrophobic hairpin of colicins was inserting only shallowly and laying more or less parallel to the plane of the membrane, in a "penknife" conformation [17,152,213]. One possibility is that the hydrophobic hairpin is in fact alternating between these two different conformations [214]. The conformation adopted by the non-transmembrane domains of the PFPs is also uncertain, and might be very different in each case. Although for several α-PFPs there is evidence that helices other than the transmembrane helices lay parallel to the membrane in the lipid headgroup region [37,175], the picture obtained from the atomic structure of the β-PFP prepore or pore is quite different, as these proteins tend to adopt a "mushroom" conformation, with the non-transmembrane regions of the protein retaining a high level of tertiary structure [85,140]. The mushroom conformation is more likely to allow strong specific interactions between the non-transmembrane domains leading to the formation of homooligomers with well-defined stoichiometries, as was indeed observed for some β-PFPs, but not for α-PFPs.

9.5.4. Oligomerization

The majority of PFPs have been shown to form membrane oligomers on their way to pore formation. In the case of α-PFPs, oligomerization is hard to separate from final pore formation, which means that these two steps may not be distinct. In the case of β-PFPs, however, a clear intermediate prepore state has been identified, where the proteins have oligomerized but where an aqueous membrane channel has not yet been formed [215].

The oligomerization process of PFPs has been discussed in terms of cooperativity. In cooperative models of oligomer formation, there is a rate-limiting nucleation step and a propagation step corresponding to the growth of the oligomer by addition of monomers, which happens at a much faster rate than the nucleation. In other words, there is a significant energy barrier present only at the nucleation step. Nucleation-dependent mechanisms seem to be the rule rather than the exception for the auto-assembly of PFPs, since cooperative oligomerization has been shown to occur both for α-PFPs such as Bax [162] or α-synuclein [131,216] and for β-PFPs such as streptolysin O [205]. It is interesting to note that one of the prototypical nucleation-dependent mechanisms is the formation of insoluble fibrils by amyloid proteins [217,218]. It should therefore come as no surprise that fibril and membrane oligomer formation are also cooperative for the amyloid-like protein α-synuclein [216], and by extension for other PFPs.

One interesting question is the nature of the nucleation step. For α-synuclein, the formation of membrane oligomers has been investigated using computer simulations, which showed that in this particular case the nucleation step corresponds to the formation of propagating dimers

[131]. Dimerization was also proposed to be the rate-limiting step in the case of streptolysin O [219]. For Bax, on the other hand, the nucleation step is thought to correspond to the activation of one monomer into an oligomerization-competent state by direct interaction with membrane-inserted tBid [46].

Although the growth of the oligomer that follows the rate-limiting nucleation step is generally thought to be rapid, in the case of large oligomers such as those formed by the cholesterol-dependent cytolysins, it is still possible to observe intermediate states with partially formed, arc-like oligomers [220].

9.5.5. Membrane Permeabilization

For α-PFPs, as mentioned in the previous section, pore formation seems to occur concurrently with oligomerization. In fact, experimental evidence indicates that many α-PFPs, such as colicins or diphtheria toxin, do not require oligomerization to form small channels in the membrane. For those α-PFPs that seemingly do require oligomerization, such as Bax, the size of the pores can be increased by addition of monomers [162,221]. These observations can be explained in the context of the "umbrella" and "penknife" models by the fact that a single protein can independently adopt two conformations in the membrane, one corresponding to a closed state (with no, or only few, transmembrane helices) and one corresponding to an open state (with the insertion of supplementary helices in the membrane) [41,212]. The gathering of several proteins in the same region of the membrane will increase the potential for the formation of large aqueous channels, as many transmembrane helices act in concert. The nature of this process is still being debated, but it is reminiscent of pore formation by the antimicrobial α-helical peptide, which is better understood [222].

For β-PFPs on the other hand, intermediate structures have been observed where the proteins are oligomerized into a prepore, that is, an insertion-competent but non-lytic precursor ring structure [20,223–226]. In the case of anthrax protective antigen, this structure has even been crystallized [87]. For those β-PFPs that form very large pores, membrane-inserted incomplete arc-shaped pore structures have been observed [220,219], which means that membrane insertion does not require the oligomer to form a complete ring structure, but does require a sufficient number of monomers to be present. The formation of the pore, rather than oligomerization itself, seems to be rate limiting for these β-PFPs. This may be because the formation of the β-barrel that constitutes the transmembrane part of the pore for these proteins involves secondary structure rearrangements. For cholesterol-dependent cytolysins, for example, two helical domains change to a β-sheet upon membrane insertion [103,154]. Membrane insertion requires the oligomerization of a significant number of monomers before it can occur, showing that the process is cooperative [224,227]. For aerolysin, for example, it was shown that the heptamer, but not the monomer or dimer, is able to insert in the membrane [20]. The requirement for oligomerization before insertion in the membrane can proceed is due to the fact that a membrane-inserted β-sheet that is not folded into a barrel will have unsatisfied hydrogen bonds. Indeed it was shown that hydrophobic peptides form β-sheets in membranes in a cooperative manner because of the prohibitive energy cost of inserting peptide bonds with unsatisfied hydrogen bonds into membranes [228]. The driving force for membrane insertion is therefore coming from the hydrophobicity generated when the proteins oligomerize. The membrane insertion of β-PFPs is essentially irreversible, as proven by the fact that non-denaturing detergents or even sodium dodecyl sulfate (SDS) cannot break up the oligomer [220,229]. This means that the free energy of the oligomer drops considerably upon membrane insertion.

9.6. SUMMARY

The membrane pores formed by pore-forming toxins and other closely structurally related proteins are a fascinating example of oligomeric structures involving both proteins and lipids. Pore-forming proteins can be separated into two categories, depending on the secondary structure adopted by their transmembrane domain in the lipid bilayer. For α-PFPs, the interaction with the membrane is mediated by an α-helical hairpin, while for β-PFPs it is mediated by a β-sheet. There are several important technical difficulties associated with the study of membrane protein oligomers, and traditional high-resolution structural methods such as X-ray diffraction and nuclear magnetic resonance need to be complemented by other biophysical methods such as electron microscopy, atomic force microscopy, and fluorescence spectroscopy and imaging. Although a detailed knowledge of the pore structures and of the molecular mechanisms leading to their formation is still missing, the current understanding is that α-PFPs tend to form lipid-lined pores with a loose structure and variable stoichiometry, while β-PFPs tend to form stable protein-lined pores with a well-defined stoichiometry. Another complicating element compared to the study of soluble molecular complexes is the pervading role played by lipids, which influence every aspect of pore formation, starting with the targeting of the initially monomeric water-soluble pore-forming proteins to the membrane. For β-PFPs, the emerging picture is that the proteins, after activation, partially insert in the lipid bilayer as monomers, after which several monomers oligomerize into a prepore, which allows subsequent collective insertions of their transmembrane domains assembled into a β-barrel. For α-PFPs, the picture is not as clear, but it is possible that after activation and membrane binding, a single monomeric protein is already able to fully insert its transmembrane domain and cause serious membrane disruption. At high protein concentrations, however, α-PFPs do associate into oligomeric structures and form long-lived pores. Several outstanding questions remain on our way to a complete characterization of the pore-formation mechanism by PFPs, and maybe the most important one is the precise structure of the pores formed by α-helical PFPs. Answering those questions should prove very rewarding, as PFPs are a simple and functionally interesting example of membrane supramolecular assembly, and the fact that they constitute a genetically encodable and modifiable hole puncher tool makes them extremely interesting for biotechnology.

ACKNOWLEDGMENTS

The authors acknowledge funding support from the Canadian Institutes of Health Research, grants NTA-71857 and MOP-86657. CF and DWA also acknowledge support from the Canada Research Chair program. The authors are grateful to Joaquin Ortega, Giuseppe Melacini, Aisha Shamas-Din, and Duncan O'Dell for their critical reading of this chapter.

PROBLEMS

In this series of related problems, we will use simple arguments to estimate the free energy necessary to form different types of pores in lipid membranes, in the presence and in the absence of pore-forming proteins. This will help us understand how pore-forming proteins can help form stable water-filled channels in biological membranes.

9.1. Free energy of a pore: line tension and surface tension.

a. Let us consider a membrane with a surface tension γ. What is the cost in free energy, E, for the creation of a pore of radius R, if the line tension along the edge of the pore is λ?

b. How does ΔE vary with R? What does this imply about the stability of the pores?

c. What is the surface tension in a biological membrane? What does it imply for pore formation?

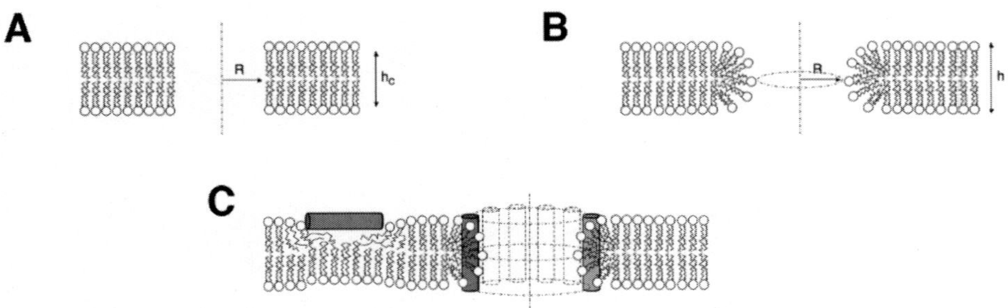

Figure 9.8. Different models of membrane pores for energy calculations. (A) Circular pore formed without any rearrangement of the lipid molecules. (B) Toroidal pore. (C) Toroidal pore stabilized by pore-forming proteins. The proteins are present both as monomers inserted in the outer leaflet and as oligomers lining the pores.

9.2. Free energy of a water-filled membrane channel. We first consider a hypothetic type of water-filled channel, where the lipids do not rearrange themselves along the edge of the pore, as illustrated in Figure 9.8A, in a relaxed membrane for which the surface tension is $\gamma = 0$.

a. Find the expression for the line tension at the edge of the pore as a function of the thickness of the hydrophobic part of the bilayer, h_C, and the surface tension between the apolar medium formed by the lipid carbon chains and the polar water solvent, γ_C.

b. The typical thickness of the hydrophobic part of a phospholipid bilayer is $h_C = 3$ nm, and the value of the surface tension between oil and water is $\gamma_C \approx 50$ mN/m. Estimate the free energy associated with the formation of a pore of radius $R = 1$ nm.

c. Can such pores form spontaneously in biological membranes?

9.3. Free energy of a toroidal pore. We next consider a more realistic model for the pores, where the lipid surface curves in order to connect the two membrane leaflets (toroidal pore), as illustrated in Figure 9.8B.

a. At the pore, what is the curvature, C_1, of the lipid surface in a plane perpendicular to the membrane? What is the curvature, C_2, in a plane parallel to the membrane? Which of these two membrane deformations is the more energetically costly?
b. By considering only the most costly type of curvature, give an expression for the line tension associated with the edge of the pore as a function of the pore radius, R, the bilayer thickness, h, and the bending modulus of the lipid monolayer, k_C.
c. The bending modulus of a lipid monolayer can be estimated to be half that of the lipid bilayer, leading to $k_C \approx 2\times10^{-20}$ J [233]. Estimate the free energy associated with the formation of a toroidal pore with radius R = 1 nm,
d. Can such pores form spontaneously in biological membranes?

9.4. Reduction of pore free energy by pore-forming proteins. We now consider the stabilization effect that PFPs have on membrane pores, by adapting a model originally developed by Huang et al. for antimicrobial peptides [222]. The premise of this model is that the pore-forming protein (present at a total surface concentration c) can exist in two different membrane states: either as a partially inserted monomer (concentration αc), or as part of an oligomeric membrane pore. Monomeric proteins increase the surface area of the membrane leaflet in which they insert (each by an amount A_m), and thus induce a positive tension in the membrane, $\gamma = \gamma_0 A_m \alpha c$. Oligomeric proteins are lining the pores, and although the pore size may vary, the linear density of protein along the edge, ρ, is constant, so that the line tension at the pore, λ, is also constant. This model is therefore better adapted to PFPs for which the pore size and stoichiometry vary with the protein concentration, typically α-PPFs.

a. If ε_m is the binding energy of a monomeric PFP and ε_0 is the pore energy per protein lining the pore, what is the free energy per unit area, F, of the system constituted by the lipids and the proteins? Use this to show that above a critical concentration of proteins in the membrane, c^*, the concentration of monomeric protein remains constant.
b. Consider that $c > c^*$, and that pores exist on the membrane at a concentration C. Write the expression for the free energy of a single pore, ΔE, as a function of its radius, R.
c. How many energy minima are there for a single pore, and which radius do they correspond to? Are those energy minima stable or unstable? Explain the origin of the pore stabilization according to this model.
d. In addition to influencing the membrane surface tension, the PFPs also reduce the line tension of the pores. How does reducing the line tension further influence the structure and/or stability of the pores?

FURTHER STUDY

Mitchell T, Godfree A, Stewart DE, eds. 1999. *Toxins, society for applied microbiology symposium*, Vol. 27. Boston: Blackwell Publishing.

Van der Goot G. 2001. Pore-forming toxins. In *Current topics in microbiology and immunology*, Vol. 257. Paris: Lavoisier Publishers.

Menestrina G, Dalla Sera M, Lazarovici P, eds. 2003. Pore-forming peptides and protein toxins. In *Cell and molecular mechanisms of toxin action*, Vol. 5. Philadelphia: Taylor & Francis.

REFERENCES

1. Gouaux E. 1997. Channel-forming toxins: tales of transformation. *Curr Opin Struct Biol* **7**:566–573.
2. Matsuzaki K. 1999. Why and how are peptide–lipid interactions utilized for self-defense? Magainins and tachyplesins as archetypes. *Biochim Biophys Acta* **1462**:1–10.
3. Lesieur C, Vecsey-Semjen B, Abrami L, Fivaz M, Gisou van der Goot F. 1997. Membrane insertion: the strategies of toxins (review). *Mol Membr Biol* **14**:45–64.
4. Walker JE, Saraste M. 1996. Membrane proteins: membrane protein structure. *Curr Opin Struct Biol* **6**:457–459.
5. Montal M. 1996. Protein folds in channel structure. *Curr Opin Struct Biol* **6**:499–510.
6. Henderson R, Baldwin JM, Ceska TA, Zemlin F, Beckmann E, Downing KH. 1990. Model for the structure of bacteriorhodopsin based on high-resolution electron cryo-microscopy. *J Mol Biol* **213**:899–929.
7. Doyle DA, Morais Cabral J, Pfuetzner RA, Kuo A, Gulbis JM, Cohen SL, Chait BT, MacKinnon R. 1998. The structure of the potassium channel: molecular basis of K+ conduction and selectivity. *Science* **280**:69–77.
8. Cramer WA, Heymann JB, Schendel SL, Deriy BN, Cohen FS, Elkins PA, Stauffacher CV. 1995. Structure–function of the channel-forming colicins. *Annu Rev Biophys Biomol Struct* **24**:611–641.
9. Lakey JH, Slatin SL. 2001. Pore-forming colicins and their relatives. *Curr Top Microbiol Immunol* **257**:131–161.
10. Cascales E, Buchanan SK, Duche D, Kleanthous C, Lloubès R, Postle K, Riley M, Slatin S, Cavard D. 2007. Colicin biology. *Microbiol Mol Biol Rev* **71**:158–229.
11. Wiener M, Freymann D, Ghosh P, Stroud RM. 1997. Crystal structure of colicin Ia. *Nature* **385**:461–464.
12. Vetter IR, Parker MW, Tucker AD, Lakey JH, Pattus F, Tsernoglou D. 1998. Crystal structure of a colicin N fragment suggests a model for toxicity. *Structure* **6**:863–874.
13. Parker MW, Pattus F, Tucker AD, Tsernoglou D. 1989. Structure of the membrane-pore-forming fragment of colicin A. *Nature* **337**:93–96.
14. Elkins P, Bunker A, Cramer WA, Stauffacher CV. 1997. A mechanism for toxin insertion into membranes is suggested by the crystal structure of the channel-forming domain of colicin E1. *Structure* **5**:443–458.
15. Muga A, Gonzalez-Manas JM, Lakey JH, Pattus F, Surewicz WK. 1993. pH-dependent stability and membrane interaction of the pore-forming domain of colicin A. *J Biol Chem* **268**:1553–1557.
16. Gonzalez-Manas JM, Lakey JH, Pattus F. 1992. Brominated phospholipids as a tool for monitoring the membrane insertion of colicin A. *Biochemistry* **31**:7294–7300.
17. Shin YK, Levinthal C, Levinthal F, Hubbell WL. 1993. Colicin E1 binding to membranes: time-resolved studies of spin-labeled mutants. *Science* **259**:960–963.
18. Lakey JH, Massotte D, Heitz F, Dasseux JL, Faucon JF, Parker MW, Pattus F. 1991. Membrane insertion of the pore-forming domain of colicin A: a spectroscopic study. *Eur J Biochem* **196**:599–607.
19. Rath P, Bousche O, Merrill AR, Cramer WA, Rothschild KJ. 1991. Fourier transform infrared evidence for a predominantly alpha-helical structure of the membrane bound channel forming COOH-terminal peptide of colicin E1. *Biophys J* **59**:516–522.
20. van der Goot FG, Didat N, Pattus F, Dowhan W, Letellier L. 1993. Role of acidic lipids in the translocation and channel activity of colicins A and N in *Escherichia coli* cells. *Eur J Biochem* **213**:217–221.
21. Slatin SL, Qiu XQ, Jakes KS, Finkelstein A. 1994. Identification of a translocated protein segment in a voltage-dependent channel. *Nature* **371**:158–161.
22. Qiu XQ, Jakes KS, Finkelstein A, Slatin SL. 1994. Site-specific biotinylation of colicin Ia: a probe for protein conformation in the membrane. *J Biol Chem* **269**:7483–7488.
23. Parker MW, Feil SC. 2005. Pore-forming protein toxins: from structure to function. *Prog Biophys Mol Biol* **88**:91–142.
24. Qiu XQ, Jakes KS, Kienker PK, Finkelstein A, Slatin SL. 1996. Major transmembrane movement associated with colicin Ia channel gating. *J Gen Physiol* **107**:313–328.

25. Jakes KS, Kienker PK, Slatin SL, Finkelstein A. 1998. Translocation of inserted foreign epitopes by a channel-forming protein. *Proc Natl Acad Sci USA* **95**:4321–4326.
26. Zhang YL, Cramer WA. 1993. Intramembrane helix-helix interactions as the basis of inhibition of the colicin E1 ion channel by its immunity protein. *J Biol Chem* **268**:10176–10184.
27. Collier RJ. 2001. Understanding the mode of action of diphtheria toxin: a perspective on progress during the 20th century. *Toxicon* **39**:1793–1803.
28. Tsuneoka M, Nakayama K, Hatsuzawa K, Komada M, Kitamura N, Mekada E. 1993. Evidence for involvement of furin in cleavage and activation of diphtheria toxin. *J Biol Chem* **268**:26461–26465.
29. Yamaizumi M, Mekada E, Uchida T, Okada Y. 1978. One molecule of diphtheria toxin fragment A introduced into a cell can kill the cell. *Cell* **15**:245–250.
30. Naglich JG, Metherall JE, Russell DW, Eidels L. 1992. Expression cloning of a diphtheria toxin receptor: identity with a heparin-binding EGF-like growth factor precursor. *Cell* **69**:1051–1061.
31. Kagan BL, Finkelstein A, Colombini M. 1981. Diphtheria toxin fragment forms large pores in phospholipid bilayer membranes. *Proc Natl Acad Sci USA* **78**:4950–4954.
32. Sharpe JC, London E. 1999. Diphtheria toxin forms pores of different sizes depending on its concentration in membranes: probable relationship to oligomerization. *J Membr Biol* **171**:209–221.
33. Choe S, Bennett MJ, Fujii G, Curmi PM, Kantardjieff KA, Collier RJ, Eisenberg D. 1992. The crystal structure of diphtheria toxin. *Nature* **357**:216–222.
34. Bennett MJ, Eisenberg D. 1994. Refined structure of monomeric diphtheria toxin at 2.3 Å resolution. *Protein Sci* **3**:1464–1475.
35. Carroll SF, Barbieri JT, Collier RJ. 1986. Dimeric form of diphtheria toxin: purification and characterization. *Biochemistry* **25**:2425–2430.
36. Chenal A, Savarin P, Nizard P, Guillain F, Gillet D, Forge V. 2002. Membrane protein insertion regulated by bringing electrostatic and hydrophobic interactions into play: a case study with the translocation domain of diphtheria toxin. *J Biol Chem* **277**:43425–43432.
37. Oh KJ, Zhan H, Cui C, Hideg K, Collier RJ, Hubbell WL. 1996. Organization of diphtheria toxin T domain in bilayers: a site-directed spin labeling study. *Science* **273**:810–812.
38. Knowles BH. 1994. Mechanism of action of *Bacillus thuringiensis* insecticidal delta-endotoxins. *Adv Insect Physiol* **24**:275–308.
39. Li JD, Carroll J, Ellar DJ. 1991. Crystal structure of insecticidal delta-endotoxin from *Bacillus thuringiensis* at 2.5-Å resolution. *Nature* **353**:815–821.
40. Grochulski P, Masson L, Borisova S, Pusztai-Carey M, Schwartz JL, Brousseau R, Cygler M. 1995. *Bacillus thuringiensis* CryIA(a) insecticidal toxin: crystal structure and channel formation. *J Mol Biol* **254**:447–464.
41. Li J, Derbyshire DJ, Promdonkoy B, Ellar DJ. 2001. Structural implications for the transformation of the *Bacillus thuringiensis* delta-endotoxins from water-soluble to membrane-inserted forms. *Biochem Soc Trans* **29**:571–577.
42. Galitsky N, Cody V, Wojtczak A, Ghosh D, Luft JR, Pangborn W, English L. 2001. Structure of the insecticidal bacterial delta-endotoxin Cry3Bb1 of *Bacillus thuringiensis*. *Acta Crystallogr D Biol Crystallogr* **57**:1101–1109.
43. Morse RJ, Yamamoto T, Stroud RM. 2001. Structure of Cry2Aa suggests an unexpected receptor binding epitope. *Structure* **9**:409–417.
44. Cory S, Huang DC, Adams JM. 2003. The Bcl-2 family: roles in cell survival and oncogenesis. *Oncogene* **22**:8590–8607.
45. Youle RJ, Strasser A. 2008. The BCL-2 protein family: opposing activities that mediate cell death. *Nat Rev Mol Cell Biol* **9**:47–59.
46. Leber B, Lin J, Andrews DW. 2007. Embedded together: the life and death consequences of interaction of the Bcl-2 family with membranes. *Apoptosis* **12**:897–911.
47. Muchmore SW, Sattler M, Liang H, Meadows RP, Harlan JE, Yoon HS, Nettesheim D, Chang BS, Thompson CB, Wong SL, Ng SL, Fesik SW. 1996. X-ray and NMR structure of human Bcl-xL, an inhibitor of programmed cell death. *Nature* **381**:335–341.
48. Aouacheria A, Brunet F, Gouy M. 2005. Phylogenomics of life-or-death switches in multicellular animals: Bcl-2, BH3-Only, and BNip families of apoptotic regulators. *Mol Biol Evol* **22**:2395–2416.
49. Petros AM, Medek A, Nettesheim DG, Kim DH, Yoon HS, Swift K, Matayoshi ED, Oltersdorf T, Fesik SW. 2001. Solution structure of the antiapoptotic protein bcl-2. *Proc Natl Acad Sci USA* **98**:3012–3017.

50. Denisov AY, Madiraju MS, Chen G, Khadir A, Beauparlant P, Attardo G, Shore GC, Gehring K. 2003. Solution structure of human BCL-w: modulation of ligand binding by the C-terminal helix. *J Biol Chem* **278**:21124–21128.
51. Woo JS, Jung JS, Ha NC, Shin J, Kim KH, Lee W, Oh BH. 2003. Unique structural features of a BCL-2 family protein CED-9 and biophysical characterization of CED-9/EGL-1 interactions. *Cell Death Differ* **10**:1310–1319.
52. Suzuki M, Youle RJ, Tjandra N. 2000. Structure of Bax: coregulation of dimer formation and intracellular localization. *Cell* **103**:645–654.
53. McDonnell JM, Fushman D, Milliman CL, Korsmeyer SJ, Cowburn D. 1999. Solution structure of the proapoptotic molecule BID: a structural basis for apoptotic agonists and antagonists. *Cell* **96**:625–634.
54. Chou JJ, Li H, Salvesen GS, Yuan J, Wagner G. 1999. Solution structure of BID, an intracellular amplifier of apoptotic signaling. *Cell* **96**:615–624.
55. Wolter KG, Hsu Y-T, Smith CL, Nechushtan A, Xi X-G, Youle RJ. 1997. Movement of Bax from the cytosol to mitochondria during apoptosis. *J Cell Biol* **139**:1281–1292.
56. Sharpe JC, Arnoult D, Youle RJ. 2004. Control of mitochondrial permeability by Bcl-2 family members. *Biochim Biophys Acta* **1644**:107–113.
57. Minn AJ, Velez P, Schendel SL, Liang H, Muchmore SW, Fesik SW, Fill M, Thompson CB. 1997. Bcl-x(L) forms an ion channel in synthetic lipid membranes. *Nature* **385**:353–357.
58. Basanez G, Sharpe JC, Galanis J, Brandt TB, Hardwick JM, Zimmerberg J. 2002. Bax-type apoptotic proteins porate pure lipid bilayers through a mechanism sensitive to intrinsic monolayer curvature. *J Biol Chem* **277**:49360–49365.
59. Antonsson B, Montessuit S, Lauper S, Eskes R, Martinou JC. 2000. Bax oligomerization is required for channel-forming activity in liposomes and to trigger cytochrome c release from mitochondria. *Biochem J* **345 Pt 2**:271–278.
60. Spillantini MG, Schmidt ML, Lee VM, Trojanowski JQ, Jakes R, Goedert M. 1997. Alpha-synuclein in Lewy bodies. *Nature* **388**:839–840.
61. Goldberg MS, Lansbury Jr PT. 2000. Is there a cause-and-effect relationship between alpha-synuclein fibrillization and Parkinson's disease? *Nat Cell Biol* **2**:E115–E119.
62. Davidson WS, Jonas A, Clayton DF, George JM. 1998. Stabilization of alpha-synuclein secondary structure upon binding to synthetic membranes. *J Biol Chem* **273**:9443–9449.
63. Jo E, McLaurin J, Yip CM, St George-Hyslop P, Fraser PE. 2000. α-Synuclein membrane interactions and lipid specificity. *J Biol Chem* **275**:34328–34334.
64. Chandra S, Chen X, Rizo J, Jahn R, Sudhof TC. 2003. A broken alpha-helix in folded alpha-Synuclein. *J Biol Chem* **278**:15313–15318.
65. Ulmer TS, Bax A, Cole NB, Nussbaum RL. 2005. Structure and dynamics of micelle-bound human alpha-synuclein. *J Biol Chem* **280**:9595–9603.
66. Zakharov SD, Hulleman JD, Dutseva EA, Antonenko YN, Rochet JC, Cramer WA. 2007. Helical alpha-synuclein forms highly conductive ion channels. *Biochemistry* **46**:14369–14379.
67. Volles MJ, Lansbury Jr PT. 2002. Vesicle permeabilization by protofibrillar alpha-synuclein is sensitive to Parkinson's disease-linked mutations and occurs by a pore-like mechanism. *Biochemistry* **41**:4595–4602.
68. Lashuel HA, Petre BM, Wall J, Simon M, Nowak RJ, Walz T, Lansbury Jr PT. 2002. Alpha-synuclein, especially the Parkinson's disease-associated mutants, forms pore-like annular and tubular protofibrils. *J Mol Biol* **322**:1089–1102.
69. Ding TT, Lee SJ, Rochet JC, Lansbury Jr PT. 2002. Annular alpha-synuclein protofibrils are produced when spherical protofibrils are incubated in solution or bound to brain-derived membranes. *Biochemistry* **41**:10209–10217.
70. Hashimoto M, Rockenstein E, Mante M, Mallory M, Masliah E. 2001. β-Synuclein inhibits alpha-synuclein aggregation: a possible role as an anti-parkinsonian factor. *Neuron* **32**:213–223.
71. Uversky VN, Li J, Souillac P, Millett IS, Doniach S, Jakes R, Goedert M, Fink AL. 2002. Biophysical properties of the synucleins and their propensities to fibrillate: inhibition of alpha-synuclein assembly by beta- and gamma-synucleins. *J Biol Chem* **277**:11970–11978.
72. Volles MJ, Lee SJ, Rochet JC, Shtilerman MD, Ding TT, Kessler JC, Lansbury Jr PT. 2001. Vesicle permeabilization by protofibrillar alpha-synuclein: implications for the pathogenesis and treatment of Parkinson's disease. *Biochemistry* **40**:7812–7819.

73. Kayed R, Sokolov Y, Edmonds B, McIntire TM, Milton SC, Hall JE, Glabe CG. 2004. Permeabilization of lipid bilayers is a common conformation-dependent activity of soluble amyloid oligomers in protein misfolding diseases. *J Biol Chem* **279**:46363–46366.
74. Lashuel HA, Lansbury Jr PT. 2006. Are amyloid diseases caused by protein aggregates that mimic bacterial pore-forming toxins? *Q Rev Biophys* **39**:167–201.
75. Murphy DD, Rueter SM, Trojanowski JQ, Lee VM. 2000. Synucleins are developmentally expressed, and alpha-synuclein regulates the size of the presynaptic vesicular pool in primary hippocampal neurons. *J Neurosci* **20**:3214–3220.
76. Lashuel HA, Hartley D, Petre BM, Walz T, Lansbury Jr PT. 2002. Neurodegenerative disease: amyloid pores from pathogenic mutations. *Nature* **418**:291.
77. Allured VS, Collier RJ, Carroll SF, McKay DB. 1986. Structure of exotoxin A of *Pseudomonas aeruginosa* at 3.0-Å resolution. *Proc Natl Acad Sci USA* **83**:1320–1324.
78. Amino R, Martins RM, Procopio J, Hirata IY, Juliano MA, Schenkman S. 2002. Trialysin, a novel pore-forming protein from saliva of hematophagous insects activated by limited proteolysis. *J Biol Chem* **277**:6207–6213.
79. Martins RM, Amino R, Daghastanli KR, Cuccovia IM, Juliano MA, Schenkman S. 2008. A short proregion of trialysin, a pore-forming protein of Triatoma infestans salivary glands, controls activity by folding the N-terminal lytic motif. *FEBS J* **275**:994–1002.
80. Anderluh G, Macek P. 2002. Cytolytic peptide and protein toxins from sea anemones (Anthozoa: Actiniaria). *Toxicon* **40**:111–124.
81. Mancheno JM, Martin-Benito J, Martinez-Ripoll M, Gavilanes JG, Hermoso JA. 2003. Crystal and electron microscopy structures of sticholysin II actinoporin reveal insights into the mechanism of membrane pore formation. *Structure* **11**:1319–1328.
82. Mancheno JM, Martinez-Ripoll M, Gavilanes JG, Hermoso JA. 2002. Crystallization and preliminary X-ray diffraction studies of the water-soluble state of the pore-forming toxin sticholysin II from the sea anemone Stichodactyla helianthus. *Acta Crystallogr D Biol Crystallogr* **58**:1229–1231.
83. Anderluh G, Krizaj I, Strukelj B, Gubensek F, Macek P, Pungercar J. 1999. Equinatoxins, pore-forming proteins from the sea anemone Actinia equina, belong to a multigene family. *Toxicon* **37**:1391–1401.
84. Parker MW. 2003. Cryptic clues as to how water-soluble protein toxins form pores in membranes. *Toxicon* **42**:1–6.
85. Song L, Hobaugh MR, Shustak C, Cheley S, Bayley H, Gouaux JE. 1996. Structure of staphylococcal alpha-hemolysin, a heptameric transmembrane pore. *Science* **274**:1859–1866.
86. Schulz GE. 1996. Porins: general to specific, native to engineered passive pores. *Curr Opin Struct Biol* **6**:485–490.
87. Petosa C, Collier RJ, Klimpel KR, Leppla SH, Liddington RC. 1997. Crystal structure of the anthrax toxin protective antigen. *Nature* **385**:833–838.
88. Menestrina G, Serra MD, Prevost G. 2001. Mode of action of beta-barrel pore-forming toxins of the staphylococcal alpha-hemolysin family. *Toxicon* **39**:1661–1672.
89. Montoya M, Gouaux E. 2003. Beta-barrel membrane protein folding and structure viewed through the lens of alpha-hemolysin. *Biochim Biophys Acta* **1609**:19–27.
90. Olson R, Nariya H, Yokota K, Kamio Y, Gouaux E. 1999. Crystal structure of staphylococcal LukF delineates conformational changes accompanying formation of a transmembrane channel. *Nat Struct Biol* **6**:134–140.
91. Miles G, Movileanu L, Bayley H. 2002. Subunit composition of a bicomponent toxin: staphylococcal leukocidin forms an octameric transmembrane pore. *Protein Sci* **11**:894–902.
92. Sugawara-Tomita N, Tomita T, Kamio Y. 2002. Stochastic assembly of two-component staphylococcal gamma-hemolysin into heteroheptameric transmembrane pores with alternate subunit arrangements in ratios of 3:4 and 4:3. *J Bacteriol* **184**:4747–4756.
93. Palmer M. 2001. The family of thiol-activated, cholesterol-binding cytolysins. *Toxicon* **39**:1681–1689.
94. Tweten RK, Parker MW, Johnson AE. 2001. The cholesterol-dependent cytolysins. *Curr Top Microbiol Immunol* **257**:15–33.
95. Gonzalez MR, Bischofberger M, Pernot L, van der Goot FG, Freche B. 2008. Bacterial pore-forming toxins: the (w)hole story? *Cell Mol Life Sci* **65**:493–507.
96. Billington SJ, Jost BH, Songer JG. 2000. Thiol-activated cytolysins: structure, function and role in pathogenesis. *FEMS Microbiol Lett* **182**:197–205.

97. Rossjohn J, Feil SC, McKinstry WJ, Tweten RK, Parker MW. 1997. Structure of a cholesterol-binding, thiol-activated cytolysin and a model of its membrane form. *Cell* **89**:685–692.
98. Rossjohn J, Polekhina G, Feil SC, Morton CJ, Tweten RK, Parker MW. 2007. Structures of perfringolysin O suggest a pathway for activation of cholesterol-dependent cytolysins. *J Mol Biol* **367**:1227–1236.
99. Sekiya K, Satoh R, Danbara H, Futaesaku Y. 1993. A ring-shaped structure with a crown formed by streptolysin O on the erythrocyte membrane. *J Bacteriol* **175**:5953–5961.
100. Olofsson A, Hebert H, Thelestam M. 1993. The projection structure of perfringolysin O (*Clostridium perfringens* theta-toxin). *FEBS Lett* **319**:125–127.
101. Morgan PJ, Hyman SC, Rowe AJ, Mitchell TJ, Andrew PW, Saibil HR. 1995. Subunit organisation and symmetry of pore-forming, oligomeric pneumolysin. *FEBS Lett* **371**:77–80.
102. Bonev BB, Gilbert RJ, Andrew PW, Byron O, Watts A. 2001. Structural analysis of the protein/lipid complexes associated with pore formation by the bacterial toxin pneumolysin. *J Biol Chem* **276**:5714–5719.
103. Shatursky O, Heuck AP, Shepard LA, Rossjohn J, Parker MW, Johnson AE, Tweten RK. 1999. The mechanism of membrane insertion for a cholesterol-dependent cytolysin: a novel paradigm for pore-forming toxins. *Cell* **99**:293–299.
104. Young JA, Collier RJ. 2007. Anthrax toxin: receptor binding, internalization, pore formation, and translocation. *Annu Rev Biochem* **76**:243–265.
105. Escuyer V, Collier RJ. 1991. Anthrax protective antigen interacts with a specific receptor on the surface of CHO-K1 cells. *Infect Immun* **59**:3381–3386.
106. Klimpel KR, Molloy SS, Thomas G, Leppla SH. 1992. Anthrax toxin protective antigen is activated by a cell surface protease with the sequence specificity and catalytic properties of furin. *Proc Natl Acad Sci USA* **89**:10277–10281.
107. Blaustein RO, Koehler TM, Collier RJ, Finkelstein A. 1989. Anthrax toxin: channel-forming activity of protective antigen in planar phospholipid bilayers. *Proc Natl Acad Sci USA* **86**:2209–2213.
108. Koehler TM, Collier RJ. 1991. Anthrax toxin protective antigen: low-pH-induced hydrophobicity and channel formation in liposomes. *Mol Microbiol* **5**:1501–1506.
109. Milne JC, Furlong D, Hanna PC, Wall JS, Collier RJ. 1994. Anthrax protective antigen forms oligomers during intoxication of mammalian cells. *J Biol Chem* **269**:20607–20612.
110. Koni PA, Ellar DJ. 1994. Biochemical characterization of *Bacillus thuringiensis* cytolytic delta-endotoxins. *Microbiology* **140**(Pt 8):1869–1880.
111. Wilmsen HU, Pattus F, Buckley JT. 1990. Aerolysin, a hemolysin from *Aeromonas hydrophila*, forms voltage-gated channels in planar lipid bilayers. *J Membr Biol* **115**:71–81.
112. Tucker AD, Parker MW, Tsernoglou D, Buckley JT. 1990. Crystallization of a proform of aerolysin, a hole-forming toxin from *Aeromonas hydrophila*. *J Mol Biol* **212**:561–562.
113. Parker MW, Buckley JT, Postma JP, Tucker AD, Leonard K, Pattus F, Tsernoglou D. 1994. Structure of the *Aeromonas* toxin proaerolysin in its water-soluble and membrane-channel states. *Nature* **367**:292–295.
114. Melton JA, Parker MW, Rossjohn J, Buckley JT, Tweten RK. 2004. The identification and structure of the membrane-spanning domain of the *Clostridium septicum* alpha toxin. *J Biol Chem* **279**:14315–14322.
115. Tweten RK. 2001. *Clostridium perfringens* beta toxin and *Clostridium septicum* alpha toxin: their mechanisms and possible role in pathogenesis. *Vet Microbiol* **82**:1–9.
116. Perelle S, Gibert M, Boquet P, Popoff MR. 1993. Characterization of *Clostridium perfringens* iota-toxin genes and expression in *Escherichia coli*. *Infect Immun* **61**:5147–5156.
117. Sliwinski-Korell A, Engelhardt H, Kampka M, Lutz F. 1999. Oligomerization and structural changes of the pore-forming *Pseudomonas aeruginosa* cytotoxin. *Eur J Biochem* **265**:221–230.
118. Xiong G, Struckmeier M, Lutz F. 1994. Pore-forming *Pseudomonas aeruginosa* cytotoxin. *Toxicology* **87**:69–83.
119. Stern HA, Feller SE. 2003. Calculation of the dielectric permittivity profile for a nonuniform system: application to a lipid bilayer simulation. *J Chem Phys* **118**:3401–3412.
120. Baumann G, Mueller P. 1974. A molecular model of membrane excitability. *J Supramol Struct* **2**:538–557.
121. Killian JA, von Heijne G. 2000. How proteins adapt to a membrane–water interface. *Trends Biochem Sci* **25**:429–434.
122. Galdiero S, Gouaux E. 2004. High-resolution crystallographic studies of alpha-hemolysin-phospholipid complexes define heptamer–lipid head group interactions: implication for understanding protein–lipid interactions. *Protein Sci* **13**:1503–1511.

123. Ludtke SJ, He K, Heller WT, Harroun TA, Yang L, Huang HW. 1996. Membrane pores induced by magainin. *Biochemistry* **35**:13723–13728.
124. Matsuzaki K, Murase O, Fujii N, Miyajima K. 1996. An antimicrobial peptide, magainin 2, induced rapid flip-flop of phospholipids coupled with pore formation and peptide translocation. *Biochemistry* **35**:11361–11368.
125. Yang L, Harroun TA, Weiss TM, Ding L, Huang HW. 2001. Barrel-stave model or toroidal model? A case study on melittin pores. *Biophys J* **81**:1475–1485.
126. Sobko AA, Kotova EA, Antonenko YN, Zakharov SD, Cramer WA. 2006. Lipid dependence of the channel properties of a colicin E1-lipid toroidal pore. *J Biol Chem* **281**:14408–14416.
127. Epand RF, Martinou JC, Montessuit S, Epand RM. 2003. Transbilayer lipid diffusion promoted by Bax: implications for apoptosis. *Biochemistry* **42**:14576–14582.
128. Valcarcel CA, Dalla Serra M, Potrich C, Bernhart I, Tejuca M, Martinez D, Pazos F, Lanio ME, Menestrina G. 2001. Effects of lipid composition on membrane permeabilization by sticholysin I and II, two cytolysins of the sea anemone Stichodactyla helianthus. *Biophys J* **80**:2761–2774.
129. Hinds MG, Zhang W, Anderluh G, Hansen PE, Norton RS. 2002. Solution structure of the eukaryotic pore-forming cytolysin equinatoxin II: implications for pore formation. *J Mol Biol* **315**:1219–1229.
130. Gesell J, Zasloff M, Opella SJ. 1997. Two-dimensional 1H NMR experiments show that the 23-residue magainin antibiotic peptide is an alpha-helix in dodecylphosphocholine micelles, sodium dodecylsulfate micelles, and trifluoroethanol/water solution. *J Biomol NMR* **9**:127–135.
131. Tsigelny IF, Bar-On P, Sharikov Y, Crews L, Hashimoto M, Miller MA, Keller SH, Platoshyn O, Yuan JX, Masliah E. 2007. Dynamics of alpha-synuclein aggregation and inhibition of pore-like oligomer development by beta-synuclein. *FEBS J* **274**:1862–1877.
132. Marassi FM, Opella SJ. 1998. NMR structural studies of membrane proteins. *Curr Opin Struct Biol* **8**:640–648.
133. Watts A, Straus SK, Grage SL, Kamihira M, Lam YH, Zhao X. 2004. Membrane protein structure determination using solid-state NMR. *Methods Mol Biol* **278**:403–473.
134. Ketchem RR, Hu W, Cross TA. 1993. High-resolution conformation of gramicidin A in a lipid bilayer by solid-state NMR. *Science* **261**:1457–1460.
135. Bechinger B, Zasloff M, Opella SJ. 1993. Structure and orientation of the antibiotic peptide magainin in membranes by solid-state nuclear magnetic resonance spectroscopy. *Protein Sci* **2**:2077–2084.
136. Bechinger B. 1999. The structure, dynamics and orientation of antimicrobial peptides in membranes by multidimensional solid-state NMR spectroscopy. *Biochim Biophys Acta* **1462**:157–183.
137. Muller SA, Aebi U, Engel A. 2008. What transmission electron microscopes can visualize now and in the future. *J Struct Biol* **163**:235–245.
138. Henderson R. 1995. The potential and limitations of neutrons, electrons and X-rays for atomic resolution microscopy of unstained biological molecules. *Q Rev Biophys* **28**:171–193.
139. Rubinstein JL. 2007. Structural analysis of membrane protein complexes by single particle electron microscopy. *Methods* **41**:409–416.
140. Tilley SJ, Orlova EV, Gilbert RJ, Andrew PW, Saibil HR. 2005. Structural basis of pore formation by the bacterial toxin pneumolysin. *Cell* **121**:247–256.
141. Wilmsen HU, Leonard KR, Tichelaar W, Buckley JT, Pattus F. 1992. The aerolysin membrane channel is formed by heptamerization of the monomer. *EMBO J* **11**:2457–2463.
142. Woodcock CL, Frado LL, Wall JS. 1980. Composition of native and reconstituted chromatin particles: direct mass determination by scanning transmission electron microscopy. *Proc Natl Acad Sci USA* **77**:4818–4822.
143. Tzokov SB, Wyborn NR, Stillman TJ, Jamieson S, Czudnochowski N, Artymiuk PJ, Green J, Bullough PA. 2006. Structure of the hemolysin E (HlyE, ClyA, and SheA) channel in its membrane-bound form. *J Biol Chem* **281**:23042–23049.
144. Eifler N, Vetsch M, Gregorini M, Ringler P, Chami M, Philippsen A, Fritz A, Müller SA, Glockshuber R, Engel A, Grauschopf U. 2006. Cytotoxin ClyA from *Escherichia coli* assembles to a 13-meric pore independent of its redox-state. *EMBO J* **25**:2652–2661.
145. Shao Z. 1999. Probing nanometer structures with atomic force microscopy. *Newsl Physiol Sci* **14**:142–149.
146. Fang Y, Cheley S, Bayley H, Yang J. 1997. The heptameric prepore of a staphylococcal alpha-hemolysin mutant in lipid bilayers imaged by atomic force microscopy. *Biochemistry* **36**:9518–9522.
147. Malghani MS, Fang Y, Cheley S, Bayley H, Yang J. 1999. Heptameric structures of two alpha-hemolysin mutants imaged with in situ atomic force microscopy. *Microsc Res Tech* **44**:353–356.

148. Czajkowsky DM, Sheng S, Shao Z. 1998. Staphylococcal alpha-hemolysin can form hexamers in phospholipid bilayers. *J Mol Biol* **276**:325–330.
149. Kostka M, Hogen T, Danzer KM, Levin J, Habeck M, Wirth A, Wagner R, Glabe CG, Finger S, Heinzelmann U, Garidel P, Duan W, Ross CA, Kretzschmar H, Giese A. 2008. Single particle characterization of iron-induced pore-forming alpha-synuclein oligomers. *J Biol Chem* **283**:10992–1003.
150. Czajkowsky DM, Hotze EM, Shao Z, Tweten RK. 2004. Vertical collapse of a cytolysin prepore moves its transmembrane beta-hairpins to the membrane. *EMBO J* **23**:3206–3215.
151. Lakey JH, Baty D, Pattus F. 1991. Fluorescence energy transfer distance measurements using site-directed single cysteine mutants: the membrane insertion of colicin A. *J Mol Biol* **218**:639–653.
152. Lakey JH, Duche D, Gonzalez-Manas JM, Baty D, Pattus F. 1993. Fluorescence energy transfer distance measurements. The hydrophobic helical hairpin of colicin A in the membrane bound state. *J Mol Biol* **230**:1055–1067.
153. Ward RJ, Palmer M, Leonard K, Bhakdi S. 1994. Identification of a putative membrane-inserted segment in the alpha-toxin of *Staphylococcus aureus*. *Biochemistry* **33**:7477–7484.
154. Shepard LA, Heuck AP, Hamman BD, Rossjohn J, Parker MW, Ryan KR, Johnson AE, Tweten RK. 1998. Identification of a membrane-spanning domain of the thiol-activated pore-forming toxin *Clostridium perfringens* perfringolysin O: an alpha-helical to beta-sheet transition identified by fluorescence spectroscopy. *Biochemistry* **37**:14563–14574.
155. Heuck AP, Johnson AE. 2002. Pore-forming protein structure analysis in membranes using multiple independent fluorescence techniques. *Cell Biochem Biophys* **36**:89–101.
156. Ulbrich MH, Isacoff EY. 2007. Subunit counting in membrane-bound proteins. *Nat Methods* **4**:319–321.
157. Somes K. Das MD, Cheley S, Wallace MI, Bayley H. 2007. Membrane protein stoichiometry determined from the step-wise photobleaching of dye-labelled subunits. *ChemBioChem* **8**:994–999.
158. Chen Y, Muller JD, So PT, Gratton E. 1999. The photon counting histogram in fluorescence fluctuation spectroscopy. *Biophys J* **77**:553–567.
159. Kask P, Palo K, Ullmann D, Gall K. 1999. Fluorescence-intensity distribution analysis and its application in biomolecular detection technology. *Proc Natl Acad Sci USA* **96**:13756–13761.
160. Chen Y, Wei LN, Muller JD. 2003. Probing protein oligomerization in living cells with fluorescence fluctuation spectroscopy. *Proc Natl Acad Sci USA* **100**:15492–15497.
161. Saffarian S, Li Y, Elson EL, Pike LJ. 2007. Oligomerization of the EGF receptor investigated by live cell fluorescence intensity distribution analysis. *Biophys J* **93**:1021–1031.
162. Satsoura D, Griffiths C, Leber B, Andrews DW, Fradin C. Activated Bax forms oligomers of varying size to permeabilize lipid membranes. In preparation.
163. Epand RF, Savage PB, Epand RM. 2007. Bacterial lipid composition and the antimicrobial efficacy of cationic steroid compounds (Ceragenins). *Biochim Biophys Acta* **1768**:2500–2509.
164. van Meer G. 1989. Lipid traffic in animal cells. *Annu Rev Cell Biol* **5**:247–275.
165. Daum G, Vance JE. 1997. Import of lipids into mitochondria. *Prog Lipid Res* **36**:103–130.
166. Ohno-Iwashita Y, Iwamoto M, Ando S, Iwashita S. 1992. Effect of lipidic factors on membrane cholesterol topology—mode of binding of theta-toxin to cholesterol in liposomes. *Biochim Biophys Acta* **1109**:81–90.
167. Ohno-Iwashita Y, Iwamoto M, Mitsui K, Ando S, Iwashita S. 1991. A cytolysin, theta-toxin, preferentially binds to membrane cholesterol surrounded by phospholipids with 18-carbon hydrocarbon chains in cholesterol-rich region. *J Biochem* **110**:369–375.
168. Waheed AA, Shimada Y, Heijnen HF, Nakamura M, Inomata M, Hayashi M, Iwashita S, Slot JW, Ohno-Iwashita Y. 2001. Selective binding of perfringolysin O derivative to cholesterol-rich membrane microdomains (rafts). *Proc Natl Acad Sci USA* **98**:4926–4931.
169. Zitzer A, Bittman R, Verbicky CA, Erukulla RK, Bhakdi S, Weis S, Valeva A, Palmer M. 2001. Coupling of cholesterol and cone-shaped lipids in bilayers augments membrane permeabilization by the cholesterol-specific toxins streptolysin O and Vibrio cholerae cytolysin. *J Biol Chem* **276**:14628–14633.
170. Johnson MK, Geoffroy C, Alouf JE. 1980. Binding of cholesterol by sulfhydryl-activated cytolysins. *Infect Immun* **27**:97–101.
171. Shimada Y, Maruya M, Iwashita S, Ohno-Iwashita Y. 2002. The C-terminal domain of perfringolysin O is an essential cholesterol-binding unit targeting to cholesterol-rich microdomains. *Eur J Biochem* **269**:6195–6203.
172. Lutter M, Fang M, Luo X, Nishijima M, Xie X, Wang X. 2000. Cardiolipin provides specificity for targeting of tBid to mitochondria. *Nat Cell Biol* **2**:754–761.

173. Thomas WE, Ellar DJ. 1983. Mechanism of action of *Bacillus thuringiensis* var *israelensis* insecticidal delta-endotoxin. *FEBS Lett* **154**:362–368.
174. Esposti MD, Erler JT, Hickman JA, Dive C. 2001. Bid, a widely expressed proapoptotic protein of the Bcl-2 family, displays lipid transfer activity. *Mol Cell Biol* **21**:7268–7276.
175. Oh KJ, Barbuto S, Meyer N, Kim RS, Collier RJ, Korsmeyer SJ. 2005. Conformational changes in BID, a pro-apoptotic BCL-2 family member, upon membrane binding: a site-directed spin labeling study. *J Biol Chem* **280**:753–767.
176. Papini E, Colonna R, Schiavo G, Cusinato F, Tomasi M, Rappuoli R, Montecucco C. 1987. Diphtheria toxin and its mutant crm 197 differ in their interaction with lipids. *FEBS Lett* **215**:73–78.
177. Seelig J. 2004. Thermodynamics of lipid-peptide interactions. *Biochim Biophys Acta* **1666**:40–50.
178. McLaughlin S. 1989. The electrostatic properties of membranes. *Annu Rev Biophys Biophys Chem* **18**:113–136.
179. Beschiaschvili G, Seelig J. 1990. Peptide binding to lipid bilayers: binding isotherms and zeta-potential of a cyclic somatostatin analogue. *Biochemistry* **29**:10995–11000.
180. Wenk MR, Seelig J. 1998. Magainin 2 amide interaction with lipid membranes: calorimetric detection of peptide binding and pore formation. *Biochemistry* **37**:3909–3916.
181. Kim J, Mosior M, Chung LA, Wu H, McLaughlin S. 1991. Binding of peptides with basic residues to membranes containing acidic phospholipids. *Biophys J* **60**:135–148.
182. McMahon HT, Gallop JL. 2005. Membrane curvature and mechanisms of dynamic cell membrane remodelling. *Nature* **438**:590–596.
183. Drin G, Casella JF, Gautier R, Boehmer T, Schwartz TU, Antonny B. 2007. A general amphipathic alpha-helical motif for sensing membrane curvature. *Nat Struct Mol Biol* **14**:138–146.
184. Nuscher B, Kamp F, Mehnert T, Odoy S, Haass C, Kahle PJ, Beyer K. 2004. Alpha-synuclein has a high affinity for packing defects in a bilayer membrane: a thermodynamics study. *J Biol Chem* **279**:21966–21975.
185. Lakey JH, Parker MW, Gonzalez-Manas JM, Duché D, Vriend G, Baty D, Pattus F. 1994. The role of electrostatic charge in the membrane insertion of colicin A: calculation and mutation. *Eur J Biochem* **220**:155–163.
186. van der Goot FG, Gonzalez-Manas JM, Lakey JH, Pattus F. 1991. A "molten-globule" membrane-insertion intermediate of the pore-forming domain of colicin A. *Nature* **354**:408–410.
187. Ulmschneider MB, Sansom MS. 2001. Amino acid distributions in integral membrane protein structures. *Biochim Biophys Acta* **1512**:1–14.
188. White SH, Wimley WC. 1998. Hydrophobic interactions of peptides with membrane interfaces. *Biochim Biophys Acta* **1376**:339–352.
189. Scarlata S, Gruner SM. 1997. Role of phosphatidylethanolamine lipids in the stabilization of protein–lipid contacts. *Biophys Chem* **67**:269–279.
190. Scarlata S, McBath H, Haspel HC. 1995. Effect of lipid packing on the conformational states of purified GLUT-1 hexose transporter. *Biochemistry* **34**:7703–7711.
191. Nyholm TK, Ozdirekcan S, Killian JA. 2007. How protein transmembrane segments sense the lipid environment. *Biochemistry* **46**:1457–1465.
192. Killian JA. 1998. Hydrophobic mismatch between proteins and lipids in membranes. *Biochim Biophys Acta* **1376**:401–415.
193. Kooijman EE, Chupin V, Fuller NL, Kozlov MM, de Kruijff B, Burger KN, Rand PR. 2005. Spontaneous curvature of phosphatidic acid and lysophosphatidic acid. *Biochemistry* **44**:2097–2102.
194. Matsuzaki K, Sugishita K, Ishibe N, Ueha M, Nakata S, Miyajima K, Epand RM. 1998. Relationship of membrane curvature to the formation of pores by magainin 2. *Biochemistry* **37**:11856–11863.
195. Malev VV, Schagina LV, Gurnev PA, Takemoto JY, Nestorovich EM, Bezrukov SM. 2002. Syringomycin E channel: a lipidic pore stabilized by lipopeptide? *Biophys J* **82**:1985–1994.
196. Epand RF, Martinou JC, Montessuit S, Epand RM. 2002. Membrane perturbations induced by the apoptotic Bax protein. *Biochem J* **367**:849–855.
197. Sobko AA, Kotova EA, Antonenko YN, Zakharov SD, Cramer WA. 2004. Effect of lipids with different spontaneous curvature on the channel activity of colicin E1: evidence in favor of a toroidal pore. *FEBS Lett* **576**:205–210.
198. Li H, Zhu H, Xu CJ, Yuan J. 1998. Cleavage of BID by caspase 8 mediates the mitochondrial damage in the Fas pathway of apoptosis. *Cell* **94**:491–501.
199. Howard SP, Buckley JT. 1985. Activation of the hole-forming toxin aerolysin by extracellular processing. *J Bacteriol* **163**:336–340.

200. Li J, Koni PA, Ellar DJ. 1996. Structure of the mosquitocidal delta-endotoxin CytB from *Bacillus thuringiensis* sp. *kyushuensis* and implications for membrane pore formation. *J Mol Biol* **257**:129–152.
201. Milne JC, Collier RJ. 1993. pH-dependent permeabilization of the plasma membrane of mammalian cells by anthrax protective antigen. *Mol Microbiol* **10**:647–653.
202. van der Goot FG, Lakey JH, Pattus F. 1992. The molten globule intermediate for protein insertion or translocation through membranes. *Trends Cell Biol* **2**:343–348.
203. Mel SF, Stroud RM. 1993. Colicin Ia inserts into negatively charged membranes at low pH with a tertiary but little secondary structural change. *Biochemistry* **32**:2082–2089.
204. Korsmeyer SJ, Wei MC, Saito M, Weiler S, Oh KJ, Schlesinger PH. 2000. Pro-apoptotic cascade activates BID, which oligomerizes BAK or BAX into pores that result in the release of cytochrome c. *Cell Death Differ* **7**:1166–1173.
205. Palmer M, Valeva A, Kehoe M, Bhakdi S. 1995. Kinetics of streptolysin O self-assembly. *Eur J Biochem* **231**:388–395.
206. Berg HC, Purcell EM. 1977. Physics of chemoreception. *Biophys J* **20**:193–219.
207. London E. 1992. How bacterial protein toxins enter cells; the role of partial unfolding in membrane translocation. *Mol Microbiol* **6**:3277–3282.
208. Yethon JA, Epand RF, Leber B, Epand RM, Andrews DW. 2003. Interaction with a membrane surface triggers a reversible conformational change in Bax normally associated with induction of apoptosis. *J Biol Chem* **278**:48935–48941.
209. Zhan H, Choe S, Huynh PD, Finkelstein A, Eisenberg D, Collier RJ. 1994. Dynamic transitions of the transmembrane domain of diphtheria toxin: disulfide trapping and fluorescence proximity studies. *Biochemistry* **33**:11254–11263.
210. Silverman JA, Mindell JA, Finkelstein A, Shen WH, Collier RJ. 1994. Mutational analysis of the helical hairpin region of diphtheria toxin transmembrane domain. *J Biol Chem* **269**:22524–22532.
211. Palmer M, Vulicevic I, Saweljew P, Valeva A, Kehoe M, Bhakdi S. 1998. Streptolysin O: a proposed model of allosteric interaction between a pore-forming protein and its target lipid bilayer. *Biochemistry* **37**:2378–2383.
212. Parker MW, Tucker AD, Tsernoglou D, Pattus F. 1990. Insights into membrane insertion based on studies of colicins. *Trends Biochem Sci* **15**:126–129.
213. Massotte D, Yamamoto M, Scianimanico S, Sorokine O, van Dorsselaer A, Nakatani Y, Ourisson G, Pattus F. 1993. Structure of the membrane-bound form of the pore-forming domain of colicin A: a partial proteolysis and mass spectrometry study. *Biochemistry* **32**:13787–13794.
214. Kienker PK, Qiu X, Slatin SL, Finkelstein A, Jakes KS. 1997. Transmembrane insertion of the colicin Ia hydrophobic hairpin. *J Membr Biol* **157**:27–37.
215. Valeva A, Palmer M, Bhakdi S. 1997. Staphylococcal alpha-toxin: formation of the heptameric pore is partially cooperative and proceeds through multiple intermediate stages. *Biochemistry* **36**:13298–13304.
216. Fink AL. 2006. The aggregation and fibrillation of alpha-synuclein. *Acc Chem Res* **39**:628–634.
217. Jarrett JT, Berger EP, Lansbury Jr PT. 1993. The carboxy terminus of the beta amyloid protein is critical for the seeding of amyloid formation: implications for the pathogenesis of Alzheimer's disease. *Biochemistry* **32**:4693–4697.
218. Come JH, Fraser PE, Lansbury Jr PT. 1993. A kinetic model for amyloid formation in the prion diseases: importance of seeding. *Proc Natl Acad Sci USA* **90**:5959–5963.
219. Palmer M, Harris R, Freytag C, Kehoe M, Tranum-Jensen J, Bhakdi S. 1998. Assembly mechanism of the oligomeric streptolysin O pore: the early membrane lesion is lined by a free edge of the lipid membrane and is extended gradually during oligomerization. *EMBO J* **17**:1598–1605.
220. Bhakdi S, Tranum-Jensen J, Sziegoleit A. 1985. Mechanism of membrane damage by streptolysin-O. *Infect Immun* **47**:52–60.
221. Saito M, Korsmeyer SJ, Schlesinger PH. 2000. BAX-dependent transport of cytochrome c reconstituted in pure liposomes. *Nat Cell Biol* **2**:553–555.
222. Huang HW, Chen FY, Lee MT. 2004. Molecular mechanism of peptide-induced pores in membranes. *Phys Rev Lett* **92**:198304.
223. Thelestam M, Olofsson A, Blomqvist L, Hebert H. 1991. Oligomerisation of cell-bound staphylococcal alpha-toxin in relation to membrane permeabilisation. *Biochim Biophys Acta* **1062**:245–254.
224. Walker B, Krishnasastry M, Zorn L, Bayley H. 1992. Assembly of the oligomeric membrane pore formed by Staphylococcal alpha-hemolysin examined by truncation mutagenesis. *J Biol Chem* **267**:21782–21786.

225. Vecsey-Semjen B, Lesieur C, Mollby R, van der Goot FG. 1997. Conformational changes due to membrane binding and channel formation by staphylococcal alpha-toxin. *J Biol Chem* **272**:5709–5717.
226. Vecsey-Semjen B, Mollby R, van der Goot FG. 1996. Partial C-terminal unfolding is required for channel formation by staphylococcal alpha-toxin. *J Biol Chem* **271**:8655–8660.
227. Valeva A, Palmer M, Hilgert K, Kehoe M, Bhakdi S. 1995. Correct oligomerization is a prerequisite for insertion of the central molecular domain of staphylococcal alpha-toxin into the lipid bilayer. *Biochim Biophys Acta* **1236**:213–218.
228. Wimley WC, Hristova K, Ladokhin AS, Silvestro L, Axelsen PH, White SH. 1998. Folding of beta-sheet membrane proteins: a hydrophobic hexapeptide model. *J Mol Biol* **277**:1091–1110.
229. Shepard LA, Shatursky O, Johnson AE, Tweten RK. 2000. The mechanism of pore assembly for a cholesterol-dependent cytolysin: formation of a large prepore complex precedes the insertion of the transmembrane beta-hairpins. *Biochemistry* **39**:10284–10293.
230. Krasilnikov OV, Da Cruz JB, Yuldasheva LN, Varanda WA, Nogueira RA. 1998. A novel approach to study the geometry of the water lumen of ion channels: colicin Ia channels in planar lipid bilayers. *J Membr Biol* **161**:83–92.
231. Athanasiadis A, Anderluh G, Macek P, Turk D. 2001. Crystal structure of the soluble form of equinatoxin II, a pore-forming toxin from the sea anemone Actinia equina. *Structure* **9**:341–346.
232. Gouaux E. 1998. α-Hemolysin from *Staphylococcus aureus*: an archetype of beta-barrel, channel-forming toxins. *J Struct Biol* **121**:110–122.
233. Sandre O, Moreaux L, Brochard-Wyart F. 1999. Dynamics of transient pores in stretched vesicles. *Proc Natl Acad Sci USA* **96**:10591–10596.
234. Evans E, Heinrich V, Ludwig F, Rawicz W. 2003. Dynamic tension spectroscopy and strength of biomembranes. *Biophys J* **85**:2342–2350.
235. Zhelev DV, Needham D. 1993. Tension-stabilized pores in giant vesicles: determination of pore size and pore line tension. *Biochim Biophys Acta* **1147**:89–104.
236. Malinin VS, Frederik P, Lentz BR. 2002. Osmotic and curvature stress affect PEG-induced fusion of lipid vesicles but not mixing of their lipids. *Biophys J* **82**:2090–2100.
237. Weaver JC, Chizmadzhev YA. 1996. Theory of electroporation: a review. *Bioelectrochem Bioenerg* **41**:135–160.
238. Moroz JD, Nelson P. 1997. Dynamically stabilized pores in bilayer membranes. *Biophys J* **72**:2211–2216.
239. Brochard-Wyart F, de Gennes PG, Sandre O. 2000. Transient pores in stretched vesicles: role of leak-out. *Physica A* **278**:32–51.
240. Loi S, Sun G, Franz V, Butt HJ. 2002. Rupture of molecular thin films observed in atomic force microscopy, II: experiment. *Phys Rev E Stat Nonlin Soft Matter Phys* **66**:031602.

10

Morphogens, Membranes and Mechanotransduction in Articular Cartilage

Shirley Motaung,[1,2] Stephanie Chan,[1] and A. Hari Reddi[1]
[1]*Department of Orthopaedic Surgery, Lawrence Ellison Center for Tissue Regeneration, University of California Davis, School of Medicine, Sacramento, California;* [2]*Faculty of Science, Department of Biomedical Sciences, Tshwane University of Technology, Pretoria, South Africa*

10.1. INTRODUCTION

The three fundamentals of developmental biology are cell growth, cellular differentiation, and morphogenesis. Morphogenesis is the process of the generation of the shape of tissues, organs, and entire organisms from various cells. During embryonic development, as cells reproduce and divide, chemical and mechanical signals induce the cell to sort and differentiate into specialized cells. Morphogenesis is the process by which these cells become distributed and organized into tissues and organs. Morphogenetic responses can be stimulated in organisms by morphogenetic proteins, hormones, and environmental cues. There are different types of molecules that play an important role during morphogenesis and include the transcription factors and morphogens themselves.

10.2. MORPHOGENS

Morphogens are soluble molecules that can diffuse and signal cellular responses by acting directly on the cell by binding to specific protein receptors [1]. Morphogens diffuse through a concentration gradient and therefore act on cells in a spatially dependent manner. Morphogens were first identified in the fruit fly and frog embryos by genetic approaches, differential

Address correspondence to A. Hari Reddi, UC Davis Medical Center, Research Building 1, Room 2000, 4635 Second Avenue, Sacramento, CA 95817, USA, 916 734-5749, <ahreddi@ucdavis.edu>.

R. Faller et al. (eds.), *Biomembrane Frontiers: Nanostructures, Models, and the Design of Life*, Handbook of Modern Biophysics, DOI: 10.1007/978-1-60761-314-5_10,

displays, subtractive hybridization, and expression cloning. These research techniques were then extended to mice and men. Alternatively, morphogens were isolated from bone, with its well-known regeneration potential. There are different types of morphogen families that have been identified, which are presented in Table 10.1.

Table 10.1. Morphogen Families

Morphogen family	Morphogenetic function
Hedgehog family (Hh)	*Drosophila* development
Sonic hedgehog (Shh)	Neural development
Indian hedgehog (Ihh)	Modulating chondrogenesis in the appendicular skeleton
Desert hedgehog (Dhh)	Testis development
Bone morphogenetic proteins (BMPs)	Induce endochondral bone morphogenesis
Wnt family member wingless (Wg)	*Drosophila* appendage development
Fibroblast growth factor (FGF)	Mesodermal cells development

10.3. CARTILAGE MORPHOGENESIS

Cartilage morphogenesis is the key rate-limiting step in the dynamics of endochondral bone development. Development of the skeleton begins with formation of cartilage that is subsequently replaced by bone. The process by which the cartilage endoskeleton grows and ossifies during development is called endochondral ossification. Endochondral ossification is a multistep process and includes extracellular matrix production, cell hypertrophy, vascular invasion, resorption of calcified cartilage, and bone deposition. The morphogenesis of cartilage determines the shape of bones, location of joint tissues (including tendons and ligaments), and formation of articular cartilage. Cartilage morphogenesis is critical for both the joint and bone. The end of bone forms articular cartilage while growth-plate cartilage is the focal center for longitudinal growth of the skeleton. The maintenance factors of articular chondrocytes include bone morphogenetic proteins (BMPs) and TGF-β isoforms. Hence all BMPs can be cartilage morphogenetic proteins.

10.4. BONE MORPHOGENETIC PROTEINS

Urist made the key discovery that new bone formation can be induced by implantation of dimineralized bone matrix [2,3]. In extraskeletal sites, characterization of the active fractions in the matrix led to the discovery, isolation, and characterization of several osteoinductive proteins, also called bone morphogenetic proteins (BMPs) [4,5]. BMPs, based on their variable functional expression, are able to initiate chondrogenesis [6] and are present at the apical ectodermal ridge in the developing limb bud. BMPs and members of the human transforming growth factor (TGF-β) superfamily induce new cartilage and bone formation in vitro and in vivo [7–10] , and regulate cell proliferation and differentiation [11,12].

10.5. STRUCTURE AND FUNCTION OF ARTICULAR CARTILAGE

Articular cartilage is critical to the normal function of human and animal joints, providing lubrication and load-bearing to allow locomotion and movement. It is a uniquely avascular, aneural, and alymphatic tissue comprised almost entirely of an extensive extracellular matrix (ECM) with very few cells. Cartilage is notoriously recalcitrant to regeneration and repair. Cartilage has a defined stratified structure composed of superficial, middle, deep, and calcified zones, each with distinct cell densities and phenotypes, molecular architecture, and mechanical properties. It is composed of extracellular matrix macromolecules such as type II collagen, aggrecan, hyaluronan, chondroitin sulfate, and decorin. Water makes up the majority of the tissue, accounting for 60–85% of the wet weight (ww), and provides fluid for lubrication. Collagen is the major organic constituent of cartilage and accounts for 15–22% ww and provides the tensile strength of cartilage. Proteoglycans are the second most abundant organic component, accounting for 4–7% ww [13]. The negative charges in the glycosaminoglycans (GAGs) on proteoglycans attract collagen, thus building the ECM. They also attract water, providing the compressive strength of cartilage as it resists the outflow of water during compression. Collagen and water content are the highest in the superficial zone and decrease through deeper zones. In contrast, proteoglycan content is lowest in the superficial zone and increases in the deeper zones of the articular cartilage.

During osteoarthritis (OA) the first changes to articular cartilage occur in the superficial zone. Superficial zone protein (SZP) is a proteoglycan that exists at the superficial zone of cartilage and functions as a lubricant during body movement. It is possible that normal accumulation of SZP is impaired during OA, which leads to progressive joint degeneration. SZP is a large proteoglycan that is synthesized by superficial zone chondrocytes and synoviocytes, and is secreted into synovial fluid [14,15]. SZP is also present in other joint-lining tissues such as the tendon [16] and meniscus [17], suggesting a major role in joint function. Radin and Swann first discovered the role for a glycoprotein fraction in joint lubrication [18]. This glycoprotein was isolated and named lubricin [19,20] in synovial fluid. Lubricin and SZP are key mediators in boundary lubrication and protect the joint from excessive wear [21–23]. Mutations in the gene encoding SZP cause a rare autosomal-recessive disorder called camptodactyl-arthropathy-coxa vara-pericarditis (CACP) syndrome in humans [24,25].

10.6. ROLE OF TRANSFORMING GROWTH FACTOR (TGF-β) IN CHONDROCYTES

Transforming growth factor beta (TGF-β) includes three related mammalian isoforms. TGF-β1, -2, and -3 are pleiotropic cytokines that regulate extracellular matrix production, wound healing, immune functions, and cell proliferation and differentiation. They belong to the large TGF-β superfamily, which also includes the activins, inhibins, bone morphogenetic proteins (BMPs), and growth and differentiation factors (GDFs). TGF-β has been identified in bone matrix, cartilage, platelets, activated lymphocytes, and other tissues [26]. TGF-β members have multiple functions, but appear to be part of the regulatory network of growth factors that maintain articular cartilage in the differentiation of cell phenotype [27].

Recent publications show that BMPs, growth factors, and cytokines such as FGF-2, IGF-1, PDGF, TGF-β1 FGF-2, IGF-1, PDGF, TGF-β1, IL-1a, IL-1b, and TNF-α in the stifle (knee) joints from 3-month-old calves affect the accumulation of articular cartilage superficial zone protein (SZP). Explants and cell culture experiments were designed to investigate the role of BMPs, growth factors, and cytokines in the accumulation of SZP. Explant cultures were treated with 300 ng/mL BMP-7 for a period of 1, 3, or 7 days. Monolayer cell cultures were treated with 100 ng/mL each of BMP-2, BMP-4, or BMP-7, and 30 ng/mL each of FGF-2, IGF-1, PDGF, or TGF-β1 for 3 days. It was found that in both explant and monolayer culture systems BMP-7 increased SZP accumulation in a dose- and time-dependent fashion. SZP accumulation also varied with different growth factor treatments in monolayer cell cultures. In addition to being stimulated by BMP-7, SZP levels also significantly increased when cells were treated with FGF-2, IGF-1, PDGF, or TGF-β1, but not BMP-2 or BMP-4. Such catabolic cytokines as IL-1a, IL-1b, and TNFα inhibit SZP levels in monolayer cell cultures. In conclusion, the results provide novel insights into the role of morphogens, especially BMP-7, growth factors, and cytokines on the accumulation of SZP in articular cartilage [28,29].

In 2007, Niikura and Reddi investigated the roles of TGF-β and BMP superfamily members in SZP accumulation in both articular chondrocytes and synoviocytes using primary cell culture. They also investigated the effects of TGF-β1, TGF-β2, TGF-β3, BMP-2, BMP-4, BMP-7, GDF-5, activin A, activin B, and activin AB on SZP accumulation and further demonstrated the critical functional role of TGF-β receptor type I (TGF–β-RI) kinase in the action of TGF-β isoforms. It was found that maximal stimulation of SZP accumulation occurred with the TGF-β isoforms (TGF-β1, TGF–β2, and TGF–β3) in both superficial zones of chondrocytes and synoviocytes. This showed the critical role of TGF-β in the regulation of SZP accumulation.

Studies have shown that BMPs and TGF-β can act as differentiation factors depending on environmental conditions. The use of growth factors in tissue engineering has been proposed as a powerful tool to induce tissue growth and regeneration [9,28]. It was well documented that these growth factors have the ability to influence engineered cartilage with respect to tissue growth, concentration, and distribution of ECM components [6,11,29]. Articular chondrocytes can be stimulated in many ways to enhance the production of extracellular matrix molecules. The most common method is the application of growth factors, which can positively affect the formation of new cartilage tissue in both explants and engineered constructs [27,28].

Several studies have indicated that BMPs and TGF-β have important roles in articular chondrocyte differentiation and production and in maintenance of the matrix. Several animal experiments on the healing of articular cartilage defects have been performed using BMPs, a repair-signaling molecule. It is well documented that the most efficacious treatment for articular cartilage damage and osteoarthritis is joint surgery. However, the potential use of growth factors and BMPs together with tissue-engineering methods to treat cartilage and bone damage is exciting, yet not fully realized, and is likely to be a future treatment strategy.

10.7. MEMBRANES AND MATRIX IN MECHANOTRANSDUCTION

Articular cartilage structure and composition is subject to regulation by mechanical forces in the skeletal system. The mechanobiology of articular cartilage is intimately linked to the cell membrane and extracellular matrix continuum. The articular cartilage consists of chondrocytes surrounded by vast expanses of extracellular matrix. The membrane extracellular matrix continuum functions at the organismic, organ, tissue, and cellular levels. It is well known that genetic factors are key in tissue morphogenesis. The roles of mechanical forces, including shear stress, are among the epigenetic factors that influence morphogenesis. Thus the constellation of morphogens and mechanical forces are integrated during mechanotransduction, which will be discussed next.

10.8. MECHANICAL PROPERTIES OF CARTILAGE

Articular cartilage is viscoelastic, exhibiting a nonlinear and time-dependent strain response to mechanical loading. The viscoelasticity of cartilage is due in part to its anisotropic and depth-dependent structure [13]. Because of this, the material properties of the tissue, such as moduli and hydraulic permeability, depend on many factors, such as thickness of the cartilage, source and age of the specimen, history of loading, and loading rate and duration.

The most common loads experienced by articular cartilage are compression and shear. Peak compressive loads in the hip joint can reach over four times body weight (BW) during a normal gait cycle and over seven times BW walking at a fast speed [29]. Peak shear forces in the hip can range from ~0.2 to 0.4 times BW during normal gait [30]. In the human knee, compressive forces of over three times BW occur during walking [31], up to six times BW during stair climbing [31], and over 6–11 times BW during running [32,33]. Shear forces in the knee can range from 0.6–0.7 times BW during walking and running [31,33], up to 1.3 BW during stair climbing [31], and up to 0.6 times BW during deep flexion [34]. Given the large forces imposed on the joint during everyday activities, articular cartilage plays a critical role in supporting the body and distributing these loads across the joint.

When an external force is applied to articular cartilage, the tissue responds by developing an internal stress and strain. In order to function properly through decades of load cycles during an average person's lifespan, the material properties of cartilage must exceed the stress levels generated by daily activities. The compressive modulus of human femoral head cartilage in confined compression has been reported to depend on the depth, increasing exponentially from 1.16 MPa in the superficial zone to 7.75 MPa in the deep zone [35]. The more compliant superficial surface allows fluid exudation from cartilage during loading, while the stiffer deep layer provides compressive resistance. Because cartilage is a viscoelastic material, it experiences creep under static loading. Furthermore, the rate at which creep occurs also varies with time. Measurements of human femoral condyles and tibial plateaus under cyclic loading showed compressive moduli between 7.5 and 19.5 MPa for regions of the condyle and 0.7–9.8 MPa for regions in the plateau [36]. It was also shown that stiffer samples experienced higher strains and lower creep rates than softer ones [36]. To provide lateral strength of the tissue during sliding contacts, articular cartilage must also have sufficient shear strength. Shear moduli have also been reported to vary with depth: from 0.18 MPa at the surface to ~3 MPa in the deep zone of human femoral condylar cartilage [37]. Others report an increase from 70 to 650 kPa (0.07–0.65 MPa) from the superficial to deeper zones, respectively, of bovine cartilage from the patellofemoral groove [38].

The shear strength of articular cartilage is crucial for preventing wear. Osteoarthritis, also known as degenerative arthritis, is the most common joint disease and is associated with excessive loading and wear of the joint. Mechanical wear mechanisms include adhesion and abrasion, and occur most commonly during solid-on-solid contact. Synovial fluid and the water content within cartilage help provide hydrodynamic, elastohydrodynamic, weeping [39], and boosted [40] lubrication. Within these modes of lubrication, the cartilage is well protected from wear. However, in many instances during extreme loading conditions, a fluid film cannot be maintained and cartilage relies on boundary lubrication. Boundary lubrication acts as a "last line of defense" against solid contact and eventual wear. In this mode, surfaces are separated by a molecular-scale lubricant film monolayer adsorbed to the tissue surface. Therefore, the molecular entities present at the articular surface are important for normal joint function. To maintain a normal ECM as well as a normal boundary lubricant film, chondrocytes must respond to mechanical loading to keep the tissue in homeostasis and provide the material compressive and shear strengths required to support daily loads.

10.9. MECHANICAL SIGNAL TRANSDUCTION

It is well established that chondrocytes respond to mechanical loading. Articular chondrocytes exist in a microenvironment involving complex biomechanical and biochemical interactions. Many of the mechanical stimuli that induce biological responses of chondrocytes have been identified, but the mechanisms by which they occur remain largely unknown. Proteoglycans are upregulated during low rates (0.001 Hz) of dynamic compression [41] and downregulated at higher loading rates (1 Hz) [42]. Dynamic shear loading increases synthesis of superficial zone protein (SZP) [43]. Thus, in addition to morphogenetic factors, biomechanical signals are critical for cartilage structure and function.

Of particular interest are the signaling pathways that allow mechanical signals to be sensed, received, and processed by chondrocytes to evoke downstream responses on synthesizing matrix and surface proteins. An important proteoglycan present at the surface of cartilage is SZP, which provides boundary lubrication to the joint. As discussed previously, it has been shown that chemical factors that regulate SZP expression include TGF-β, BMP-7, and oncostatin M (OSM), which increase SZP synthesis, and interleukin 1 (IL-1) and tumor necrosis factor α (TNF-α), which decrease SZP synthesis [44–46]. Recent work has shown that the TGF-β/Smad-2, 3 signaling pathway plays a key role in the mechanotransduction of SZP expression. It was found that shear loading of bovine femoral condylar cartilage upregulated SZP in the absence of TGF-β inhibitor SB431542, but effected no change in the presence of inhibitor [47]. Furthermore, a regional dependence of SZP regulation was found, having a greater effect in areas of the joint that are highly loaded (anterior medial region of the condyle), and a smaller effect in lightly loaded areas (posterior region of the condyle) [47] (Fig. 10.1). Therefore, growth factor pathways appear to be one mechanism by which transduction of mechanical stimuli occurs in cartilage. It is likely that other possible mechanisms, including integrin signaling and autocrine signaling pathways [48,49], also contribute to regulation of cartilage proteins.

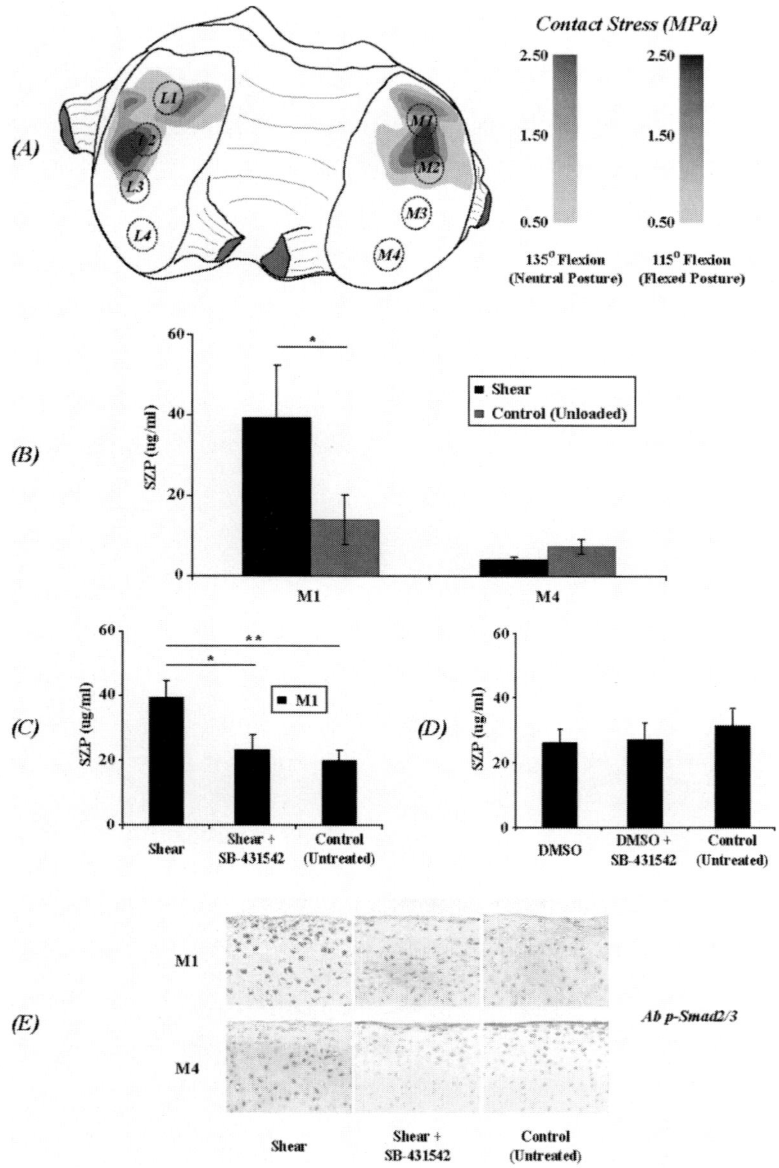

Figure 10.1. Determination of SZP expression by contact stress distribution on femoral condyles through transforming growth factor (TGF)-β-mediated signaling pathways. (A) Representative pressure distributions during compressive loading and simulated in-vivo postures, determined using pressure-sensitive film and a load application system for joint testing. For a normal load equal to 1, 2, and 3 times body weight in the 135° neutral posture and 115° flexed posture, regions of contact stress were consistently located in the anterior regions of the joint, corresponding to SZP expression patterns. (B) Significant increase in SZP expression levels during shear loading of explants from location M1, but not from location M4, compared to unloaded controls. (C) Decrease in SZP expression levels at location M1 to control levels during shear loading of explants in the presence of SB431542, a specific inhibitor of TGF-β receptor type I kinase activity. (D) SZP expression levels at location M1 in unloaded control samples, samples treated with DMSO (the carrier vehicle for SB431542), and samples treated with DMSO plus SB431542. (E) Immunostaining for phospho-Smad2/3. Staining was particularly enhanced in superficial zone chondrocytes from location M1 compared with untreated controls or explants treated with SB431542. Please visit http://www.springer.com/series/7845 to view a high-resolution full-color version of this illustration.

10.10. APPLICATIONS OF BMPs

BMPs have a wide range of applications and potential in regenerative medicine and surgery. BMPs can be applied in the following clinical areas:

- Othopaedic surgery
- Fracture treatment
- Spine fusion
- Articular cartilage repair
- Dentistry and oral surgery
- Periodontal surgery
- Craniofacial surgery
- Plastic surgery

10.11. SUMMARY

Articular cartilage consists of four different types of zones: superficial, middle, deep, and calcified. This zonal arrangement is an indication of the different functional capabilities of the respective zones that act together to provide normal movement and locomotion through the joint. The superficial zone is strong in tension because of the collagen fibers and resists the shear applied to cartilage during articulation. The middle zone consists of proteoglycans that impart compressive strength to the tissue. The deep zone consists of collagen fibers that are continuous through the tidemark to mechanically integrate cartilage with the subchondral bone. The structure and function relationships of these zones are unique and different from each other. This indicates that in order to tissue-engineer articular cartilage it may be necessary to replicate the zonal arrangement of articular cartilage. Culturing superficial, middle, and deep zone cells separately may allow one to find different chemical stimuli that optimize cellular biosynthesis for each subpopulation of chondrocytes. Articular chondrocytes can be stimulated to enhance the production of extracellular matrix molecules. The most common method is the application of growth factors and BMPs, which can stimulate the formation of new cartilage tissue in both explants and engineered constructs. The current standard of treatment for articular cartilage damage and osteoarthritis is joint surgery, but the potential use of growth factors and BMPs together with tissue-engineering methods to treat cartilage and bone damage is promising and exciting.

ACKNOWLEDGMENTS

We thank the Lawrence Ellison Endowment for research support. This work was supported by grants from the NIH and the Fulbright Fellowship from the Institute for International Education. We are also grateful to Dr. Corey Neu for allowing us to reprint the graphic in Figure 10.1.

PROBLEMS

10.1. **Case Study of Human BMP 2 and Fracture Healing**. Patients (n = 450) with an open tibial fracture were involved in randomized, single-blind study. The study was conducted at 49 centers in 11 countries. Patients were prospectively randomized to one of three groups. Patients who met the study eligibility had provided informed consent and had been properly randomized irrespective of whether they received treatment:

Group 1 — received standard of care (SOC) (intramedullary nail fixation), which was the control group

Group 2 — received rhBMP-2 (0. 75 mg/ml)

Group 3 — received intramedullary nail fixation and BMP 2.

Treatments were assigned to each group. They were followed after the treatment for the following lengths of time: 6, 10, 14, 20, 26, 39, and 52 weeks after treatment.

A fracture was considered to be healed when there was radiographic evidence of fracture union and met all the clinical criteria. The primary outcome measure was the recommendation of secondary intervention because of delayed union or nonunion within 12 months postoperatively. All interventions were classified according to the degree of invasiveness.

Which of the three groups will report a shorter time to fracture union and why?

FURTHER STUDY

Reddi AH 2003. Cartilage morphogenetic proteins: role in joint development, homoeostasis, and regeneration. *Ann Rheum Dis* **62**(Suppl 2)73–78.

Reddi AH. 1998. Role of morphogenetic proteins in skeletal tissue engineering and regeneration. *Nat Biotechnol* **16**(3):247–252.

Khouri RK, Koudsi B, Reddi AH. 1991. Tissue transformation into bone in vivo. *JAMA* 266:1953–1955.

REFERENCES

1. Fata JE, Werb Z, Bissell MJ. 2004. Regulation of mammary gland branching morphogenesis by the extracellular matrix and its remodeling enzymes. *Breast Cancer Res* **6**(1):1–11.
2. Reddi AH, Huggins C. 1972. Biochemical sequences in the transformation of normal fibroblasts in adolescent rats. *Proc Natl Acad Sci USA* **69**(6):1601–1605.
3. Urist MR. 1965. Bone: formation by autoinduction. *Science* **150**(698):893–899.
4. Luyten FP, Chen P, Paralkar V, Reddi AH. 1994 Recombinant bone morphogenetic protein-4, transforming growth factor-beta 1, and activin A enhance the cartilage phenotype of articular chondrocytes in vitro. *Exp Cell Res* **210**(2):224–229.
5. Sampath TK, Coughlin JE, Whetstone RM, Banach D, Corbett C, Ridge RJ, Ozkaynak E, Oppermann H, Rueger DC. 1990. Bovine osteogenic protein is composed of dimers of OP-1 and BMP-2A, two members of the transforming growth factor-beta superfamily. *J Biol Chem* **265**(22):13198–131205.
6. Chen P, Carrington JL, Hammonds RG, Reddi AH. 1991 Stimulation of chondrogenesis in limb bud mesoderm cells by recombinant human bone morphogenetic protein 2B (BMP-2B) and modulation by transforming growth factor beta 1 and beta 2. *Exp Cell Res* **195**(2):509–515.
7. Pecina M, Jelic M, Martinovic S, Haspl M, Vukicevic S. 2002. Articular cartilage repair: the role of bone morphogenetic proteins. *Int Orthop* **26**(3):131–136.

8. Reddi AH. 2003. Cartilage morphogenetic proteins: role in joint development, homoeostasis, and regeneration. *Ann Rheum Dis* **62**(Suppl 2):73–78.
9. Reddi AH. 1998. Role of morphogenetic proteins in skeletal tissue engineering and regeneration. *Nat Biotechnol* **16**(3):247–252.
10. Sampath TK, DeSimone DP, Reddi AH. 1982. Extracellular bone matrix-derived growth factor. *Exp Cell Res* **142**(2):460–464.
11. Asahina I, Sampath TK, Nishimura I, Hauschka PV. 1993. Human osteogenic protein-1 induces both chondroblastic and osteoblastic differentiation of osteoprogenitor cells derived from newborn rat calvaria. *J Cell Biol* **123**(4):921–933.
12. Vukicevic S, Luyten FP, Reddi AH. 1989. Stimulation of the expression of osteogenic and chondrogenic phenotypes in vitro by osteogenin. *Proc Natl Acad Sci USA* **86**(22):8793–8797.
13. Mow VC, Ratcliffe A, Poole AR. 1992. Cartilage and diarthrodial joints as paradigms for hierarchical materials and structures. *Biomaterials* **13**(2):67–97.
14. Schumacher BL, Block JA, Schmid TM, Aydelotte MB, Kuettner KE. 1994. A novel proteoglycan synthesized and secreted by chondrocytes of the superficial zone of articular cartilage. *Arch Biochem Biophys* **311**(1):144–152.
15. Schumacher BL, Hughes CE, Kuettner KE, Caterson B, Aydelotte MB. 1999. Immunodetection and partial cDNA sequence of the proteoglycan, superficial zone protein, synthesized by cells lining synovial joints. *J Orthop Res* **17**(1):110–120.
16. Rees SG, Davies, JR, Tudor D, Flannery CR, Hughes CE, Dent CM, Caterson B. 2002. Immunolocalisation and expression of proteoglycan 4 (cartilage superficial zone proteoglycan) in tendon. *Matrix Biol* **21**(7):593–602.
17. Schumacher BL, Schmidt TA, Voegtline MS, Chen AC, Sah RL. 2005. Proteoglycan 4 (PRG4) synthesis and immunolocalization in bovine meniscus. *J Orthop Res* **23**(3):562–568.
18. Radin E L, Swann DA, Weisser PA. 1970 Separation of a hyaluronate-free lubricating fraction from synovial fluid. *Nature* **228**(5269):377–378.
19. Swann DA Silver FH, Slayter HS, Stafford W, Shore E. 1985. The molecular structure and lubricating activity of lubricin isolated from bovine and human synovial fluids. *Biochem J* **225**(1):195–201.
20. Swann DA, Slayter HS, Silver FH. 1981. The molecular structure of lubricating glycoprotein-I, the boundary lubricant for articular cartilage. *J Biol Chem* **256**(11):5921–5925.
21. Jay GD, Britt DE, Cha CJ. 2000. Lubricin is a product of megakaryocyte stimulating factor gene expression by human synovial fibroblasts. *J Rheumatol* **27**(3):594–600.
22. Jay GD, Haberstroh K, Cha CJ. 1998. Comparison of the boundary-lubricating ability of bovine synovial fluid, lubricin, and Healon. *J Biomed Mater Res* **40**(3):414–418.
23. Jay GD, Tantravahi U, Britt DE, Barrach HJ, Cha CJ. 2001. Homology of lubricin and superficial zone protein (SZP): products of megakaryocyte stimulating factor (MSF) gene expression by human synovial fibroblasts and articular chondrocytes localized to chromosome 1q25. *J Orthop Res* **19**(4):677–687.
24. Bahabri SA, Suwairi WM, Laxer RM, Polinkovsky A, Dalaan AA, Warman ML. 1998. The camptodactyly-arthropathy-coxa vara-pericarditis syndrome: clinical features and genetic mapping to human chromosome 1. *Arthr Rheum* **41**(4):730–735.
25. Marcelino J, Carpten JD, Suwairi WM, Gutierrez OM, Schwartz S, Robbins C, Sood R, Makalowska I, Baxevanis A, Johnstone B, Laxer RM, Zemel L, Kim CA, Herd JK, Ihle J, Williams C, Johnson M, Raman V, Alonso LG, Brunoni D, Gerstein A, Papadopoulos N, Bahabri SA, Trent JM, Warman ML. 1999. CACP, encoding a secreted proteoglycan, is mutated in camptodactyly-arthropathy-coxa vara-pericarditis syndrome. *Nat Genet* **23**(3):319–322.
26. Trippel SB. 2004. Growth factor inhibition: potential role in the etiopathogenesis of osteoarthritis. *Clin Orthop Relat Res* **427**(Suppl):S47–S52.
27. O'Connor WJ, Botti T, Khan SN, Lane JM. 2000. The use of growth factors in cartilage repair. *Orthop Clin North Am* **31**(3):399–410.
28. Khalafi A, Schmid TM, Neu C, Reddi AH. 2007. Increased accumulation of superficial zone protein (SZP) in articular cartilage in response to bone morphogenetic protein-7 and growth factors. *J Orthop Res* **225**(3):293–303.
29. Niikura T, Reddi AH. 2007. Differential regulation of lubricin/superficial zone protein by transforming growth factor beta/bone morphogenetic protein superfamily members in articular chondrocytes and synoviocytes. *Arthr Rheum* **56**(7):2312–2321.

30. Reddi AH. 2005. BMPs: from bone morphogenetic proetin to body morphogenetic proteins. *Cytokine Growth Factor Rev* **16**:249–250.
31. Paul JP. 1976. Force actions transmitted by joints in the human body. *Proc R Soc Lond B Biol Sci* **192**(1107): 163–172.
32. Wu G, Millon D. 2008. Joint kinetics during Tai Chi gait and normal walking gait in young and elderly Tai Chi Chuan practitioners. *Clin Biomech* **23**(6):787–795.
33. Taylor WR, Heller MO, Bergmann G, Duda GN. 2004. Tibio-femoral loading during human gait and stair climbing. *J Orthop Res* **22**(3):625–632.
34. Flynn TW, Soutas-Little RW. 1995. Patellofemoral joint compressive forces in forward and backward running. *J Orthop Sports Phys Ther* **21**(5):277–282.
35. Scott SH, Winter DA. 1990. Internal forces of chronic running injury sites. *Med Sci Sports Exerc* **22**(3):357–369.
36. Nagura T, Dyrby CO, Alexander EJ, Andriacchi TP. 2002. Mechanical loads at the knee joint during deep flexion. *J Orthop Res* **20**(4):881–886.
37. Chen SS, Falcovitz YH, Schneiderman R, Maroudas A, Sah RL. 2001. Depth-dependent compressive properties of normal aged human femoral head articular cartilage: relationship to fixed charge density. *Osteoarthritis Cartilage* **9**(6):561–569.
38. Barker MK, Seedhom BB. 2001. The relationship of the compressive modulus of articular cartilage with its deformation response to cyclic loading: does cartilage optimize its modulus so as to minimize the strains arising in it due to the prevalent loading regime? *Rheumatology (Oxford)* **40**(3):274–284.
39. Wong BL, Bae WC, Chun J, Gratz KR, Lotz M, Sah RL. 2008. Biomechanics of cartilage articulation: effects of lubrication and degeneration on shear deformation. *Arthr Rheum* **58**(7):2065–2074.
40. Buckley MR, Gleghorn JP, Bonassar LJ, Cohen I. 2008. Mapping the depth dependence of shear properties in articular cartilage. *J Biomech* **41**(11):2430–2437.
41. Lewis PR, McCutchen CW. 1959. Mechanism of animal joints. *Nature* **184**:1285.
42. Walker PS, Dowson D, Longfield MD, Wright V. 1968. "Boosted lubrication" in synovial joints by fluid entrapment and enrichment. *Ann Rheum Dis* **27**(6):512–520.
43. Kim YJ, Sah RL, Grodzinsky AJ, Plaas AH, Sandy JD. 1994. Mechanical regulation of cartilage biosynthetic behavior: physical stimuli. *Arch Biochem Biophys* **311**(1):1–12.
44. Buschmann MD, Kim YJ, Wong M, Frank E, Hunziker EB, Grodzinsky AJ. 1999. Stimulation of aggrecan synthesis in cartilage explants by cyclic loading is localized to regions of high interstitial fluid flow. *Arch Biochem Biophys* **366**(1):1–7.
45. Nugent GE, Aneloski NM, Schmidt TA, Schumacher BL, Voegtline MS, Sah RL. 2006. Dynamic shear stimulation of bovine cartilage biosynthesis of proteoglycan 4. *Arthr Rheum* **54**(6):1888–1896.
46. Jones AR, Flannery CR. 2007. Bioregulation of lubricin expression by growth factors and cytokines. *Eur Cell Mater* **13**:40–45.
47. Lee SY, Niikura T, Reddi AH. 2008. Superficial zone protein (lubricin) in the different tissue compartments of the knee joint: modulation by transforming growth factor beta 1 and interleukin-1 beta. *Tissue Eng Part A* **14**(11):1799–1808.
48. Neu CP, Khalafi A, Komvopoulos K, Schmid TM, Reddi AH. 2007. Mechanotransduction of bovine articular cartilage superficial zone protein by transforming growth factor beta signaling. *Arthr Rheum* **56**(11):3706–3714.
49. Millward-Sadler SJ, Salter DM. 2004. Integrin-dependent signal cascades in chondrocyte mechanotransduction. *Ann Biomed Eng* **32**(3):435–446.
50. Tschumperlin D J, Dai G, Maly IV, Kikuchi T, Laiho LH, McVittie, AK, Haley KJ, Lilly CM, So PT, Lauffenburger DA, Kamm RD, Drazen JM. 2004. Mechanotransduction through growth-factor shedding into the extracellular space. *Nature* **429**(6987):83–86.

11

LIFECYCLE OF A LIPOPROTEIN FROM A BIOPHYSICAL PERSPECTIVE

John C. Rutledge,[1,3] Thomas Huser,[1,3] John Voss,[1]
James Chan,[3] and Atul Parikh[2,,3]
[1]*School of Medicine,* [2]*Department of Applied Physics, and* [3]*Center for Biophotonics Science and Technology, University of California Davis*

11.1. INTRODUCTION

The goal of our project was to understand how lipids and lipoproteins interact with cell membranes. This chapter will present the five major areas in which we have focused our attention on understanding how lipids and lipoproteins interact with cell membranes (Fig. 11.1): (1) triglycerides and vascular injury, (2) single lipoprotein analysis, (3) apolipoprotein E (apoE) conformation changes in the postprandial state, (4) triglyceride-rich lipoproteins (TGRLs) and endothelial cell inflammation, and (5) TGRL lipolysis products and monocyte activation.

For over a hundred years, Western civilization has questioned how the food we eat translates into disease, and specifically atherosclerotic cardiovascular disease. Although most information indicates that this basic pathophysiological process is mediated through consumption of excess saturated fats, much remains unknown.

After humans eat a meal, there is an elevation of triglycerides in the blood in the postprandial state. In normal individuals, triglycerides can rise after a meal by 50 to 100%. This has been documented many times in the past, including a paper by Hyson et al, (1998) [1]. In that study, normal healthy individuals were given a 40%-fat meal. Plasma triglycerides, which were modestly elevated initially, rose about 60% higher three to four hours after ingestion of the meal. Subsequently plasma triglycerides fell to baseline levels six hours after the meal. Even in these healthy individuals, a significant elevation of triglycerides was noted after ingestion of a moderately high-fat meal.

Address correspondence to John C. Rutledge, Department of Medicine, School of Medicine, University of California Davis, 5404 GBSF, Davis, CA 95618, USA, 530 752-6778, 530 752-3470 (fax), <jcrutledge@ucdavis.edu>, <jcrutledge@gmail.com>.

R. Faller et al. (eds.), *Biomembrane Frontiers: Nanostructures, Models, and the Design of Life*,
Handbook of Modern Biophysics, DOI: 10.1007/978-1-60761-314-5_11,

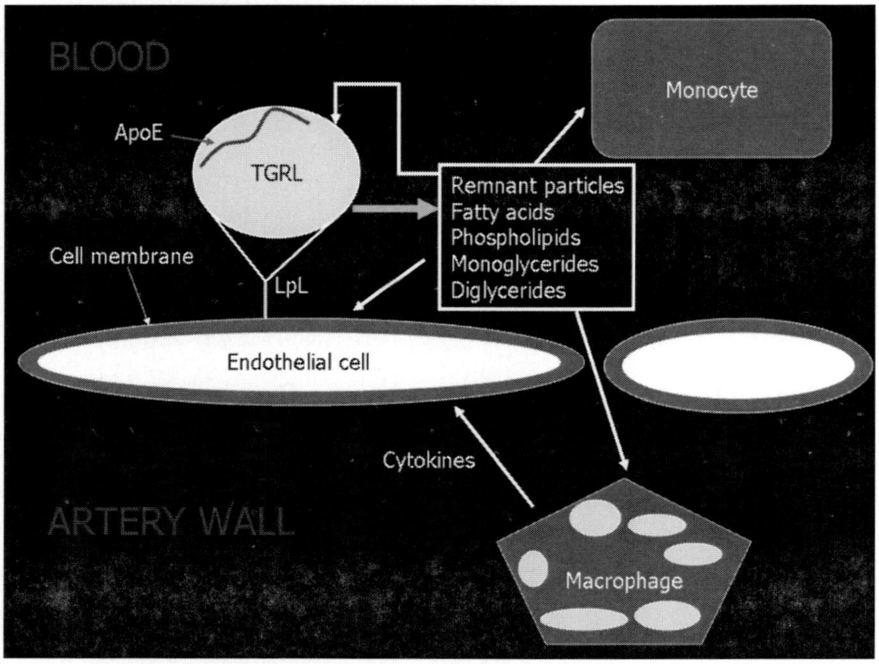

Figure 11.1. Pathophysiological model of TGRL-induced vascular inflammation. Please visit http://www.springer.com/series/7845 to view a high-resolution full-color version of this illustration.

11.2. LASER TRAPPING RAMAN SPECTROSCOPY FOR ANALYSIS OF SINGLE LIPOPROTEINS

So then how do plasma triglycerides cause a vascular inflammatory response? In a recent paper from our group, we tested the response of healthy individuals to a moderately high-fat meal [2,3]. In this study, we measured monocyte tumor necrosis factor-alpha (TNFα) before and after the meal using flow cytometry. A low level of monocytes with TNFα was noted before the meal. However, after ingestion of the meal, monocyte TNFα rose four- to fivefold. Further, although triglycerides at six hours post-consumption had returned to their control values, monocyte TNFα remained elevated.

We have developed a model of vascular inflammation that involves triglyceride-rich lipoproteins, endothelial cells, monocytes, and macrophages (Fig. 11.1). Endothelial cells line the blood vessel walls. On the surface of endothelial cells, lipoprotein lipase is anchored to the plasma membrane of the endothelial cell. Lipoprotein lipase hydrolyzes TGRL to generate remnant particles, fatty acids, phospholipids, monoglycerides, and diglycerides. In addition, on the surface of triglyceride-rich lipoproteins, apolipoproteins assume a variety of conformations. The remnant lipoprotein particles, fatty acids, phospholipids, monoglycerides, and diglycerides can have effects on monocytes, endothelial cells, macrophages, and arterial smooth muscle cells. Lipolysis products also affect or remodel other lipoproteins. Thus, there are many possible interactions of triglyceride-rich lipoprotein lipolysis products with other lipoproteins and the cells in the blood and vascular wall.

Triglyceride-rich lipoproteins are spherical lipid and protein particles. As the name implies, these lipoproteins have a high triglyceride content. Cholesterol ester and triglycerides are in the core of the particle, and unesterified cholesterol, phospholipid, and apolipoproteins reside on the surface of the particle. This is the basic structure of all lipoproteins in blood, including chylomicrons, very-low-density (VLDL), low-density (LDL), and high-density (HDL) lipoproteins.

In a recent paper [4], we investigated the changes in individual triglyceride-rich lipoproteins by combining laser-trapping with Raman spectroscopy. TGRLs were isolated by ultracentrifugation. A 30-μL drop containing TGRL is placed on a glass cover slip. A 633-nm continuous light wave laser beam is focused on a single-trapped TGRL. The lipoprotein particle is viewed through a 100×-power, 1.3-NA oil immersion objective. The lipoproteins are imaged easily above 30 nm in diameter using this equipment. A Raman spectrum can be generated by illuminating the lipoprotein particle. Saturated bond peaks can be detected, as well as unsaturated bond peaks. Thus, single lipoproteins can be imaged and analyzed in real time.

Recent advances in biophysical and biochemical analysis techniques now enable the nondestructive analysis of cellular biochemistry in living cells. Raman spectroscopy is a particularly powerful nondestructive analysis technique that probes molecular bond vibrations by inelastic scattering of light. At the tissue and organ levels, Raman spectroscopy is being translated as a clinical technique for rapid analysis of diseased tissue during surgery, etc. Its strengths are high sensitivity and chemical specificity without the need for optical labels (e.g., H&E stains or fluorescent probes). Our group and others have recently advanced the sensitivity of this technique and demonstrated that the favorable properties of Raman spectroscopy are maintained at the single-cell and even the single-lipoprotein levels, enabling the label-free analysis of individual cells and TGRLs.

In [4] we showed that the lipoproteins in the fasting state, when compared with the postprandial state, change dramatically in terms of saturated and unsaturated fats. We showed that in individuals given a high-glycemic diet and meal that lipoproteins became more saturated after the meal. This potentially has implications for uptake by vascular cells and atherosclerotic cardiovascular disease.

Additionally, our studies have shown that laser trapping Raman spectroscopy can be used to perform experiments on single lipoproteins. We also showed that lipoprotein lipase has a characteristic spectrum. In addition, fasting VLDL generated a spectrum characteristic of the meal previously ingested. However, if we added lipoprotein lipase to the VLDL, a dramatic change in the Raman spectra was seen indicating greater content of saturated fats. As far as we know, this is the first experiment on a single lipoprotein that has ever been performed.

Additionally, we showed that specific fatty acids gave characteristic Raman spectra. For example, the saturated fatty acids palmitic acid and stearic acid demonstrate dramatically different spectra that allow us to identify specific lipids in lipoproteins (Fig. 11.2).

We conclude from these studies that the saturated-to-unsaturated content of individual TGRL can be assessed by laser trapping Raman spectroscopy. Future studies are designed to better understand how VLDL remodeling by lipolysis products affects uptake by cells, and ultimately atherogenesis.

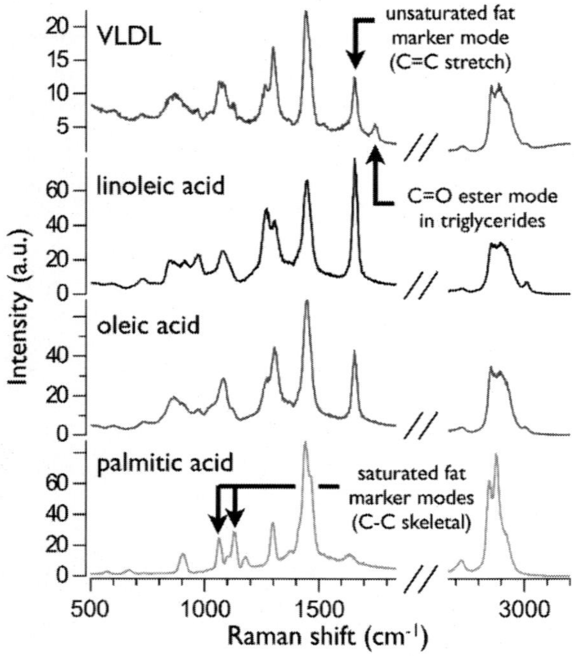

Figure 11.2. Raman spectra of fatty acids and lipoproteins. Comparison of the different Raman spectra obtained from saturated and unsaturated fatty acids in comparison to the one obtained from a single VLDL lipoprotein. Note the sharp structural peaks in the 1000–1150 cm^{-1} range in the saturated lipid palmitic acid, which are also partially reflected in the VLDL spectrum, representing its heterogeneous fat content. Please visit http://www.springer.com/series/7845 to view a high-resolution full-color version of this illustration.

11.3. APOLIPOPROTEIN E CONFORMATIONAL CHANGES IN THE POSTPRANDIAL STATE

Apolipoprotein E3 (apoE3) is associated with protection from atherosclerotic cardiovascular disease. ApoE4 is associated with a tenfold greater risk of Alzheimer's disease and a fourfold greater risk of atherosclerosis. ApoE3 and apoE4 have dramatic conformational differences, despite only a single amino acid difference. We used electron paramagnetic resonance (EPR) spectroscopy of site-directed spin labels to analyze these conformational differences, where spin-labeled sidechains in apoE report on the local dynamics, polarity, and proximity to other spin labels [5–8]. This technique therefore reports on the structural adaptation of apolipoproteins upon binding to different lipoprotein species, as well as structural changes that occur during the lifecycle of the lipoprotein particle.

Tetali et al (2006) [9] showed that apoE4 had dramatic conformational changes in the postprandial state. However, apoE3 was little changed in the postprandial state. The implications for these dramatic changes in apoE4 are significant. As noted above, apoE4 indicates a much greater risk of Alzheimer's disease and atherosclerotic cardiovascular disease.

In brief, our study showed that conformational changes in apoE4 associated with decreased protein–protein interactions are observed in the postprandial state. Additionally, further studies have shown changes in lipoprotein fluidity increased dramatically in VLDL as compared to low-density and high-density lipoproteins. Further, VLDL lipolysis products dramatically increase VLDL fluidity when using 12-doxyl stearic acid probes. In conclusion, VLDL lipolysis products increase VLDL fluidity and apoE4 conformational changes.

11.4. TGRL LIPOLYSIS PRODUCTS AND VASCULAR INFLAMMATION

Triglyceride-rich lipoproteins, including chylomicrons and VLDLs, generate lipolysis products consisting of remnant particles, fatty acids, phospholipids, monoglycerides, and diglycerides. These lipolysis products can act on endothelial cells to increase endothelial cell injury and endothelial layer permeability. Rutledge, Mullick, Gardner, and Goldberg (2000) [10] showed that VLDL lipolysis products strongly deposit these products in and on the endothelial cell layer. Not only are lipolysis products posited on the endothelial layer, but a dramatic increase in endothelial layer permeability was observed.

In a recent study, Wang, Butti, Parikh, and Rutledge (2008) [11] showed that endothelial cell plasma membrane lipid rafts were dramatically altered when treated with VLDL lipolysis products as compared with VLDLs only. VLDL protein was labeled with tetramethylrhodamine isothiocyanate and lipid rafts were labeled with cholera toxin B and imaged using confocal fluorescence microscopy. Colocalization of the lipid rafts and the VLDL remnant particles was observed on the endothelial surface. These studies conform to previous studies using model synthetic membranes [12–14] (Fig. 11.3).

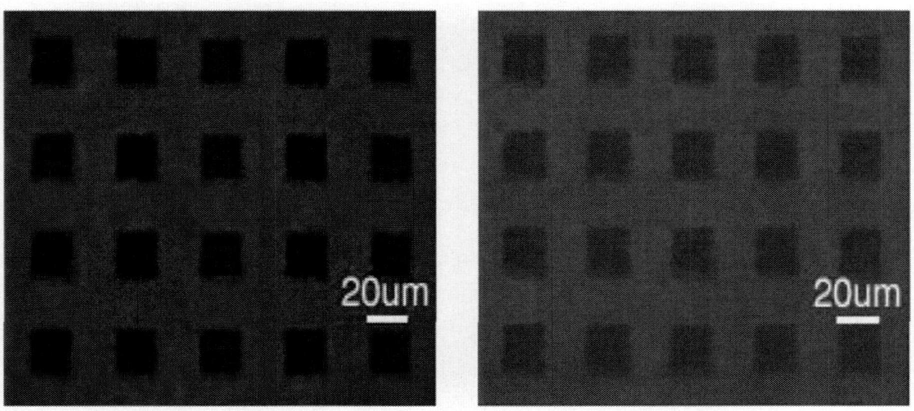

Figure 11.3. Effects of TGRL on synthetic lipid rafts. The two-color images in the left panel reveal the patterned fluid POPC bilayer labeled with TR-DHPE stably coralling a cholesterol/sphingomyelin/POPC (1:1:1) raft mixture labeled with Marina Blue-DHPE. The two-color image on the right reveals colocalization of atto-520 labeled TGRL (green) with raft-like domains. Please visit http://www.springer.com/series/7845 to view a high-resolution full-color version of this illustration.

11.5. CELL-FREE MEMBRANE-MIMETIC MODEL BILAYERS TO STUDY INTERACTION OF TGRL WITH RAFT-LIKE MICROENVIRONMENTS

Supported phospholipid membranes displaying an engineered array of cholesterol-rich raft-like microcompositions were incubated with postprandial plasma and their interactions probed in real time. The plasma was modified by adding TGRLs fluorescently labeled with Atto 520 dye, which binds to surface apolipoproteins. A striking selectivity between cholesterol-rich raft-like domains and TGRL was observed. Figure 11.3 shows two-color (false colors correspond to emission properties of the probes used) epifluorescence images of raft micropatterns before and after incubation with atto-520-labeled TGRL for 30 min. The image in the left panel reveals that the raft microdomains (blue) remain stably confined within the fluid POPC matrix (labeled pink). The image shown in the right panel reveals an almost exclusive preference by the TGRL particles (green) to associate with the cholesterol-rich raft-like microdomains. Current studies are focused on characterization of the selectivity of TGRL association in cholesterol-enriched, raft-like domains by incubation with methyl-β-cyclodextrin that selectively extracts cholesterols from plasma membranes. Control samples are those with free fatty acids that do not show this selectivity. These model experiments will be useful in correlating lipoproteins and lipids, and raft physical properties (e.g., sizes, densities, and distributions) with endothelial layer function, e.g., permeability.

11.6. TGRL LIPOLYSIS PRODUCTS AND INCREASED ENDOTHELIAL LAYER PERMEABILITY

Eiselein, Wilson, Lame, and Rutledge (2007) [15] examined the effect of junctional proteins on endothelial layer permeability. When junctional proteins ZO-1, occludin, and VE-cadherin were examined before and after treatment with TGRL lipolysis products, a dramatic rearrangement of ZO-1 and occludin on junctional protein morphology was observed.

11.7. MONOCYTES AND TGRL LIPOLYSIS PRODUCTS

Triglyceride-rich lipoproteins have been shown to regulate monocytes. Monocytes are known important cell types in the atherosclerotic process. We have observed that when circulating monocytes are exposed to triglyceride-rich lipoprotein lipolysis products, monocytes become activated and lipid droplets form in the monocytes (Fig. 11.4).

To interrogate monocytes, individual cells were captured with a tightly focused laser trap that suspends the cell. A characteristic Raman spectrum from in the monocyte is illuminated by the laser. In one experiment, we imaged monocytes with light microscopy. In addition, we developed Raman spectra of monocytes in media (Fig. 11.5). Next we treated the monocytes with lipopolysaccharide and noted a morphological change in the monocyte and a modest change in the Raman spectra. Finally, VLDL was incubated with LpL and the monocyte was exposed to the VLDL lipolysis products. A dramatic increase in lipid droplet formation was detected in monocytes. In addition, a dramatic change in the Raman spectra was observed.

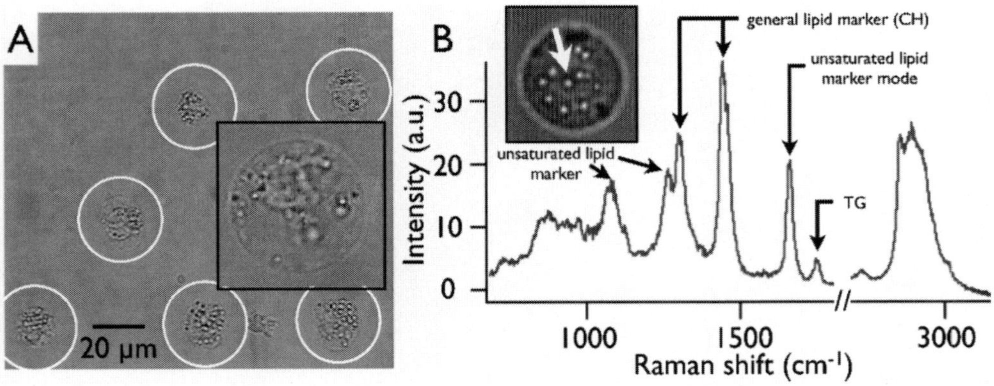

Figure 11.4. Monocytes form lipid vesicles in their cytosol after treatment with VLDL lipolysis products. (A) THP1 cells 3 hours after treatment with lipolysis products showing lipid vesicles. Cells are highlighted by white circles for better visibility. The inset shows a close-up of a monocyte with lipid vesicles. (B) Raman spectrum of the lipid vesicle highlighted by an arrow in the white-light confocal micrograph in the inset. Please visit http://www.springer.com/series/7845 to view a high-resolution full-color version of this illustration.

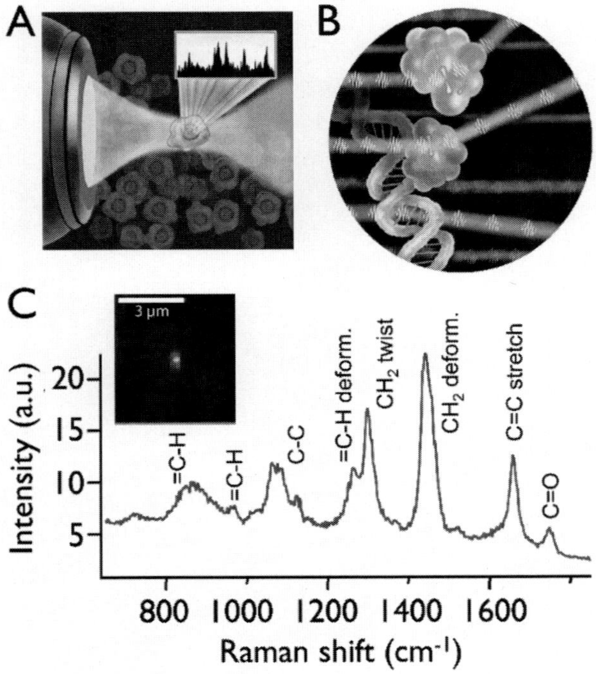

Figure 11.5. Laser tweezers Raman spectroscopy. (A) Cartoon of a single, optically trapped monocyte immobilized by a tightly focused laser beam. (B) Cartoon of the process of Raman scattering. A fraction of the incoming photons (left) can scatter off of molecular bonds and change their energy. (C) Raman spectrum of an optically trapped TGRL. Major Raman peaks are labeled with their respective molecular bonds. The inset shows a back-reflected light image of an individual optically trapped VLDL (~50 nm in diameter). Please visit http://www.springer.com/series/7845 to view a high-resolution full-color version of this illustration.

Coherent anti-Stokes Raman scattering (CARS) microscopy was first demonstrated in the 1980s, but was severely limited by the short-pulsed laser excitation sources in the visible part of the optical spectrum available at that time. Not until 1999 was this type of label-free microscopy rediscovered. CARS is a four-wave mixing process, where two short pulsed laser beams are closely overlapped in time and in space to probe specific Raman modes in a sample. Simplistically speaking, one laser source serves as the Raman excitation source and generates Raman-scattered photons, while the second laser source is overlapped with a specific Raman transition (e.g., a lipid CH vibration) relative to the excitation source and creates a coherent Raman signal on the anti-Stokes, i.e., blue-shifted, side of the spectrum. This process is nonlinear in that it requires the simultaneous scattering of two photons from the excitation laser source and one photon from the probe laser source. This creates an inherent confocal effect similar to multiphoton fluorescence excitation, because the signal is only generated in the very focus, where all laser beams overlap temporally and spatially. This form of microscopy provides chemical maps of microscopic samples at speeds and signal intensities similar to confocal fluorescence microscopy and has recently even been extended to in-vivo imaging in animal models [16–18].

In addition, we used CARS to image the lipid-filled monocytes. Numerous 1-μm and less lipid droplets were formed in the monocytes after exposure to the TGRL lipolysis products. In comparison, TGRL can cause the formation of lipid droplets; however, they are much less frequent and the Raman spectra are very different. The origin and functional significance of the lipid droplets in the monocytes are unknown at the present time.

These studies with monocytes show the utility, flexibility, and importance of these new biophysical techniques. We showed that individual monocytes can be analyzed by laser trapping Raman spectroscopy, and also that lipolysis products injure monocytes with subsequent formation of lipid droplets.

11.8. SUMMARY

The lipid saturated-to-unsaturated ratio of individual TGRLs can be assessed by laser trapping Raman spectroscopy.

VLDL lipolysis products increase remnant particle fluidity and induce apoE4 conformational changes.

VLDL lipolysis products injure endothelial cells and cause rearrangement of junctional proteins.

VLDL remnant particles colocalized with remodeled endothelial cell lipid rafts after treatment with lipolysis products.

Individual monocytes can be analyzed by laser trapping Raman spectroscopy, and lipolysis products injure monocytes and induce lipid droplet formation.

11.9. CONCLUSIONS

Our studies demonstrate how a collaboration of biologists/medical scientists, and biophysicists and bioengineers can provide a constructive and productive interaction of multiple points of view. Our studies show at the single molecule and cell levels that lipids and lipoproteins have dramatic effects on lipoproteins and a variety of vascular cells. Some of the changes that we had observed potentially can be eliminated by medical therapies and potentially provide biomarkers of occult disease in the future.

ACKNOWLEDGMENTS

NIH HL78615 and HL55067. This work was also supported by funding from the National Science Foundation. The Center for Biophotonics, an NSF Science and Technology Center, is managed by the University of California Davis under Cooperative Agreement No. PHY 0120999.

PROBLEM

11.1. How can apoE conformation be used clinically to detect vascular inflammation and atherosclerosis?

FURTHER STUDY

Hyson DA, Paglieroni TG, Wun T, Rutledge JC. 2002. Postprandial lipemia is associated with platelet and monocyte activation and increased monocyte cytokine expression in normolipemic men. *Clin Appl Thromb Hemost* **8**(2):147–155.

Chan JW, Motton D, Rutledge JC, Keim NL, Huser T. 2005. Raman spectroscopic analysis of biochemical changes in individual triglyceride-rich lipoproteins in the pre- and postprandial state. *Anal Chem* **77**(18):5870–5876.

Tetali SD, Budamagunta MS, Voss JC, Rutledge JC. 2006. C-termini interactions of apolipoprotein E4 respond to the postprandial state. J Lipid Res **47**(7):1358–1365.

REFERENCES

1. Redard CL, Davis PA, Schneeman BO. 1990. Dietary fiber and gender: effect on postprandial lipemia. *Am J Clin Nutr* **52**(5):837–845.
2. Hyson D, Rutledge JC, Berglund L. 2003. Postprandial lipemia and cardiovascular disease. *Curr Atheroscler Rep* **5**(6):437–444.
3. Hyson DA, Paglieroni TG, Wun T, Rutledge JC. 2002. Postprandial lipemia is associated with platelet and monocyte activation and increased monocyte cytokine expression in normolipemic men. *Clin Appl Thromb Hemost* **8**(2):147–155.
4. Chan JW, Motton D, Rutledge JC, Keim NL, Huser T. 2005. Raman spectroscopic analysis of biochemical changes in individual triglyceride-rich lipoproteins in the pre- and postprandial state. *Anal Chem* **77**(18):5870–5876.
5. Martin DD, Budamagunta MS, Ryan RO, Voss JC, Oda MN. 2006. Apolipoprotein A-I assumes a looped belt conformation on reconstituted high density lipoprotein. *J Biol Chem* **281**(29):20418–20426
6. Shao B, Bergt C, Fu X, Green P, Voss JC, Oda MN, Oram JF, Heinecke JW. 2005. Tyrosine 192 in apolipoprotein A-I is the major site of nitration and chlorination by myeloperoxidase, but only chlorination markedly impairs ABCA1-dependent cholesterol transport. *J Biol Chem* **280**(7):5983–5993.
7. Hatters DM, Budamagunta MS, Voss JC, Weisgraber KH. 2005. Modulation of apolipoprotein E structure by domain interaction: differences in lipid-bound and lipid-free forms. *J Biol Chem* **280**(40):34288–34295.
8. Oda MN, Forte TM, Ryan RO, Voss JC. 2003. The C-terminal domain of apolipoprotein A-I contains a lipid-sensitive conformational trigger. *Nat Struct Biol* **10**(6):455–460.
9. Tetali SD, Budamagunta MS, Voss JC, Rutledge JC. 2006. C-termini interactions of apolipoprotein E4 respond to the postprandial state. *J Lipid Res* **47**(7):1358–1365.

10. Rutledge JC, Mullick AE, Gardner G, Goldberg IJ. 2000. Direct visualization of lipid deposition and reverse lipid transport in a perfused artery: roles of VLDL and HDL. *Circ Res* **86**(7):768–773.
11. Wang L, Sapuri-Butti AR, Aung HH, Parikh AN, Rutledge JC. 2008. Triglyceride-rich lipoprotein lipolysis increases aggregation of endothelial cell membrane microdomains and produces reactive oxygen species. *Am J Physiol* **295**(1):H237–H244.
12. Vinchurkar MS, Bricarello DA, Lagerstedt JO, Buban JP, Stahlberg H, Oda MN, Voss JC, Parikh AN. 2008. Bridging across length scales: multi-scale ordering of supported lipid bilayers via lipoprotein self-assembly and surface patterning. *J Am Chem Soc* **130**(33):11164–11169.
13. Oliver AE, Kendall EL, Howland MC, Sanii B, Shreve AP, Parikh AN. 2008. Protecting, patterning, and scaffolding supported lipid membranes using carbohydrate glasses. *Lab Chip* **8**(6):892–897.
14. Sanii B, Smith AM, Butti R, Brozell AM, Parikh AN. 2008. Bending membranes on demand: fluid phospholipid bilayers on topographically deformable substrates. *Nano Lett* **8**(3):866–871.
15. Eiselein L, Wilson DW, Lame MW, Rutledge JC. 2007. Lipolysis products from triglyceride-rich lipoproteins increase endothelial permeability, perturb zonula occludens-1 and F-actin, and induce apoptosis. *Am J Physiol* **292**(6):H2745–H2753.
16. Chan JW, Taylor DS, Lane SM, Zwerdling T, Tuscano J, Huser T. 2008. Nondestructive identification of individual leukemia cells by laser trapping Raman spectroscopy. *Anal Chem* **80**(6):2180–2187.
17. Chan JW, Winhold H, Corzett MH, Ulloa JM, Cosman M, Balhorn R, Huser T. 2007. Monitoring dynamic protein expression in living *E. coli* bacterial cells by laser tweezers Raman spectroscopy. *Cytometry A* **71**(7):468–474.
18. Chan JW, Taylor DS, Zwerdling T, Lane SM, Ihara K, Huser T. 2006. Micro-Raman spectroscopy detects individual neoplastic and normal hematopoietic cells. *Biophys J* **90**(2):648–656.

12

TARGETING APOLIPOPROTEINS IN MAGNETIC RESONANCE IMAGING

Renuka Sriram,[1] Jens O. Lagerstedt,[1] Haris Samardzic,[1] Ulrike Kreutzer,[1] Jitka Petrolova,[1] Hongtao Xie,[1] George A. Kaysen,[1] John C. Voss,[1] Jean F. Desreux,[2] and Thomas Jue[1]
[1]*Biochemistry and Molecular Medicine, University of California Davis;*
[2]*Coordination and Radiochemistry, University of Liège, Belgium*

12.1. INTRODUCTION

Maintaining normal physiological homeostasis depends upon a coordinated metabolism of both water-soluble and -insoluble substrates. In humans the body derives these molecules — such as glucose, amino acids, and fatty acids — from complex food matter. Water-soluble substrates can circulate readily in blood, while water-insoluble molecules — such as fatty acid, triacylglycerol, and cholesterol — require ampiphathic carriers to transport them from the site of biosynthesis (liver and intestine) to the target tissue. For fatty acid, albumin serves as the major transporter. For triacylglycerol and cholesterol, however, macromolecular complexes aggregate the hydrophobic molecules into the core and cover the surface with amphiphatic proteins and phospholipids to solubilize the particles in the lymphatic and circulatory systems. These macromolecules belong to a class of proteins, plasma lipoproteins, with specific functions and cellular targets. In the clinic these lipoproteins prognosticate the risk of cardiovascular disease (CVD).

Lipoproteins divide usually into five major types: chylomicron, very-low-density lipoprotein (VLDL), intermediate-density lipoprotein (IDL), low-density lipoprotein (LDL), and high-density lipoprotein (HDL). Each lipoprotein type exhibits characteristic density, size, and composition. As implied in the name, the density varies from the low-density chylomicron (<0.95 g/ml) to the high-density HDL (1.2 g/ml). Size also varies. The chylomicron has the

Address correspondence to Thomas Jue, PhD, Biochemistry and Molecular Medicine, University of California Davis, Davis, CA 95616-8635, 530 752-4569, 530 752-3516 (fax), <<tjue@ucdavis.edu>>.

R. Faller et al. (eds.), *Biomembrane Frontiers: Nanostructures, Models, and the Design of Life*,
Handbook of Modern Biophysics, DOI: 10.1007/978-1-60761-314-5_12,

largest diameter (75–1,200 nm), and HDL has the smallest (5–12 nm). The physical property variation arises from each lipoprotein's distinct composition. In a chylomicron, cholesterol, triacylglycerol, and phospholipid predominate and constitute about 90% of the particle. Protein constitutes only about 10%. In contrast, the smaller HDL has less cholesterol, triacylglycerol, and phospholipid (65% of the particle) but more protein (over 30%).

Even though lipoproteins contain a high fraction of triacylglycerol and cholesterol, these hydrophobic molecules do not determine the physiological function. Instead, function depends heavily upon the action of the associated proteins. Each lipoprotein contains a unique set of apoproteins or apolipoproteins. (Proteins separated from a lipid–protein complex have the designation of apoproteins, derived from the contraction of the two terms "apodized protein"). In LDL, apoB-100 (apolipoprotein B-100) predominates, whereas in HDL apoA-I (apolipoprotein A1) predominates. Indeed, these markedly different proteins, apoB-100 (513 kD, (kilodalton)) and apoA-I (29 kD), confer lipoprotein targeting and functional specificity [1,2].

12.2. CHYLOMICRONS AND TRIACYLGLYCEROL

The function of chylomicron and LDL illustrates the distinct physiological roles. Chylomicrons play a major role in transporting triacylglycerol from the intestine after a meal. Normal plasma lipid, cholesterol, and triacylglycerol levels range from 3.6–6.8, 1.3–2.6, and 0.8–2.4 g/l, respectively [2]. In the postprandial state, triglycerides can increase 50–100%. Depending upon the fat content of a meal, plasma triglyceride levels can remain high for up to 4 hours, even in healthy individuals [3].

Chylomicrons carry the triacylglycerol from the intestine into the lymphatic system and then into the circulatory system via the thoracic duct. As a consequence, the lungs and heart receive the initial flow of triacylglycerol-laden chylomicrons from the intestine after a meal. Indeed, the heart prefers fatty acid over glucose as a fuel source. In addition, chylomicrons contain apoB-48 (240,000 kD), apoA-IV (44,000 kD), apoC-II (8,837 kD), apoC-III (8,751 kD), and apoE (34,145 kD). ApoC-II activates lipoprotein lipase in the capillary endothelium of adipose tissue, heart, skeletal muscle, and lactating mammary glands. The lipase hydrolyzes the triacylglycerol to release free fatty acid to the target tissue, where the cell can utilize it as energy source or a fuel storage. The remnant chylomicrons, rich in cholesterol, return to the liver. In the liver, which synthesizes de novo both triacylglycerol and cholesterol, any excess triacylglycerol and cholesterol gets repackaged and re-exported to the circulation as VLDL [4].

12.3. LDL AND CHOLESTEROL TRANSPORT

In contrast, LDL does not play any significant role in transporting triacylglycerol after a meal. It has a major function in transporting cholesterol and regulating its metabolism. Moreover, releasing cholesterol from LDL does not require the action of any lipase or protease. The apoB-100 (513 kD) associated with LDL binds to the target cell surface receptor. Both the receptor and the LDL get endocytosed. Inside the cell, the LDL receptor fuses with lysosomes, where enzymes catalyze the release of the unesterifed cholesterol for membrane biosynthesis. Alternatively, ACAT (acyl-CoA-choleserol acyl transferase) can re-esterify cholesterol for storage [5]. The receptor then recycles to the cell surface.

Typically, the liver, adrenals, and adipose tissue remove about 50% of the total plasma LDL pool in one day. However, patients with familial hypercholesteremia (FH) have a deficiency in synthesizing functional LDL receptors. Homozygotes have almost no LDL receptors. Heterozygotes have less than 50% of the normal LDL receptors. Consequently, plasma LDL with its cholesterol cargo cannot endocytose into the cell for metabolism or reprocessing. Instead, LDL remains in the plasma and elevates the serum cholesterol level. A normal individual has about 1.3–2.6 g/l of cholesterol in the plasma. FH heterozygotes have about 50% increased cholesterol level. FH homozygotes have even higher levels: 6.5–10 g/l. Such an elevated cholesterol level promotes the formation of atherosclerotic plaques, which narrow the arterial vessels to impede blood flow. Indeed, FH subjects suffer severely from atherosclerosis and cardiovascular disease.

Reducing the level of LDL cholesterol can improve significantly the risk of atherosclerotic plaque formation. Since saturated, monounsaturated, and polyunsaturated fatty acids in the diet contribute to the rise in cholesterol, lipid, and lipoprotein levels, simply moderating the diet will reduce cholesterol level. Palmitic acid alone accounts for about 60% of the ingested saturated fatty acid. Moreover, saturated fatty acids (12:0; lauric acid), 14 (14:0; myristic acid), and 16 (16:0; palmitic acid)) can actually promote the formation of LDL cholesterol. Accordingly, individuals with high LDL cholesterol should reduce the intake of 16:0 saturated fatty acid. Ingested saturated fat should provide <10% of total energy need. Total fat should meet about 30% of total energy demand. Cholesterol intake should not exceed 300 mg/day [6,7].

12.4. HDL AND CHOLESTEROL SCAVENGING

HDL does not serve to deliver either triacylglycerol or cholesterol to the peripheral tissue. Its composition differs from LDL and the chylomicron and hints at a distinct function. A chylomicron has by weight 2% protein, 3% cholesteryl ester, and 85% triacylglycerol. In contrast, LDL has 23% protein, 37% cholesteryl ester, and 10% triacylglycerol. With HDL, the composition shifts dramatically toward the protein fraction: 55% protein, 15% cholesterol ester, and 4% triacylglycerol [8]. Indeed, HDL has a major function as a bioscavenger of cholesterol.

HDL contains apoA-I, apoC-I, apoC-II, and the enzyme LCAT (lecithin cholesterol acyl transferase), which catalyzes the esterification of cholesterol by lecithin. HDL biosynthesis starts in the liver or the intestine, and the assembly requires a cell surface enzyme — ABCA1 (ATP-binding cassette transporter A1) — to mediate the transfer of cellular phospholipids and cholesterol to extracellular lipid-poor apoA-I. The nascent, discoidal shaped HDL particle contains proteins but does not have a significant amount of cholesterol and no cholesterol ester. Without the cholesterol ester and triacylglycerol, the HDL lipoprotein cannot form a particle core. With the apoproteins and LCAT on the surface, however, the nascent HDL particle begins to take up cholesterol from chylomicrons, VLDL remnants, and extrahepatic tissues. HDL converts cholesterol to cholesteryl ester, which along with the uptake of triacylglycerol forms the lipoprotein core. The interaction of HDL with cholesterol scavenger receptor class B type I (SR-BI) transfers more cholesterol to the growing particles, which transforms the nascent discoidal HDL to the mature spherical HDL particle. Alternatively, the CETP (cholesteryl ester transfer protein) enzyme can also mediate the transfer of cholesterol from HDL to VLDL, IDL, and LDL in exchange for triglycerides [10,11]. HDL delivers its scavenged cholesterol cargo to the liver for excretion in the bile or feces. HDL has a pivotal role in the reverse cholesterol pathway [9].

12.5. LIPOPROTEINS AS INDICES OF CVD RISK

Because of the role of LDL in transporting cholesterol and regulating its metabolism, the clinical treatment of dyslipidemia has focused on reducing LDL cholesterol in order to lower plasma cholesterol levels [12–14]. Indeed, the widely used statin drugs inhibit cholesterol biosynthesis and increase available LDL receptors. Statins belong to a class of drugs that act on the committed step in cholesterol biosynthesis involving HMG-CoA (3 hydroxy-3 methyl glutaryl CoA) reductase. HMG-CoA reductase catalyzes the conversion of HMG-CoA to mevalonate. The statin molecules resemble HMG-CoA and therefore competitively inhibit HMG-CoA reductase activity. Statins therefore decrease cholesterol biosynthesis. As a result, the cells now have an increased number of cholesterol-free LDL receptors available to bind cholesterol. Plasma cholesterol levels also fall.

Prospective risk studies also relate the HDL cholesterol level to a reduced risk of myocardial infarction [15]. Other studies have followed male and female patients from 35 to 70 years for 5–20 years and have reached the same general conclusion: the higher the HDL cholesterol level, the lower the risk of CVD [10]. The CVD risk becomes 2.5 times higher at low HDL cholesterol values of 0.9 mmol/l for men and 1.2 mmol/l for women than at high HDL cholesterol values of 1.9 mmol/l for men and 2.3 mmol/l for women. Each 1 mg/dl increase in HDL cholesterol level decreases CVD risk by about 2–3%.

HDL and LDL levels reflect then the overall state of cholesterol transport/metabolism. Simply measuring the overall cholesterol, HDL, and LDL levels, however, would yield an ambiguous assessment, since LDL level correlates directly with an atherogenic disposition, while HDL level correlates with an anti-atherogenic condition. The HDL:LDL ratio presents a much better indicator of CVD risk. These ratios include cholesterol:HDL cholesterol, LDL cholesterol:HDL cholesterol, and non-HDL cholesterol:HDL cholesterol [16].

12.6. APOLIPOPROTEINS AND CVD RISK

Instead of lipoprotein levels or ratios, some researchers have proposed that apolipoproteins may serve as more accurate biomarkers of CVD risk [17,18]. Because the LDL particle (as well as VLDL and IDL particles) contains only one apoB molecule and the HDL particle has predominantly apoA-I, the apoB/apoA-I ratio reflects the "cholesterol balance" between atherogenic (apoB) and anti-atherogenic (apoA-I) conditions. Apolipoprotein ratios appear to track better than LDL cholesterol, lipids, and lipid ratios the risk of myocardial infarction and stroke [18,19].

A rationale for using apolipoprotein ratio predicates on a concern about lipoprotein vs. apoprotein assay accuracy and international standardization [20]. A clinical laboratory usually achieves a coefficient of variation of less than 5% for the lipoprotein assay. Between laboratories, however, the variation increases by as much as 5–15%. But standardization for HDL cholesterol determination also varies, leading to more uncertainty. In contrast, a well-established international standardization exists for apoA-I and apoB assays [21,22]. The international reference data facilitate immensely comparative analysis, leading to a more accurate indexing of apolipoprotein level and CVD risk. For these reasons, the use of the apoB/apoA-I ratio appears to have advantages over lipoprotein ratios such as total cholesterol/HDL cholesterol and LDL cholesterol/HDL cholesterol [19,21,22].

12.7. APOA-I AND HDL

HDL particles appear in a distribution of composition, shape, and sizes and have a function of maintaining cellular cholesterol homeostasis by binding to cholesterol from peripheral tissues and transporting it to the liver for clearance in the bile or feces. This reverse cholesterol process depends upon the bidirectional flux of free cholesterol via free diffusion or SRB-I (scavenger receptor class-B type I) receptor in the direction of the concentration gradient. In contrast, the ABCA1 (ATP-binding cassette transporter AI) mediates efflux of both cellular cholesterol and phospholipids unidirectionally from cell to HDL. The largest particle, HDL2b, contains the most cholesterol. Other HDL particles include the intermediate sized HDL2a, HDL3a, and HDL3b as well as the small HDL3c.

In addition to managing the cholesterol homeostasis, HDL particles influence antioxidant and anti-inflammatory interaction, stimulate endothelial nitric oxide formation, and have anti-thrombotic potential. These interactions depend upon the action of apoA-I. In fact, HDL particles can subdivide into three classes: (1) with only apoA-I (70% of apolipoprotein in HDL), (2) with only apoA-II (20% of apolipoprotein in HDL), and (3) with both apoA-I and apoA-II. Once synthesized in either liver or intestine, apoA-I associates almost completely with HDL, follows a complex metabolic pathway, and helps determine HDL particle size distribution and function [11,23].

12.8. APOA-I THERAPY

ApoA-I provides another target for therapeutic intervention to reduce CVD risk [14]. ApoA-I level by itself already correlates inversely with CVD risk, because it facilitates free cholesterol efflux from the peripheral tissue. In particular, it facilitates the efflux of cholesterol from macrophages via the ABC (adenosine triphosphate-binding cassette) A1 transporter [24]. In addition, apoA-I acts putatively as an anti-oxidative and anti-inflammatory agent.

Despite the inverse correlation of ApoA-I with CVD risk, the efficacy of an ApoA-I directed therapy remains unclear. The case of ApoA-I Milano illustrates this. In 1974 physicians noted that Valerio Dagnoli from the northern Italian town of Limone sul Garda had vanishingly low levels of HDL. Some of the other villagers (about 40, which comprised 3.5% of the population) also had minimal HDL. Given the role of HDL as a cholesterol bioscavenger, the reduced HDL level should have correlated with a high incidence of CVD. Surprisingly, none exhibited any signs of CVD. However, these villagers had a mutant apoA-I. Each carried a single point mutation of apoA-I (R173C) or apoA-I Milano, traceable to a single man who lived in the village during the 1700s [28–30]. Some scientists postulated that perhaps the variant apoA-I (R173C) provided an unusual cardioprotection. Indeed, when investigators injected a synthetic version of this apoA-I Milano HDL into rabbits and mice, they noted a reduction of plaque buildup and inflammation.

The therapeutic potential of apoA-I Milano to reduce atheroma stimulated the formation of a venture capital company — Esperion Therapeutics of Ann Arbor, Michigan — which held the property rights. Their initial intravascular ultrasound studies showed that apoA-I Milano can decrease atheroma volume in the coronary arteries, However, over several years of effort, Esperion managed to produce only enough apoA-I to partially treat only 30 out of the 45 people in the randomized trial, once weekly for a total of 5 weeks. Nevertheless, in 2004 Pfizer placed a $1.3 billion bet on the market potential of apoA-I Milano and its analogues. It purchased and internalized the Esperion company shortly before publication of the initial apoA-1 Milano (clinically designated ETC-216) clinical trial results. ETC 216 development, however, stalled. Moreover, animal studies showed that a bolus injection of wild type apoA-I reduced cholesterol level and atherosclerotic plaque formation [25–27]. Other HDL-directed therapy experiments with a much larger subject group (111 instead of 45) yielded conflicting results and showed no coronary atheroma reduction.

In 2006 a phase III clinical trial was stopped when several patients died after administration of torcetrapib (a CETP inhibitor) and atorvastatin. Pfizer pulled back on apoA-1 therapy development and jettisoned Esperion, but still kept ETC 216 within its drug portfolio [31,32].

12.9. IMAGING LIPOPROTEIN IN VIVO

The apoA-I Milano saga underscores an inadequate understanding of the biochemical and physiological mechanisms underlying the cardioprotective nature of HDL and apoA-I. A real-time in-vivo measurement of lipoprotein and apolipoprotein biodistribution and metabolism would go a long way to improve our understanding and to establish a solid basis for developing therapies that reduce CVD risk.

Using magnetic resonance imaging (MRI) with paramagnetic contrast agents, such as gadolinium-diethylenetriaminepentaacetic acid (Gd-DTPA) or other Gd chelates presents an approach to follow in real-time apoA-I biodistribution and metabolism in vivo [33]. The Gd^{3+}-based paramagnetic agents increase water relaxivity and thereby enhance tissue contrast. Unfortunately, most Gd-based contrast agents act nonspecifically. Although recent "smart" contrast agents can target specific tissue, they must still overcome the very low numbers of cell epitopes or binding sites [34–36]. To improve contrast agent sensitivity, researchers have experimented with nanoparticles delivering multiple Gd chelates, which increases relaxivity. These nanoparticles with attached antigen-like moieties include dendrimers, silica-coated micelles, polymeric and ceramic nanoparticles, perfluorocarbon emulsions, and crosslinked liposomes [37–43]. Although many nanoparticles have shown good targeting and relaxivity, their size limits the use to vascular compartment imaging, while their synthetic property often encounters biocompatibility, biodegradability, and toxicity hurdles.

12.10. SPECIFIC APOA-I CONTRAST AGENT

A recently synthesized magnetic resonance image (MRI) contrast agent and a method to substitute Cys at different apoA-I sites now open an opportunity to image in real time apoA-I biodistribution in vivo. Gd-methanethiosulfonate or Gd[MTS-ADO3A], an MRI contrast agent, has a

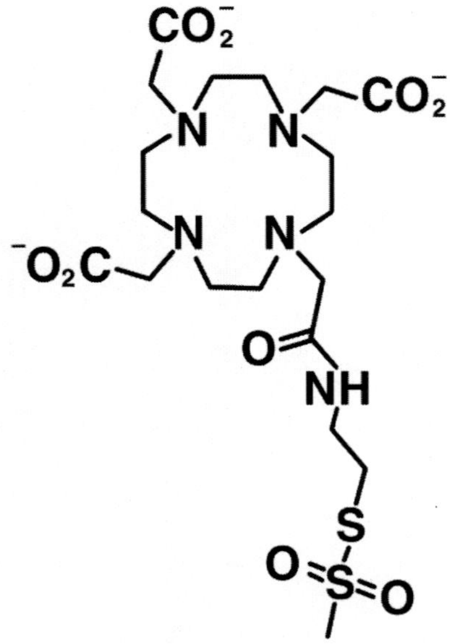

Figure 12.1. Molecular structure of Gd[MTS-ADO3A].

sulfur group that can link to a Cys amino acid residue (Fig. 12.1) [44]. For apoA-I, a biosynthetic method exists to create a cysteine mutation of apoA-I by using a bacterial expression system comprised of pNFXex plasmid in *E. coli* BL21(DE3) pLysS cells [45]. A Cys mutation would permit Gd[MTS-ADO3A] to link to the apoA-I protein at a specific site. ApoA-I then becomes an image contrast agent. Protein structure studies indicate that the Ser to Cys mutation in position 55 would provide an effective interaction of Gd[MTS-ADO3A] with water. According to molecular models, the amino acid residue at position 55 faces the aqueous face in the apoA-I–HDL complex (Fig. 12.2) [45,46]. Once Gd[MTS-ADO3A]apoA-I-S55C enters the bloodstream, it should bind tightly to HDL and target both liver and kidney. The tissue image should reflect the apoA-I biodistribution and activity. In contrast, the nonspecific gadodiamide (Omniscan) contrast agent, [GdDTPA-BMA], should produce only an image enhancement in the kidney, where the molecule gets filtered and cleared.

12.11. RELAXIVITY

In order to effect image enhancement, the contrast agent must change the image intensity in the region of interest relative to the surrounding tissue. It does so by increasing the relaxation rate of tissue water. The magnetization recovery rate after an inversion pulse depends upon a characteristic time constant, T_1, and follows the equation

$$I = I_0\left(1 - Ae^{-\tau/T_1^{\mathrm{Obs}}}\right), \tag{12.1}$$

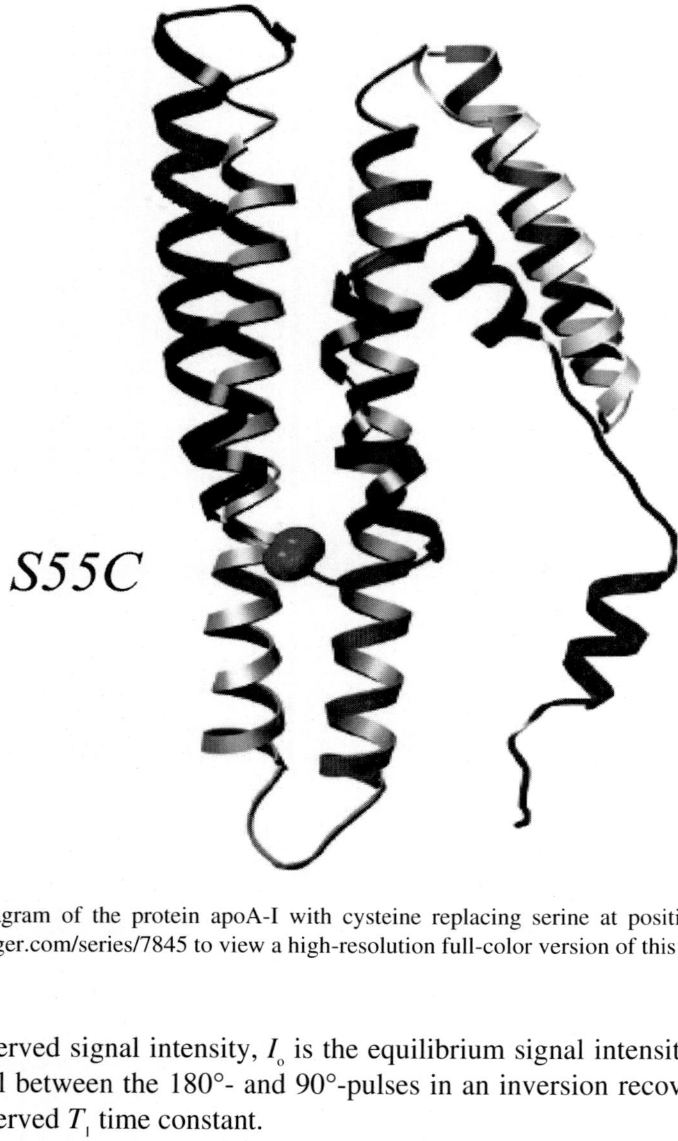

Figure 12.2. Diagram of the protein apoA-I with cysteine replacing serine at position 55. Please visit http://www.springer.com/series/7845 to view a high-resolution full-color version of this illustration.

where I is the observed signal intensity, I_0 is the equilibrium signal intensity, A is a constant, τ is the time interval between the 180°- and 90°-pulses in an inversion recovery pulse sequence, and T_1^{Obs} is the observed T_1 time constant.

In the presence of a contrast agent, the relaxation rate of water will increase in proportion to concentration of the agent, as shown in the equation

$$\frac{1}{T_1^{\mathrm{Obs}}} = \frac{1}{T_1^{\mathrm{H_2O}}} + r[\mathrm{Gd}], \qquad (12.2)$$

where $\frac{1}{T_1^{\mathrm{Obs}}}$ is the observed relaxation rate, $\frac{1}{T_1^{\mathrm{H_2O}}}$ is the intrinsic H_2O relaxation rate without any contrast agent, r is a relaxivity constant, and [Gd] is the concentration of Gd^{3+} in the Gd complexes. Because $\frac{1}{T_1^{\mathrm{Obs}}} > \frac{1}{T_1^{\mathrm{H_2O}}}$, the signal in the presence of a contrast agent will recover faster than in the absence of an agent following a T_1 pulse sequence. An appropriate choice of τ and recycle time will weight the image, so that the water magnetization in contact with the relaxation agent will return rapidly to the equilibrium state after each pulse train. In contrast, the water signal from the surrounding tissue will not return to equilibrium and will exhibit a decreased signal intensity because of partial saturation [47]. Thus the localized region with the Gd contrast agent areas will have a higher signal intensity than the surrounding regions.

At an NMR field strength of 8.9 T, water relaxes with a T_1 rate of about 0.4 s^{-1}. In the presence of Gd[MTS-ADO3A] alone, the water relaxivity increases sharply to about 8 $mM^{-1}s^{-1}$. Linking the Gd[MTS-ADO3A] to apoAI S55C in the absence and presence of rHDL changes the relaxivity to 2.0 and 12 $mM^{-1}s^{-1}$.

12.12. IMAGING APOA-I IN VIVO

Figure 12.3 shows the MRI mouse images after infusion of Gd[MTS-ADO3A]apoAI-S55C and gadodiamide. When B6 mice receive a bolus injection of gadodiamide, image enhancement appears only in the kidney, consistent with literature reports [48]. Relative to the control intensity, gadodiamide enhances kidney image intensity by 100% within 5 min of infusion. To the contrary, the infusion of Gd[MTS-ADO3A]apoAI-S55C enhances image intensity in both liver and kidney. In kidney the image intensity increases about 100%. In liver the image intensity increases also about 100% within the first 5 min.

Because Gd[MTS-ADO3A] and gadodiamide ([GdDTPA-BMA]) have similar chemical properties, any cleavage of the bond between Gd[MTS-ADO3A] and apoA-I-S55C would yield only an image enhancement in kidney. Indeed, preliminary ICPMS (inductive coupled plasma mass spectrometry) analysis, which can assess quantitatively the amount of trace elements, has confirmed that the observed image contrast originates from Gd^{3+} in the protein complex Gd[MTS-ADO3A]apoAI–S55C.

12.13. SUMMARY

The availability of Gd[MTS-ADO3A] and specific Cys mutant variants of apoA-I presents an opportunity to create an apoA-I contrast agent that can track apoA-I and HDL biodistribution and metabolism in vivo. Given Gd[MTS-ADO3A]apoAI–S55C and other potential apoA-I Cys variants, an imaging methodology now emerges to give insight into the underlying mechanism regulating HDL and apoA-I metabolism in vivo.

ACKNOWLEDGMENTS

We gratefully acknowledge funding support from NIH GM 58688 (TJ), NIH grant AG029246 (JCV), American Heart Association Western States Affiliate 0265319Y (UK), the research service of the Department of Veterans Affairs, NCHCS (Mather, CA), and the invaluable technical assistance of Drs. Jeff Walton and Jeff de Ropp.

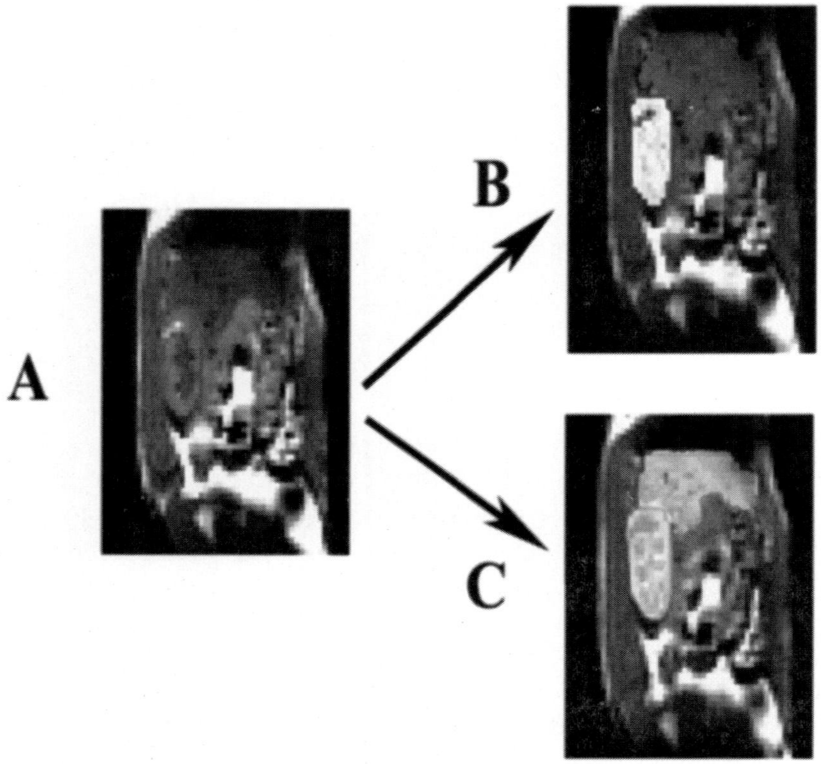

Figure 12.3. Saggital view of a set of mouse liver and kidney images acquired with T_1-weighted multi-slice spin echo sequence: (A) during control period, (B) 5 min after infusion of Omniscan (gadodiamide), (C) 5 min after infusion of Gd[MTS-ADO3A]apo AI-S55C. Please visit http://www.springer.com/series/7845 to view a high-resolution full-color version of this illustration.

PROBLEMS

1. Statins have helped patients lower their plasma cholesterol levels. Would you prescribe statins to someone with familial hypercholesterolemia to reduce his/her plasma cholesterol level?

2. How does immature HDL lead to cardiovascular disease, a major cause of death in patients with chronic kidney disease (CKD)?

FURTHER STUDY

Merbach AE, Toth E. 2001. *The chemistry of contrast agents in medical magnetic resonance imaging*. New York: John Wiley & Sons.

Dass CR, Jessup W. 1999. Apolipoprotein A-I, cylodextrins, and liposomes as potential drugs for reversal of atherosclerosis. *J Pharm Pharmacol* **52**:731–761.

Oram JF. 2003. HDL apolipoproteins and ABCA1: partners in the removal of excess cellular cholesterol. *Arterioscler Thromb Vasc Biol* **23**(5):720–727.

Martin DDO, Budamagunta MS, Ryan RO, Voss JC, Oda MN. 2006. Apolipoprotein A-I assumes a "looped belt" conformation on reconstituted high density lipoprotein. *J Biol Chem* **281**(29):20418–20426.

Oda MN, Forte TM, Ryan RO, Voss JC. 2003. The C-terminal domain of apolipoprotein A-I contains a lipid-sensitive conformational trigger. *Nat Struct Biol* **10**:455–460.

REFERENCES

1. Fielding CJ, Fielding PE. 1985. Metabolism of cholesterol and lipoproteins. In *Biochemistry of lipids and membranes*, pp. 404–474. Ed DE Vance, JE Vance. Menlo Park: Benjamin/Cummings.
2. Zubay G, Parson W, Vance D. 2006. *Principles of biochemistry*. New York: William C. Brown.
3. Rutledge JC, Huser T, Voss J, Chan J, Parikh A. 2009. Lifecycle of a lipoprotein from a biophysical perspective. In *Biomembrane frontiers: nanostructures, models, and design of life*. Ed R Faller, T Jue, M Longo, S Risbud. New York: Springer.
4. Goldberg IJ, Kako Y, Lutz EP. 2000. Responses to eating: lipoproteins, lipolytic products, and atherosclerosis. *Curr Opin Lipidol* **11**:235–241.
5. Brown MS, Goldstein JL. 1984. How LDL receptors influence cholesterol and atherosclerosis. *Sci Am* **251**: 58–66.
6. Cater NB, Garg A. 1997. Serum low-density lipoprotein cholesterol response to modification of saturated fat intake: recent insights. *Curr Opin Lipidol* **8**:332–336.
7. Stalenhoef AFH, de Graaf J. 2009. Association of fasting and nonfasting serum triglycerides with cardiovascular disease and the role of remnant-like lipoproteins and small dense LDL. *Curr Opin Lipidol* **19**:355–361.
8. Kritchevsky D. 1986. Atherosclerosis and nutrition. *Nutr Int* **2**:290–297.
9. Glomset JA. 1968. The plasma lecithins:cholesterol acyltransferase reaction. *J Lipid Res* **9**:155–167.
10. Rosenson RS. 2005. Low HDL-C: a secondary target of dyslipidemia therapy. *Am J Med* **118**:1067–1077.
11. Packard CJ, Rader DJ. 2006. Lipids and atherosclerosis: advances in translational medical science. Abingdon: Taylor & Francis.
12. Baigent C, Keech A, Kearney P, Blackwell L, Buck G, Pollicino C, Kirby A, Sourjina T, Peto R, Collins R, Simes R; Cholesterol Treatment Trialists' (CTT) Collaborators. 2005. Efficacy and safety of cholesterol-lowering treatment: prospective meta-analysis of data from 90,056 participants in 14 randomised trials of statins. *Lancet* **366**:1267–1278.
13. Third Report of the National Cholesterol Education Program (NCEP) Expert Panel on Detection, Evaluation, and Treatment of High Blood Cholesterol in Adults (Adult Treatment Panel III). 2002. Final report. *Circulation* **106**:3144–3421.
14. Luc G, Bard JM, Ferrières J, Evans A, Amouyel P, Arveiler D, Fruchart JC, Ducimetière P. 2002. Value of HDL cholesterol, apolipoprotein A-I, lipoprotein A-I, and lipoprotein A-I/A-II in prediction of coronary heart disease: The PRIME Study. *Arterioscler Thromb Vasc Biol* **22**:1155–1161.
15. Miller GJ, Miller NE. 1975. Plasma-high-density-lipoprotein concentration and development of ischaemic heart-disease. *Lancet* **1**:16–19.
16. Kinosian B, Glick H, Garland G. 1994. Cholesterol and coronary heart disease:predicting risks by levels and ratios. *Ann Intern Med* **121**:641–647.
17. Walldius G, Jungner I. 2006. The apoB/apoA-I ratio: a strong, new risk factor for cardiovascular disease and a target for lipid-lowering therapy: a review of the evidence. *J Intern Med* **259**:493–519.
18. Sniderman A, Furberg C, Keech A, Roeters van Lennep JE, Frohlich J, Jungner I, Walldius G. 2003. Apolipoproteins versus lipids as indices of coronary risk and as targets for statin treatment. *Lancet* **361**:777–780.
19. Walldius G, Jungner I, Aastveit AH, Holme I, Furberg CD, Sniderman AD. 2004. The apoB/apoA-I ratio is better than the cholesterol ratios to estimate the balance between plasma proatherogenic and antiatherogenic lipoproteins and to predict coronary risk. *Clin Chem Lab Med* **42**:1355–1363.
20. Marcovina S, Packard CJ. 2006. Measurement and meaning of apolipoprotein AI and apolipoprotein B plasma levels. *J Intern Med* **259**:437–446.

21. Sniderman AD, Jungner I, Holme I, Aastveit A, Walldius G. 2006. Errors that result from using the TC/HDL-C ratio rather than apoB/apoA-I ratio to define the lipoprotein related risk of vascular disease. *J Intern Med* **259**:455–461.
22. Marcovina SM, Albers JJ, Henderson LO, Hannon WH. 1993. International Federation of Clinical Chemistry standardization project for measurements of apolipoproteins A-I and B, III: comparability of apolipoprotein A-I values by use of international reference material. *Clin Chem* **39**:773–781.
23. Rader DJ. 2006. Molecular regulation of HDL metabolism and function: implications for novel therapies. *J Clin Invest* **116**:3090–3100.
24. Brewer HB. 2004. High-density lipoproteins: a new potential therapeutic target for the prevention of cardiovascular disease. *Arterioscler Thromb Vasc Biol* **24**:387–391.
25. Shah PK, Nilsson KSCB. 2001. Exploiting the vascular protective effects of high-density lipoprotein and its apolipoproteins: an idea whose time for testing is coming, part I. *Circulation* **104**(2376):2383.
26. Navab M, Anantharamaiah GM, Reddy ST, Hama S, Hough G, Grijalva VR, Wagner AC, Frank JS, Datta G, Garber D, Fogelman AM. 2004. Oral D-4F causes formation of pre-beta high-density lipoprotein and improves high-density lipoprotein-mediated cholesterol efflux and reverse cholesterol transport from macrophages in apolipoprotein E-null mice. *Circulation* **109**:3215–3220.
27. Shah PK, Nilsson KSCB. 2001. Exploiting the vascular protective effects of high-density lipoprotein and its apolipoproteins: an idea whose time for testing is coming, part II. *Circulation* **104**(2498):2502.
28. Franceschini G, Sirtori CR, Capurso A, Weisgraber KH, Mahley RW. 1980. A-I Milano apoprotein. Decreased high density lipoprotein cholesterol levels with significant lipoprotein modifications and without clinical atherosclerosis in an Italian family. *J Clin Invest* **66**(5):892–900.
29. Gualandri V, Franceschini G, Sirtori CR, Gianfranceschi G, Orsini GB, Cerrone A, Menotti A. 1985. AI Milano apoprotein identification of the complete kindred and evidence of a dominant genetic transmission. *Am J Hum Genet* **37**(6):1083–1097.
30. Weisgraber KH, Rall SC, Bersot TP, Mahley RW, Franceschini G, Sirtori CR. 1983. Apolipoprotein A-IMilano: detection of normal A-I in affected subjects and evidence for a cysteine for arginine substitution in the variant A-I. *J Biol Chem* **258**(4):2508–2513.
31. Avorn J. 2009. Torcetraib and atorvastatin—should marketing drive the research agenda? *N Engl J Med* **352**:2573–2576.
32. Shaffer C. 2009. Pfizer jettisons Esperion. *Nat Biotech* **26**:724–725.
33. Gilles RJ. 1994. *NMR in physiology and biomedicine*. San Diego: Academic Press.
34. Aime S, Barge A, Cabella C, Crich SG, Gianolio E. 2004. Targeting cells with MR imaging probes based on paramagnetic Gd(III) chelates. *Curr Pharm Biotechnol* **5**:509–518.
35. Artemov D, Bhujwalla ZM, Bulte JW. 2004. Magnetic resonance imaging of cell surface receptors using targeted contrast agents. *Curr Pharm Biotechnol* **5**:485–494.
36. Lanza GM, Winter P, Caruthers S, Schmeider A, Crowder K, Morawski A, Morawski A, Zhang H, Scott MJ, Wickline SA. 2004. Novel paramagnetic contrast agents for molecular imaging and targeted drug delivery. *Curr Pharm Biotechnol* **5**:495–507.
37. Anderson SA, Rader RK, Westlin WF, Null C, Jackson D, Lanza CM, Wickline SA, Kotyk JJ. 2000. Magnetic resonance contrast enhancement of neovasculature with alpha(v)beta(3)-targeted nanoparticles. *Magn Reson Med* **44**(3):433–439.
38. Kasili PM, Song JM, Vo-Dinh T. 2004. Optical sensor for the detection of caspase-9 activity in a single cell. *J Am Chem Soc* **126**(9):2799–2806.
39. Quintana A, Raczka E, Piehler L, Lee I, Myc A, Majoros I, Patri AK, Thomas T, Mulé J, Baker Jr JR. 2002. Design and function of a dendrimer-based therapeutic nanodevice targeted to tumor cells through the folate receptor. *Pharm Res* **19**(9):1310–1316.
40. Roy I, Ohulchanskyy TY, Pudavar HE, Bergey EJ, Oseroff AR, Morgan J, Dougherty TJ, Prasad PN. 2003. Ceramic-based nanoparticles entrapping water-insoluble photosensitizing anticancer drugs: a novel drug-carrier system for photodynamic therapy. *J Am Chem Soc* **125**(26):7860–7865.
41. Ruoslahti E. 2002. Antiangiogenics meet nanotechnology. *Cancer Cell* **2**(2):97–98.
42. Hood JD, Bednarski M, Frausto R, Guccione S, Reisfeld RA, Xiang R, Cheresh DA. 2002. Tumor regression by targeted gene delivery to the neovasculature. *Science* **296**(5577):2404–2407.
43. Corbin IR, Li H, Chen J, Lund-Katz S, Zhou R, Glickson JD, Zheng G. 2006. Low-density lipoprotein nanoparticles asmagnetic resonance imaging contrast agents. *Neoplasia* **8**:488–498.

44. Thonon D, Jacques V, Desreux JF. 2007. A gadolinium triacetic monamide DOTA derivative with a methanethiosulfonate anchor groiup: relaxivity properties and conjugation with albumin and thiolated particles. *Contrast Media Mol Imaging* **2**:24–34.
45. Lagerstedt JO, Budamagunta MS, Oda MN, Voss JC. 2007. Electron paramagnetic resonance spectroscopy of site-directed spin labels reveals the structural heterogeneity in the N-terminal domain of ApoA-I in solution. *J Biol Chem* **282**(12):9143–9149.
46. Martin DD, Budamagunta MS, Ryan RO, Voss JC, Oda MN. 2006. Apolipoprotein A-I assumes a "looped belt" conformation on reconstituted high density lipoprotein. *J Biol Chem* **281**(29):20418–20426.
47. Haacke EM, Brown RW, Venkatesan R, Thompson MR. 2006. *Magnetic resonance imaging: physical principles and sequence design*. New York: Wiley-Liss.
48. Bourasset F, Dencausse A, Bourrinet P, Ducret M, Corot C. 2001. Comparison of plasma and peritoneal concentrations of various categories of MRI blood pool agents in a murine experimental pharmacokinetic model. *MAGMA* **12**(2–3):82–87.

PROBLEM SOLUTIONS

CHAPTER 1

1.1. As we have discussed, a Fluids-DFT is typically formulated in an open ensemble in which the state variables are the temperature *T*, volume *V*, and chemical potentials of all the species, μ_α. Thus for a single-component membrane system consisting of lipids and solvent, we must specify four variables: μ_s, μ_L, *V*, and *T*. In the CMS-DFT, the chemical potentials are determined by the densities in the bulk reservoir. One way to set these is to fix the total bulk site density $\rho_b = \rho_L + \rho_s$ and then to consider bilayer properties as a function of the solvent number fraction $x_s = \rho_s / \rho_b$. Note that at fixed *T*, *V*, and ρ_b, the additional constraint of $\gamma = 0$ uniquely determines x_s, so that there is only one bilayer for this set of conditions. We identify this unique bilayer solution as the physical one for a bilayer at that particular temperature (this assumes our choice of the overall density is a realistic one, i.e., resulting in realistic overall fluid densities).

What happens if we now add a second component to the bilayer, such as the alcohols discussed in the text? We now have three chemical potentials: μ_s, μ_L, and μ_a for the alcohol. We still need to maintain the constraint $\gamma = 0$. What is the dimensionality of the set of possible membrane solutions at zero tension and fixed temperature and volume? How might we choose the chemical potential variables to obtain physically realistic results?

Answer

We now have five variables that describe the thermodynamic state of the system. At fixed *T*, *V*, and $\gamma = 0$, we have three constraints, so the set of possible bilayers under these conditions forms a two-dimensional surface in phase space, which could be parameterized by any two of the three chemical potentials (or any two combinations of these). If we follow our previous work and again keep ρ_b fixed, we now have a one-dimensional "line" of possible bilayers. There are various possible ways one could set, e.g., x_s and $x_a = \rho_a / \rho_b$ to choose a specific bilayer from this line. For a discussion, see the appendix in [16].

1.2. Consider the CMS-DFT described in the text (§1.3.1). The free energy is given by a functional Taylor expansion about the bulk reference state up to second order in the density as expressed in Eq. (10). Explain why this implies that the CMS-DFT cannot support a fluid/vapor phase transition or interface.

Answer

A fluid/vapor phase transition occurs in a single-component fluid. From Eq. (1.18), the grand free energy for a bulk density $\rho(r) \equiv \rho$ (different than the reference density) is given by

$$\Delta\Omega = -\frac{V}{N}(\rho - \rho_r) + \frac{1}{2}\int drc_{\alpha\beta}(r)\left[\rho^2 - \rho_r^2\right],$$

The free energy is thus quadratic in *r*. A quadratic function cannot have two minima, which is required for liquid/vapor coexistence (two minima in the free energy, one corresponding to the bulk liquid phase and the other to the bulk vapor phase). Thus, the CMS-DFT cannot describe, e.g., free liquid interfaces or wetting phenomena where there is a vapor phase. Also see the explanation beginning on p. 103 in Evans R. 1992. Density functionals in the theory of nonuniform fluids. In *Fundamentals of inhomogeneous fluids.* Ed D Henderson. New York: Marcel Dekker.

CHAPTER 2

2.1. Prove Ricci's lemma, namely, that both $\nabla_a g_{bc} = 0$ and $\nabla_a g = 0$.

Answer

Using the definition of the covariant derivative and the Christoffel symbols, we get

$$\begin{aligned}
\nabla_a g_{bc} &= \partial_a g_{bc} - \Gamma^d_{ab} g_{dc} - \Gamma^d_{ac} g_{bd} \\
&= \partial_a g_{bc} - \tfrac{1}{2} g^{de}\left(\partial_a g_{be} + \partial_b g_{ea} - \partial_e g_{ab}\right) g_{dc} - \tfrac{1}{2} g^{de}\left(\partial_a g_{ce} + \partial_c g_{ea} - \partial_e g_{ac}\right) g_{bd} \\
&= \partial_a g_{bc} - \tfrac{1}{2}\delta^e_c\left(\partial_a g_{be} + \partial_b g_{ea} - \partial_e g_{ab}\right) - \tfrac{1}{2}\delta^e_b\left(\partial_a g_{ce} + \partial_c g_{ea} - \partial_e g_{ac}\right) \\
&= \partial_a g_{bc} - \tfrac{1}{2}\left(\partial_a g_{bc} + \partial_b g_{ca} - \partial_c g_{ab}\right) - \tfrac{1}{2}\left(\partial_a g_{cb} + \partial_c g_{ba} - \partial_b g_{ac}\right) \\
&= 0.
\end{aligned}$$

We now use the fact that we can write the metric determinant as $g = \frac{1}{2}\varepsilon^{ab}\varepsilon^{cd} g_{ac} g_{bd}$. Here, ε^{ab} is the antisymmetric ε-symbol (which is 1 if *ab* is an even permutation of 12 and 0 otherwise). This is nothing but a fancy way of writing the Leibniz formula for determinants. But now we readily see that $\nabla_a g = 0$ immediately follows from $\nabla_a g_{bc} = 0$ and this equation, since the epsilon tensor is of course also a constant under differentiation.

2.2. Prove that the Gaussian curvature can be written as $K_G = \frac{1}{2}\left(K^2 - K_{ab}K^{ab}\right)$.

Answer

Since K^2 is the *square of the trace* of the curvature tensor, while $K_{ab}K^{ab}$ is the *trace of the square* of the curvature tensor, their difference can be written in terms of the eigenvalues c_1 and c_2 in the following way: $K^2 - K_{ab}K^{ab} = \left(c_1 + c_2\right)^2 - \left(c_1^2 + c_2^2\right) = 2c_1c_2$. Half of this is evidently the Gaussian curvature — the *product* of the two eigenvalues (which is identical to the *determinant* of the curvature tensor).

2.3. Show that the metric determinant is indeed given by the square of the modulus of the cross-product between the two tangent vectors, i.e., $g = \left|e_1 \times e_2\right|^2$.

Answer

Let α be the angle between $\mathbf{e}_1$ and $\mathbf{e}_2$. By using the definition of the metric as the matrix of scalar products between these vectors, we find

$$\begin{aligned}|\mathbf{e}_1\times\mathbf{e}_2|^2 &= |\mathbf{e}_1|^2|\mathbf{e}_2|^2\sin^2\alpha = |\mathbf{e}_1|^2|\mathbf{e}_2|^2(1-\cos^2\alpha) = |\mathbf{e}_1|^2|\mathbf{e}_2|^2 - (|\mathbf{e}_1||\mathbf{e}_2|\cos\alpha)^2 \\ &= |\mathbf{e}_1|^2|\mathbf{e}_2|^2 - (\mathbf{e}_1\cdot\mathbf{e}_2)^2 = g_{11}g_{22} - g_{12}g_{21} = g.\end{aligned}$$

2.4. Why is it true that $\mathbf{e}_a\cdot\partial_b\mathbf{n} = -\mathbf{n}\cdot\partial_b\mathbf{e}_a$?

Answer

Since the tangent vectors $\mathbf{e}_a$ are perpendicular to the normal vector $\mathbf{n}$, their scalar product vanishes: $\mathbf{e}_a\cdot\mathbf{n}=0$. Differentiating this orthogonality condition with respect to u^b (i.e., performing ∂_b on both sides) immediately leads to the equation above.

2.5. Given that the stretching modulus of typical phospholipid bilayers is $K_{\text{stretch}}\approx 250\text{ mN/m}$, what value for Young's modulus of the membrane interior would this imply, if we make the simplifying assumptions that such an identification is permissible? Is that value reasonable?

Answer

Using $K_{\text{stretch}} = Yh$, where Y is Young's modulus and where $h\approx 4\text{ nm}$ is the membrane thickness, we arrive at $Y\approx 60\text{ MPa}$. This is a typical modulus for rubber. Given that rubbers are often dense polymer chains with characteristic length scales in the nanometer range, this modulus is indeed plausible.

2.6. Show that an incompressible material has a Poisson's ratio of $\nu=\frac{1}{2}$.

Answer

Recall that Poisson's ratio is the negative ratio between the relative transverse material deformation and the concomitant relative longitudinal deformation. Think now of a block of material of length L and square cross-section a^2. Its volume $V=La^2$ is supposed to remain constant under stretching. Hence we must have

$$0 = dV = d(La^2) = dLa^2 + L2a\,da = V\left(\frac{dL}{L}+2\frac{da}{a}\right) = V\frac{dL}{L}\left(1+2\frac{da/a}{dL/L}\right) = V\frac{dL}{L}(1-2\nu)\ .$$

From this follows that $\nu=\frac{1}{2}$ must hold if the material is incompressible.

CHAPTER 3

3.1. Which lipid constitutes a better surfactant: triglyceride or DPPC?

Answer

Compare the hydrophobicity of lipids.

3.2. Depending on the temperature and surface pressure, lipid monolayers can form a variety of phases — e.g., gas, liquid-expanded, liquid-condensed, and condensed. At a fixed temperature and surface pressure, which monolayer has a higher molecular density: DPPC or POPC?

Answer
Compare the main phase transition temperatures of the bilayers.

3.3. The slope of the pressure–area isotherm gradually increases with increasing surface pressure. What can explain the decrease in isotherm slope at high pressures often observed experimentally?

Answer
Partial loss of lipids from the interface.

3.4. In simulations, molecular force fields often underestimate the surface tension at the water/vacuum interface. What is the possible result of a simulation, if the surface tension applied to the monolayer is larger than the surface tension at the water/vacuum interface (and smaller than the surface tension at the air/water interface in real systems).

Answer
Formation of pores in the monolayer; monolayer disruption if the applied tension is not achievable.

CHAPTER 4

4.1. What is the main driving force behind the concept of "Multiscale Modeling"?

Answer
Highly detailed models cannot describe large enough scales for collective motion. Even if possible, atomistic simulations would lead to information overkill. See further reading on multiscale modeling for more detailed information.

4.2. Which characteristics of supported bilayers are different or similar to free bilayers?

Answer
In general, the distal leaflet is very similar to a free bilayer, whereas the structure of the proximal leaflet is strongly influenced by the support. In comparison to real biomembranes, we always have to keep in mind that model membranes contain only a very limited subset of molecules.

4.3. Why are supported bilayers such an widely used model system?

Answer

Many experimental techniques are very difficult to directly apply to free bilayers (e.g., AFM). In addition, mechanical stability makes things much easier. Real membranes contain many constituents that make the analysis very complicated.

CHAPTER 5

5.1. In the chapter we discussed the overall process of pore-mediated lipid flip-flops and illustrated it for a dimyristoylphosphatidylcholine (DMPC) bilayer at $T = 323$ K. Make qualitative predictions regarding changes in the flip-flop rate when (a) temperature is decreased to e.g. the physiological value ($T = 310$ K); (b) DMPC lipids are replaced by dipalmitoylphosphatidylcholine (DPPC) lipids whose acyl chains are two hydrocarbons longer.

Answer

a. As there are no external forces applied to the bilayer system and the transmembrane lipid translocation across the bilayer is purely diffusive (driven exclusively by thermal fluctuations), one can expect that decreasing temperature should make the flip-flops slower and lead to a drop in the overall number of flip-flops, thereby decreasing the flip-flop rate. In contrast, increasing temperature will speed up flip-flop activity.
b. Longer hydrocarbon chains of DPPC lipids imply that a DPPC bilayer has a larger hydrophobic core than a DMPC does. This has two implications: (i) the energy required for pore formation and lipid desorption out of the monolayer (when a pore has been formed) is larger in the case of DPPC, and (ii) the pathway for lipid translocation is longer in a DPPC bilayer. Overall, both factors will lower the flip-flop rate for a DPPC bilayer as compared to the DMPC case.

   We note that the effects of temperature and acyl chain length on flip-flop activity were considered in detail in a very recent computational study of chemically induced lipid translocations across biological membranes [54].

5.2. Assuming that the lipid lateral diffusion coefficient is $D = 1 \times 10^{-7}$ cm^2/s, compute the time that is needed to diffuse from one side of a cell to the opposing one, assuming the cell radius to be R = 10 micrometers.

Answer

For diffusion in a plane, dimensional analysis yields $t = \ell_D^2 / D$, where ℓ_D is the length scale to be crossed. Assuming spherical cells, $\ell_D = \pi R$. Then the diffusion time asked is $t \approx 100$ s. The intriguing relationship here is the dependence of t on ℓ_D, that is, $t \propto \ell_D^2$. If the cell size was increased by a factor of 10, then the diffusion time would increase by a factor of 100, rendering diffusion of mass inefficient over a length scale of a cell. The fact that cells have a size of about tens of microns and the lipid diffusion coefficient is of the order of 10^{-7} cm^2/s implies mass diffusion to be efficient.

CHAPTER 6

6.1. Derive the cantilever amplification factor in the AFM detection system,

$$\Delta z = \frac{3L}{l} \Delta y \,, \tag{6.11}$$

where Δz is deflection of the laser spot when reaching the photodiode and Δy is the physical deflection of the cantilever end (see also §6.9).

Answer

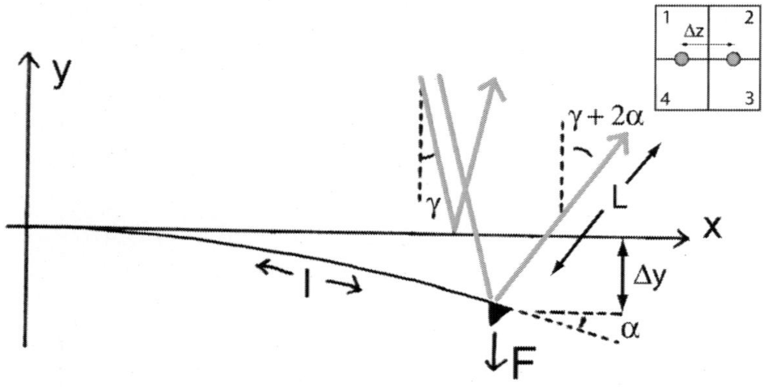

Figure 6.12.

Deflection of the laser spot is related to the cantilever bending angle as: $\Delta z = 2L\alpha$. The cantilever shape is given as a relation: $y(x)$. For small deflections ($\alpha \approx 0$) we have $\frac{dy}{dx} = \tan\alpha \simeq \alpha$, which means that

$$\Delta z = 2L\left[\frac{dy}{dx}\right]_{\text{end}} . \tag{6.12}$$

We consider the situation were a force F is applied to the tip at the cantilever end. The condition for static mechanical equilibrium is that for any value of x the torque $\tau(x)$ from internal stresses is equal to the torque due to force F:

$$\tau(x) = -F(l-x) = -\frac{EI}{R(x)} \Rightarrow \frac{1}{R(x)} = \frac{F(l-x)}{EI} \tag{6.13}$$

Where E is Young's modulus of the cantilever material, I is the *area moment of inertia* of the cantilever, and $R(x)$ is the curvature radius of the cantilever. When deflection α is small, the curvature radius can be written as

$$\frac{1}{R} = \frac{d^2y}{dx^2}.$$

We therefore obtain

$$\frac{d^2y}{dx^2} = \frac{F}{EI}(l - x). \tag{6.14}$$

When this differential equation is integrated with boundary conditions $[dy/dx]_{x=0} = 0$ and $y(x = 0) = 0$, we get the following equation for the cantilever shape, valid for small bending angles:

$$y(x) = \frac{F}{EI}\left(\frac{1}{6}x^3 - \frac{1}{2}lx^2\right). \tag{6.15}$$

Specifically, we can find the deflection Δy at the cantilever end:

$$\Delta y = y(l) = -\frac{Fl^3}{3EI}. \tag{6.16}$$

The deflection Δz of the reflected laser spot now becomes

$$\Delta z = 2L\left[\frac{dy}{dx}\right]_{x=l} = 2L\left[-\frac{Fl^2}{2EI}\right] = -\frac{Fl^2L}{EI} = \frac{3L}{l}\Delta y. \tag{6.17}$$

From the last equality, we can conclude that vertical deflections of the cantilever are amplified by a factor $3L/l$ when measured in the photodiode detector.

6.2. Consider a situation where a fluorescence image is recorded of a double-membrane system with resolvable and superimposed domain features from membranes 1 and 2. The difference in partitioning of the fluorescence probe between the membrane domains generates contrast and gives rise to two intensities — high (H) and low (L) — from the two domains types, respectively. If we are only interested in analyzing the domain pattern in membrane 2, there is a potential problem in detecting these domain features without interference from membrane 1 domains. We will consider if and how this problem can be resolved.

a. Assume that the fluorescent intensity of the two domains is the same (H.L) in membrane 1 and 2. Will it be possible to detect domains in membrane 2 by simple threshold filtering (intensity cut-off)?

Figure 6.11.

b. Assume that the fluorescent intensity of the two domains is different (H_1,L_1,H_2,L_2) in membranes 1 and 2. Will it be possible to detect domains in membrane 2 by simple threshold filtering (intensity cutoff)? Will it be possible to detect the contour of islands in membrane 2? If yes, write down the restrictions on H_1,L_1,H_2,L_2 that enable these detections.
c. Discuss alternative experimental or computational tricks to improve domain detection in the above case.

Answer

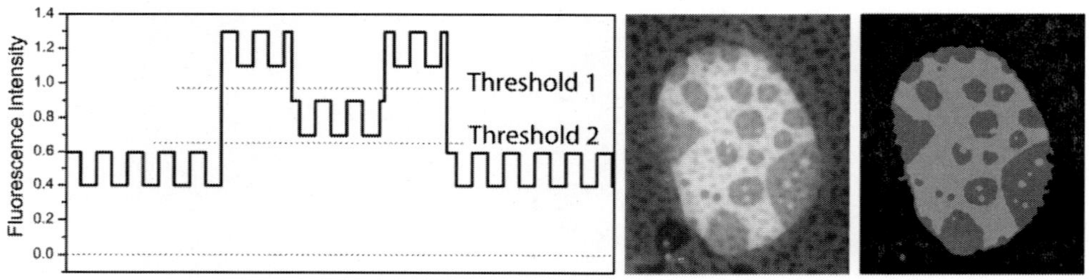

Figure 6.13. Please visit http://www.springer.com/series/7845 to view a high-resolution full-color version of this illustration.

Figure 6.13 shows schematically the fluorescence intensity of an image line crossing an island of membrane 2 on top of membrane 1. From the Figure 6.13 it can be seen that separation of domains within island 2, using threshold 1, implies the following inequality:

$$L_2 + H_1 < H_2 + L_1 \Rightarrow H_2 - L_2 > H_1 - L_1 . \tag{6.18}$$

Similarly, separation of the contour of island 2 from the background, using threshold 2, requires

$$L_1 + L_2 > H_1 \Rightarrow H_1 - L_1 < L_2 . \tag{6.19}$$

As an example, the figure illustrates the fulfillment of these conditions by the intensities (H_1,L_1,H_2,L_2) = (0.6, 0.4, 0.7, 0.3). Regarding Problem 6.2a, it is clear that domain detection by thresholding is not possible if $H_1 = H_2$ and $L_1 = L_2$ (but detection of island contour is). Simply speaking, this result shows that domain detection in membrane 2 is possible if the intensity contrast is larger than that in membrane 1. In certain cases [34], we have found that membrane–solid interactions in membrane 1 are sufficient to alter domain contrast compared to membrane 2 and enable domain detection, as described above. An example of this is shown in Figure 6.13, demonstrating threshold detection of the island boundary and of domains within the island without interference from domain features in the underlying membrane.

CHAPTER 7

7.1. Why is ionic current reduced when a single-stranded nucleic acid molecule traverses a nanopore?

Answer

The open-channel ionic current is carried through the nanopore by the electrolyte ions present in the solution. In a typical configuration using Ag/AgCl electrodes and a KCl solution, the current is carried by K^+ and Cl^- ions. When a single-stranded nucleic acid molecule enters and traverses the pore, it reduces the volume of electrolyte available for the flux of ions, thus reducing the ionic current. In addition, if the macromolecule is negatively charged, as is the case with nucleic acids, it introduces a significant negative charge density into the pore that largely prevents the flux of Cl- ions. The ionic current blockade is also related to the salt concentration used to carry the current, and it is possible to find conditions in which the translocation of nucleic acids actually increases the ionic current during translocation. This has been observed in solid-state nanopores at low salt concentrations. In this case, current increases are due to the counterions that screen the charge of the DNA backbone [40].

7.2. A homopolymer molecule such as polydeoxycytidylic acid, 100 bases in length, requires an average of 0.2 ms to traverse the pore.

a. How much time does an individual base spend in the pore?
b. Given that this time is too short for identification of individual bases, what can be done to slow down the translocation velocity?

Answer

a. About 2 μs [5].
b. A number of strategies can be used to reduce the translocation speed of macromolecules through nanopores. These include increasing the viscosity of the medium (e.g., by using glycerol), reducing the applied voltage, and reducing the temperature [41]. The most dramatic reduction was achieved when a nucleic acid processing enzyme such as an exonuclease or polymerase binds to the DNA and is drawn to the pore as a complex [26,27].

CHAPTER 8

8.1. In an FRAP experiment, the scaled fluorescence intensities are $F_i = 1$, $F_0 = 0.4$, and $F_\infty = 0.8$. Assume the bleaching spot is a disk with radius 1 μm and the half time of recovery (τ_D) is 0.2 s.

a. Find the diffusion coefficient for this protein. (Assume a uniform circular laser profile.)

b. Find the mobile and immobile fractions of this protein.

Answer

a. From Eq. (8.23), $D = 0.22\frac{1^2}{0.2} = 1.1\ \mu\text{m}^2/\text{s}$.

b. From Eq. (8.23):

$$\text{MF} = \frac{F_\infty - F_0}{F_i - F_0} = \frac{0.8-0.4}{1-0.4} = \frac{2}{3}, \quad \text{IF} = \frac{F_i - F_\infty}{F_i - F_0} = \frac{1-0.8}{1-0.4} = \frac{1}{3}.$$

8.2. Two FRAP experiments were performed for the same protein using circular ROIs of different size. The half times of recovery (τ_D) were 0.5 and 1 s. What is the ratio of the ROI radii in two experiments? (Assume a uniform circular laser profile.)

Answer

From Eq. (8.23), $0.22\frac{\omega_1^2}{0.5} = 0.22\frac{\omega_2^2}{1} \Rightarrow \frac{\omega_1}{\omega_2} = \sqrt{\frac{0.5}{1}} \simeq 0.71$.

8.3. If the diffusion coefficient of a protein X is approximately 40 μm²/s, what is the expected half time of recovery for a circular ROI with a 10-pixel diameter? (Assume a uniform circular laser profile and 1 pixel = 0.11 μm).

Answer

From Eq. (8.23), $40 = 0.22\frac{(5\times 0.11)^2}{\tau_{1/2}} \Rightarrow \tau_{1/2} = 0.22\frac{(5\times 0.11)^2}{40}$.

8.4. For two soluble proteins X and Y with molecular weights W_x and W_y and diffusion coefficients D_x and D_y, respectively, derive the following relation from the Stokes-Einstein equation. (Assume the shapes of these proteins are spherical.)

a. $D_y = D_x\sqrt[3]{\frac{W_x}{W_y}}$.

b. If the molecular weight of X is 27 kDa, and the diffusion coefficient of X is 10 μm²/s, what is the expected diffusion coefficient of a protein with a molecular weight of 40 kDa?

Answer

a. From Eq. (8.13), $D=\dfrac{\kappa_B T}{6\pi\eta r}$. If we assume that the shapes of proteins are spherical and that the molecular weights are directly proportional to the volumes of proteins, we get $r_x=\sqrt[3]{W_x}$. Therefore,

$$\frac{Dx}{Dy}=\frac{r_y}{r_x}=\frac{\sqrt[3]{W_y}}{\sqrt[3]{W_x}}\Rightarrow Dx=Dy\sqrt[3]{W_y/W_x}.$$

b. From (a), $Dx=Dy\sqrt[3]{W_y/W_x}\Rightarrow Dx=10\sqrt[3]{27/40}$.

8.5. For Eq. (8.24), assume $D_2\neq 0$ and $k_{on}\gg D_1/(4\omega^2),\ D_2/(4\omega^2)$. D_2 may not necessarily be smaller than D_1.

a. Show that Eq. (8.24) can be approximated by the effective diffusion model with an effective diffusion coefficient

$$D_{eff}=\frac{k_{off}D_1+k_{on}D_2}{k_{on}+k_{off}}.$$

b. From this show that the partition coefficient (k_p = k_{on}/k_{off}) is given by $k_p=\dfrac{D_1-D_{eff}}{D_{eff}-D_2}$.

Answer

a. From

$$\begin{cases}\dfrac{\partial u}{\partial t}=D_1\nabla^2u-k_{on}u+k_{off}c\\[2mm]\dfrac{\partial c}{\partial t}=D_2\nabla^2c+k_{on}u-k_{off}c\end{cases},$$

adding the two we get, $\dfrac{\partial}{\partial t}(u+c)=D_1\nabla^2u+D_2\nabla^2c$. Because binding is fast compared to diffusion, a local chemical equilibrium $c=\dfrac{k_{on}}{k_{off}}u$ is expected, i.e.,

$$\frac{\partial}{\partial t}\left(u+\frac{k_{on}}{k_{off}}u\right)=D_1\nabla^2u+\frac{k_{on}}{k_{off}}D_2\nabla^2u\Rightarrow\left(1+\frac{k_{on}}{k_{off}}\right)\frac{\partial}{\partial t}=\left(D_1+\frac{k_{on}}{k_{off}}D_2\right)\nabla^2u$$

$$\Rightarrow\frac{\partial u}{\partial t}=\left(\frac{k_{on}D_1+k_{off}D_2}{k_{on}+k_{off}}\right)\nabla^2u.$$

b. This can be done by solving $D_{eff}=\dfrac{k_{off}D_1+k_{on}D_2}{k_{on}+k_{off}}=\dfrac{D_1+k_pD_2}{1+k_p}$ for k_p.

8.6. If Eq. (8.24) is in the reaction-dominant regime, the concentration of fluorescence intensity is represented in terms of binding complex concentration, $c(x,y,t) \simeq c_i\left(1-e^{\kappa_{\text{off}}t'}\right)$. From this and employing $F(t) = q\varepsilon \iint_{\mathbb{R}^2} I(x,y)C(x,y,t)\,dx\,dy$, derive the corresponding FRAP formula.

Answer

$$\begin{aligned} F(t) &= q\varepsilon \iint_{\mathbb{R}^2} I(x,y)C(x,y,t)\,dx\,dy \\ &= q\varepsilon \iint_{\mathbb{R}^2} I(x,y)\left[C_i\left(1-e^{k_{\text{off}}t}\right)\right] dx\,dy \\ &= \left[q\varepsilon C_i\left(1-e^{k_{\text{off}}t}\right)\right] \iint_{\mathbb{R}^2} I(x,y)\,dx\,dy \\ &= F_i\left(1-e^{k_{\text{off}}t}\right). \end{aligned}$$

CHAPTER 9

9.1. Free energy of a pore: line tension and surface tension.

a. Let us consider a membrane with a surface tension γ. What is the cost in free energy, E, for the creation of a pore of radius R, if the line tension along the edge of the pore is λ?

b. How does ΔE vary with R? What does this imply about the stability of the pores?

c. What is the surface tension in a biological membrane? What does it imply for pore formation?

Answers

a, The formation of a pore corresponds to a reduction in the membrane surface area, $\Delta A = -\pi R^2$, and to the creation of a membrane edge of length $\Delta L = 2\pi R$. Therefore the free energy associated with the formation of the pore is

$$\Delta E(R) = -\pi R^2 \gamma + 2\pi R\lambda\,.$$

The existence of a positive surface tension favors pore formation, whereas the existence of a line tension discourages pore formation.

b. Assuming the surface and line tensions are constant, the free energy of the pore is maximum for a pore radius $R_0 = \lambda/\gamma$. Thus, in this particular situation, pores are never stable: small pores ($R < R_0$) close rapidly whereas large pores ($R > R_0$) keep increasing in size indefinitely.

c. In general, a relaxed biological membrane has a negligible surface tension. According to the equation above, pore formation in a relaxed membrane requires an input of energy of at least $2\pi R\lambda$. If this energy is larger than the thermal energy kT, pores will not appear spontaneously in this type of membranes. On the other hand, if a tension is imposed on the membrane (e.g., using micropipette aspiration [234,235], osmotic stress [236], or electroporation [237]), the spontaneous formation of pores becomes possible.

9.2. Free energy of a water-filled membrane channel. We first consider a hypothetic type of water-filled channel, where the lipids do not rearrange themselves along the edge of the pore, as illustrated in Figure 9.8A, in a relaxed membrane for which the surface tension is $\gamma = 0$.

a. Find the expression for the line tension at the edge of the pore as a function of the thickness of the hydrophobic part of the bilayer, h_C, and the surface tension between the apolar medium formed by the lipid carbon chains and the polar water solvent, γ_C.
b. The typical thickness of the hydrophobic part of a phospholipid bilayer is h_C = 3 nm, and the value of the surface tension between oil and water is $\gamma_C \approx$ 50 mN/m. Estimate the free energy associated with the formation of a pore of radius R = 1 nm.
c. Can such pores form spontaneously in biological membranes?

Answers

a. The main energy cost associated with the creation of a water-filled channel such as the one depicted in Figure 9.8A comes from the formation of a contact area between the apolar lipid hydrocarbon chains and the polar water solvent. The line tension for such a pore therefore corresponds to the surface energy of a polar/apolar interface with height h_C and with unit length:

$$\lambda = h_C \gamma_C .$$

b. Given h_C = 3 nm, and γ_C = 50 mN/m, we find a line tension for this type of pore of λ = 150 pN. The formation of a pore of radius R = 1 nm will therefore cost $\Delta E = 2\pi R\lambda \sim 10^{-18}$ J.
c. If we compare ΔE to the thermal energy at room temperature, $k_BT \sim 4 \times 10^{-21}$ J, we see that $\Delta E \sim 250\ k_BT$. Therefore,, spontaneous pore formation is very unlikely.

9.3. Free energy of a toroidal pore. We next consider a more realistic model for the pores, where the lipid surface curves in order to connect the two membrane leaflets (toroidal pore), as illustrated in Figure 9.8B.

a. At the pore, what is the curvature, C_1, of the lipid surface in a plane perpendicular to the membrane? What is the curvature, C_2, in a plane parallel to the membrane? Which of these two membrane deformations is the more energetically costly?
b. By considering only the most costly type of curvature, give an expression for the line tension associated with the edge of the pore as a function of the pore radius, R, the bilayer thickness, h, and the bending modulus of the lipid monolayer, k_C.
c. The bending modulus of a lipid monolayer can be estimated to be half that of the lipid bilayer, leading to $k_C \approx 2\times10^{-20}$ J [233]. Estimate the free energy associated with the formation of a toroidal pore with radius R = 1 nm,
d. Can such pores form spontaneously in biological membranes?

Answers

a. In a toroidal pore, the lipid surface both has a positive curvature (in planes perpendicular to the plane of the membrane) and a negative curvature (in planes parallel to the plane of the membrane. The positive curvature is constant, $C_1 = 2/h$ (where h is the total membrane thickness). The negative curvature on the other hand, goes from 0 at the top and bottom of the pore, to $C_2 = -1/R$ in its center. Since phospholipids have a bulky hydrophobic tail, a positive curvature of the lipid surface is usually more costly than a negative curvature.

b. Considering only the cost of the positive curvature of the membrane, we see that the curved surface, S, corresponds to half the surface of a torus, thus, $S = \pi^2(R + h/2)h$. The line tension corresponds to the bending energy of a unit length stretch of the pore edge:

$$\lambda = \frac{1}{2\pi R}\frac{1}{2}k_C C_1^2 S = \pi k_C \frac{R + h/2}{Rh}.$$

c. For $R = 1$ nm, $h \sim 3$ nm, and $k_C \sim 2 \times 10^{-20}$ J, we get $\lambda = 50$ pN, and in the limit of large pore diameters we obtain $\lambda = \pi k_C/h = 20$ pN. This value is in reasonable agreement with the experimental values for the line tension of membrane pores, which range from 3 to 30 pN [235,238–240]. The corresponding energy (for $R =$ 1 nm) is $\Delta E = 2\pi R\lambda \sim 3 \times 10^{-19}$ J $\sim 80\ k_BT$.

d. The energy cost to form a toroidal pore is still too high for these pores to form spontaneously in a relaxed membrane.

9.4. Reduction of pore free energy by pore-forming proteins. We now consider the stabilization effect that PFPs have on membrane pores, by adapting a model originally developed by Huang et al. for antimicrobial peptides [222]. The premise of this model is that the pore-forming protein (present at a total surface concentration c) can exist in two different membrane states: either as a partially inserted monomer (concentration αc), or as part of an oligomeric membrane pore. Monomeric proteins both increase the surface area of the membrane leaflet in which they insert (each by an amount A_m), and induce a positive tension in the membrane, $\gamma = \gamma_0 A_m \alpha c$. Oligomeric proteins are lining the pores, and although the pore size may vary, the linear density of protein along the edge, ρ, is constant, so that the line tension at the pore, λ, is also constant. This model is therefore better adapted to PFPs for which the pore size and stoichiometry vary with the protein concentration, typically α-PPFs.

a. If ε_m is the binding energy of a monomeric PFP and ε_0 is the pore energy per protein lining the pore, what is the free energy per unit area, F, of the system constituted by the lipids and the proteins? Use this to show that above a critical concentration of proteins in the membrane, c^*, the concentration of monomeric protein remains constant.

b. Consider that $c > c^*$, and that pores exist on the membrane at a concentration C. Write the expression for the free energy of a single pore, ΔE, as a function of its radius, R.

c. How many energy minima are there for a single pore, and which radius do they correspond to? Are those energy minima stable or unstable? Explain the origin of the pore stabilization according to this model.

d. In addition to influencing the membrane surface tension, the PFPs also reduce the line tension of the pores. How does reducing the line tension further influence the structure and/or stability of the pores?

Answers

a. In the absence of a tension other than that produced by the insertion of the monomeric protein, the free energy per unit area of the system formed by the lipids and the proteins is

$$F = \varepsilon_{\mathrm{m}} \alpha c + \varepsilon_0 (1-\alpha) c + \frac{1}{2} \gamma_0 (A_{\mathrm{m}} \alpha c)^2 .$$

The first term corresponds to the binding energy of the fraction of proteins that are monomeric and the second term to the energy of the pores. The third term is the increase in membrane energy after the membrane surface has increased by the relative amount $\delta A/A = A_{\mathrm{m}} \alpha c$, while acquiring a positive tension $\gamma = \gamma_0 A_{\mathrm{m}} \alpha c$.

Deriving F with respect to α shows that the minimum value of this free energy is obtained for

$$\alpha c = \frac{\varepsilon_0 - \varepsilon_{\mathrm{m}}}{\gamma_0 A_{\mathrm{m}}^2} .$$

If $\varepsilon_0 < \varepsilon_{\mathrm{m}}$, this condition cannot be fulfilled (as α cannot be negative), and we simply have $\alpha = 0$: all the proteins are part of a pore. On the other hand, if $\varepsilon_0 > \varepsilon_{\mathrm{m}}$, then this condition can be fulfilled as long as $\varepsilon_0 - \varepsilon_{\mathrm{m}} < \gamma_0 A_{\mathrm{m}}^2 c$ (as α cannot exceed 1). We define c^* as

$$c^* = \frac{\varepsilon_0 - \varepsilon_{\mathrm{m}}}{\gamma_0 A_{\mathrm{m}}^2} .$$

If $c < c^*$, then $\alpha = 1$, meaning that all the proteins on the membrane are found as monomers. If $c > c^*$, then $\alpha = c^*/c$, meaning that the concentration of monomeric proteins on the membrane is constant (and equal to c^*), while the concentration of proteins that are part of a pore is equal to $c - c^*$.

b. The energy of a pore, ΔE, is the sum of two contributions, just as in Problem 9.1a, but this time the tension in the membrane depends on the fraction of monomeric protein, $\gamma = \gamma_0 A_{\mathrm{m}} \alpha c$. And since $\alpha c = c - (1-\alpha) c = c - 2\pi R \rho C$, the membrane tension is related to the concentration, C, and radius, R, of the pores:

$$\gamma = \gamma_0 A_{\mathrm{m}} c - 2\pi \gamma_0 A_{\mathrm{m}} \rho C R .$$

Since we cannot assume that the surface tension is constant, in order to calculate ΔE we first need to consider the small change in energy, δE, associated with a small change in pore radius, δR:

$$\delta E = \lambda 2\pi\delta R - \gamma 2\pi R\delta R = 2\pi\lambda\delta R - 2\pi\gamma_0 A_m cR\delta R + 4\pi^2\gamma_0 A_m \rho CR^2\delta R.$$

Integrating this expression, we obtain the energy of a single pore of radius *R*:

$$E(R) = 2\pi\lambda R - \pi\gamma_0 A_m cR^2 + \frac{4}{3}\pi^2\gamma_0 A_m \rho CR^3.$$

c. As it is third-order in *R*, the energy *E*(*R*) has two extrema. The first corresponds to a local maximum for *R* equal to

$$R_M = \frac{c}{4\pi\rho C}\left[1 - \sqrt{1 - \frac{8\pi\rho C}{\gamma_0 A_m c^2}\lambda}\right].$$

The second corresponds to a local minimum for *R* equal to

$$R_m = \frac{c}{4\pi\rho C}\left[1 + \sqrt{1 - \frac{8\pi\rho C}{\gamma_0 A_m c^2}\lambda}\right].$$

This local minimum corresponds to a stable finite value of the pore radius. Given that *R* cannot be negative, $R = 0$ (no pore) also corresponds to a stable state.

According to this model, creation of the pores is due to the presence of monomeric proteins in the membrane, which create a positive surface tension. The stabilization of the pores comes from the fact that when the pores grow the concentration of monomers in the membrane decreases, and so does the membrane tension, which creates a negative feedback on the size of the pores.

d. For R_m to be a real number, the line tension of the pore needs to be small enough:

$$\lambda \le \lambda^* = \frac{\gamma_0 A_m c^2}{8\pi\rho C}.$$

Therefore, reducing the surface tension of the pores is another important mechanism by which PFPs can promote pore formation: If $\lambda > \lambda^*$ there is no stable solution for the diameter of the pore other than $R = 0$. The line tension also directly influences the structure of the pores: if λ decreases, R_m increases.

CHAPTER 10

10.1. **Case Study of Human BMP 2 and Fracture Healing**. Patients ($n = 450$) with an open tibial fracture were involved in randomized, single-blind study. The study was conducted at 49 centers in 11 countries. Patients were prospectively randomized to one of three groups. Patients who met the study eligibility had provided informed consent and had been properly randomized irrespective of whether they received treatment:

Group 1 — received standard of care (SOC) (intramedullary nail fixation), which was the control group

Group 2 — received rhBMP-2 (0. 75 mg/ml)

Group 3 — received intramedullary nail fixation and BMP 2.

Treatments were assigned to each group. They were followed after the treatment for the following lengths of time: 6, 10, 14, 20, 26, 39, and 52 weeks after treatment.

A fracture was considered to be healed when there was radiographic evidence of fracture union and met all the clinical criteria. The primary outcome measure was the recommendation of secondary intervention because of delayed union or nonunion within 12 months postoperatively. All interventions were classified according to the degree of invasiveness.

Which of the three groups will report a shorter time to fracture union and why?

Answer

Consult the following: Govender S. 2002. Recombinant human bone morphogenetic protein-2 for treatment of open tibial fractures: a prospective, controlled, randomized study of four hundred and fifty patients by the BMP-2 evaluation in surgery for tibial trauma. *J Bone Joint Surg* **84**:2123–2134.

Control Group 1: fractures were united after 6 months.

Groups 2 and 3: there was a consistent increase in the rate of healing starting at 10 weeks after definitive wound closure.

At 6 months the healing rate observed between the control group and Groups 2 and 3 was 21% higher in Groups 2 and 3. The rate of fracture healing was accelerated in Groups 2 and 3 as compared to the control group.

CHAPTER 11

11.1. How can apoE conformation be used clinically to detect vascular inflammation and atherosclerosis?

Answer

Apolipoprotein E4 reports on the local lipid environment to which it is associated. Conformation of apoE4 also could signify likelihood of binding to and injure of vascular cells. Thus, apoE4 conformation could become a biomarker for prediction of risk for atherosclerotic cardiovascular disease (see [9]).

CHAPTER 12

1. Patients with familial hypercholesterolemia (FH) have a defective biosynthesis of LDL receptors or dysfunctional LDL receptors. As a consequence, cholesterol levels rise dramatically in the serum, and FH patients suffer from cardiovascular diseases at an early

age. Statins reduce serum cholesterol by competitively inhibiting hepatic HMG-CoA reductase, the enzyme catalyzing the committed step in cholesterol biosynthesis in the liver. By decreasing cholesterol synthesis, a non-FH cell would have more available LDL receptors to reduce serum cholesterol level. But FH patients have no LDL or dysfunctional receptors. A statin drug, such as atorvastatin, would not alter the number of available LDL receptors by inhibiting HMG-CoA reductatse activity and therefore would not reduce an FH patient's serum cholesterol level.

2. See the following articles:

Kaysen GA 2006. Dyslipidemia in chronic kidney disease: causes and consequences. *Kidney Int* **70**:555–558.

Hovingh GK, de Groot E, van der SW, Boekholdt SM, Hutten BA, Kuivenhoven JA, Kastelein JJ. 2005. Inherited disorders of HDL metabolism and atherosclerosis. *Curr Opin Lipidol* **16**(2):139–145.

INDEX